# Legisprudence Library

Studies on the Theory and Practice of Legislation

Volume 4

The objective of the *Legisprudence Library* is to publish excellent research on legislation and related areas (such as regulation and policy-making) from the standpoint of legal theory. This series' title points to an emerging, comprehensive conception of lawmaking which focuses on the justification of laws and the overarching principles which should guide legislation and norm-giving altogether, with the rationality, the reasonableness and the quality of legislation being its major concerns. Taking on legal theory as its pivotal perspective, the series attempts to fill a significant gap in the field of legislative studies, where political science and sociological approaches remain dominant through date. Inasmuch as it fosters legal-theoretical research in lawmaking, it also contributes to widen the scope of standard jurisprudence, which has been up to recent times overwhelmingly centred on the judicial application and the interpretation of law, thereby underestimating the central role of lawmakers within the legal system.

Contributions preferably address topics connected to legislation theory, including (but not limited to) legislative rationality, legislative technique, legistics, legislative effectiveness and social compliance of laws, legislative efficiency and lawmaking economics, evaluation, legislative and regulative impact assessment, regulation management, legislative implementation, public access to legislation, democratic legitimacy of legislation, codification, legislative reasoning and argumentation, science and expertise within lawmaking, legislative language, symbolic legislation, legal policy analysis, lawmaking and adjudication, or judicial review of legislation and legislative process. Comparative and system transcending approaches are encouraged. Purely dogmatic descriptions of positive law or legislative proceedings are not taken into consideration though connections with legislative and legal practice are welcomed. The series welcomes monographs and edited volumes.

More information about this series at http://www.springer.com/series/11058

Bart van Klink • Britta van Beers
Lonneke Poort
Editors

# Symbolic Legislation Theory and Developments in Biolaw

Springer

*Editors*
Bart van Klink
Department of Legal Theory
and Legal History
Vrije Universiteit Amsterdam
Amsterdam, the Netherlands

Britta van Beers
Department of Legal Theory
and Legal History
Vrije Universiteit Amsterdam
Amsterdam, the Netherlands

Lonneke Poort
Department of Sociology, Theory
and Methodology
Erasmus University Rotterdam
Erasmus School of Law
Rotterdam, the Netherlands

ISSN 2213-2813 ISSN 2213-2856 (electronic)
Legisprudence Library
ISBN 978-3-319-33363-2 ISBN 978-3-319-33365-6 (eBook)
DOI 10.1007/978-3-319-33365-6

Library of Congress Control Number: 2016949409



Printed on acid-free paper

This Springer imprint is published by Springer Nature
The registered company is Springer International Publishing AG Switzerland

# Preface: New Challenges

In 2014, we founded the research group 'Biolaw and Symbolic Interaction' (BioSI) as part of the general research program 'Boundaries of Law' at Vrije Universiteit Amsterdam. Its underlying purpose was to bring together scholars from a variety of disciplines with a shared interest in both the regulation of bio-ethical matters and legislation theory from, in particular, a symbolic perspective. The first project we launched was the present book volume. Earlier, Luc Wintgens contacted one of us, Bart, to ask him whether he would be willing to make a book on symbolic legislation for the renowned Springer series *Legisprudence Library*. In 1998, Bart published his PhD thesis on this topic, *De wet als symbool* (*Law as a Symbol*) (Van Klink 1998). After that, he participated on several occasions in discussions on the communicative approach which he had developed together with his supervisor Willem Witteveen (see, for instance, Zeegers et al. 2005). Although he appreciated Luc's invitation very much, Bart did not want to repeat himself and previous discussions. He was looking for new theoretical challenges. Therefore, he invited Britta and Lonneke to join the book project. In 2009, Britta published her PhD thesis on the regulation of bio-ethical issues, *Persoon en lichaam in het recht* (*Person and Body in the Law*) in which she discusses, among other things, the symbolic dimensions of biolaw, building on the notion of human dignity (Van Beers 2009). Four years later, in her PhD thesis *Consensus and Controversies in Animal Biotechnology*, Lonneke developed further an interactive legislative approach – which is closely connected to (though not identical with) the communicative approach as advocated by Witteveen and Van Klink – and applied it in a comparative study to the subject of animal biotechnology regulation (Poort 2013). So the idea was to combine our mutual interests and expertise in the fields of symbolic legislation theory and bio-regulation.

In order to promote a truly multidisciplinary exchange of ideas from which both fields could profit, we invited scholars within the international community whom we expected to be interested in this topic to participate in this project from a wide array of disciplines: law, sociology of law, legal and political philosophy, and ethics. We were happy that they all responded very positively to our request. Most of their papers (in early draft versions) we discussed at the conference 'Symbolic Dimensions of Biolaw' on 23 and 24 October 2014 in Amsterdam were supported by the Royal Netherlands Academy of Arts and Sciences (KNAW). We thank the participants to the conference for presenting and discussing their viewpoints and sharing their thoughts on the other participants' papers and the general project. Moreover, we would very much like to thank all authors for their thought-provoking contributions to this volume. We also thank Esther Oldekamp for helping us to organize the conference and to analyze the notion of the symbolic and symbolic legislation in the various papers (which is by no means an easy task!) and Siebe Bakker for his assistance with the preparation of the manuscript. We are very grateful for the possibility to publish our volume in the *Legisprudence Library* series, given by the series editors Luc J. Wintgens and A. Daniel Oliver-Lalana. Two anonymous reviewers who commented on the manuscript helped us especially to strengthen the overall coherence of the volume and to clarify its central notions. Finally, we would like to thank the publisher for making this book project possible, in particular Neil Oliver and Diana Nijenhuijzen, also for the smooth cooperation and communication.

Unfortunately, we have to end our preface with a sad note. During the production of this volume, Willem Witteveen, one of the main sources of inspiration for this book, died at the MH17 crash on 17 July 2014. We regret it very much that he is no longer with us and cannot discuss the outcomes of the book with us. However, as this book testifies, his ideas live on and continue to inspire the scientific community.

Amsterdam, the Netherlands
February 2016

Bart van Klink
Britta van Beers
Lonneke Poort

## References

Poort, Lonneke. 2013. *Consensus and controversies in animal biotechnology. An interactive legislative approach to animal biotechnology in Denmark, Switzerland and the Netherlands*. The Hague: Eleven.

Van Beers, Britta. 2009. *Persoon en lichaam in het recht. Menselijke waardigheid en zelfbeschikking in het tijdperk van de medische biotechnologie* (PhD thesis, Vrije Universiteit Amsterdam, the Netherlands). The Hague: Boom juridische uitgevers.

Van Klink, Bart. 1998. *De wet als symbool. Over wetttelijke communicatie en de Wet gelijke behandeling van mannen en vrouwen bij de arbeid* (PhD thesis, Tilburg, the Netherlands). Deventer: W.E.J. Tjeenk Willink.

Zeegers, Nicolle, Willem Witteveen, and Bart van Klink (eds.). 2005. *Social and symbolic effects of legislation under the rule of law*. Lewiston: The Edwin Mellen Press.

# Contents

# Contributors

**Bernice Bovenkerk** is an Assistant Professor at the Philosophy Group at Wageningen University, the Netherlands. Her main publication in the field is *The Biotechnology Debate. Democracy in the Face of Intractable Disagreement* (Springer 2012).

**Julian Cockbain** is a Consultant European Patent Attorney based in Gent, Belgium, and Oxford, United Kingdom. He has written numerous articles on patent-related matters, and the book *Exclusions from Patentability: How Far Has the European Patent Office Eroded Boundaries?* (Cambridge University Press, 2012, co-authored with Sigrid Sterckx). Julian is also a frequently invited lecturer on patent law and innovation policies.

**Herman De Dijn** is an Emeritus Professor in Modern Philosophy at the Institute of Philosophy, Catholic University of Leuven (Louvain), Belgium. He is a member of the Royal Academy of Belgium for Science and the Arts. He has published several books, among which *Spinoza: The Way to Wisdom* (Purdue University Press 1996) and *Modernité et tradition* (Peeters/Vrin 2004) and has written many essays (in Dutch) on Spinoza, contemporary culture, ethics, and religion.

**Jonathan Herring** is a Professor of Law at Exeter College, University of Oxford, United Kingdom. He is the author of fifty books including *Caring and the Law* (Hart 2013); *Relational Autonomy and Family Law* (Springer 2014); *Older People in Law and Society* (OUP 2009), and *European Human Rights and Family Law* (with Shazia Choudhry) (Hart 2010). He has written popular textbooks on medical law and ethics, family law, criminal law, and legal ethics. He is currently working on a book on vulnerability in the law and (with Michelle Madden Dempsey) a book on when sex is justified.

**Klaus Hoeyer** is a Professor of Medical Science and Technology Studies at the University of Copenhagen, Centre for Medical Science and Technology Studies, Denmark. His background is in social anthropology and his research interests include regulatory science, ethics as policy work, and the social organization of tissue banks. He has published in a variety of journals and is the author of *Exchanging Human Bodily Material: Rethinking Bodies and Markets* (Dordrecht: Springer).

**Robert G. Lee** is a Professor of Law at the University of Birmingham and Director of the Centre for Professional Legal Education and Research, United Kingdom. He pursues a wide agenda of research on issues of regulation and governance and has co-authored two books on the regulation of biomedicine and a book on food regulation.

**Oliver W. Lembcke** Ph.D., M.A. Phil., is a Professor of Political Science at Jena University, Friedrich Schiller University, Germany. His main research interests lie in the field of democratic theory, legal philosophy, comparative politics, and the history of political thought. Some of his recent publications are together with Claudia Ritzi & Gary S. Schaal (eds.), *Zeitgenössische Demokratietheorie* (2 volumes), Wiesbaden: Springer VS, 2016; and together with Florian Weber (eds.), *Emmanuel Joseph Sieyès. The Essential Political Writings*, Leiden, Boston: Brill, 2014.

**Dorien Pessers** is a Professor of Legal Theory and the Foundations of Private Life at the University of Amsterdam and the Free University. She has published on a wide range of subjects, among them medical biotechnology, assisted reproduction and family law. Her current research is oriented to the legal-philosophical meaning of the protection of the home as a human right.

**Lonneke Poort** is an Assistant Professor at Erasmus School of Law in Rotterdam, the Netherlands. She is also member of the subcommittee on social and ethical aspects of genetic modification of the Netherlands Commission on Genetic Modification (COGEM). She published several articles and a book on the interactive approach to law among which her dissertation *Consensus & Controversies in Animal Biotechnology. An Interactive Legislative Approach to Animal Biotechnology in Denmark, Switzerland, and the Netherlands* (Den Haag: Eleven International Publishing 2013).

**Jiří Přibáň** is a Professor of Legal Philosophy and Social Theory at Cardiff University in the United Kingdom. He graduated from Charles University in Prague in 1989 and became Professor of Law in 2002. He was appointed Professor at Cardiff Law School in 2006 and held visiting professorships and fellowships at a number of academic institutions, such as UC Berkeley, New York University, European University Institute, University of Pretoria, and UNSW in Sydney. He is author and editor of many books in Czech and English, such as *Sovereignty in Post-Sovereign Society* (2015), *Legal Symbolism* (2007), *Dissidents of Law* (2002), *Liquid Society and Its Law* (edited in 2007), *Law's New Boundaries* (co-edited with D. Nelken in 2001), and *The Rule of Law in Central Europe* (co-edited with J. Young in 1999). He regularly contributes to the international and Czech media.

**Rob Schwitters** is an Associate Professor of Sociology of Law and connected to the Paul Scholten Centre at the University of Amsterdam, the Netherlands. He publishes on tort law, responsibility and liability, the welfare state, compliance, and methodological issues.

**Sigrid Sterckx** PhD, is a Professor of Ethics and Political and Social Philosophy at the Department of Philosophy and Moral Sciences of Ghent University, Belgium. She is a founding member of the Bioethics Institute Ghent and a founding member of the Ghent Centre for Global Studies. Her current research projects focus on global justice (with particular attention to climate change); human tissue research and biobanking; patenting in biomedicine and genomics; organ transplantation; neuroethics; and end-of-life decisions. She has published more than 120 books, book chapters, and articles in international academic journals on these issues.

**Elen Stokes** is the City Solicitors' Educational Trust Senior Lecturer in Property Law at Cardiff Law School, and a Research Associate of the Sustainable Places Research Institute, Cardiff University, United Kingdom. Her research focuses on the legal regulation of new and emerging technologies, such as nanotechnology and synthetic biology. More recently, her work has examined the challenges involved in regulating hydraulic fracturing ('fracking') for unconventional oil and gas in the UK and EU.

**Britta van Beers** is an Associate Professor at the Department of Legal Theory at Vrije Universiteit Amsterdam, the Netherlands. Her research involves the legal-philosophical aspects of the regulation of medical biotechnology. Her recent publications include the edited volume *Humanity across International Law and Biolaw* (with Luigi Corrias and Wouter Werner, Cambridge University Press 2014) and "Is Europe 'Giving in to Baby Markets'? Reproductive Tourism in Europe and the Gradual Erosion of Existing Legal Limits to Reproductive Markets" (*Medical Law Review* 2015).

**Wibren van der Burg** is a Professor of Legal Philosophy and Jurisprudence at Erasmus University Rotterdam, the Netherlands. He recently published *The Dynamics of Law and Morality. A Pluralist Account of Legal Interactionism* (Ashgate 2014).

**Bart van Klink** is a Professor of Legal Methodology at Vrije Universiteit Amsterdam, the Netherlands. He has published several articles and books on symbolic legislation theory, among which his dissertation *De wet als symbool* (*Law as a Symbol* 1998) and, together with Nicolle Zeegers and Willem Witteveen, the edited volume *Social and Symbolic Effects of Legislation Under the Rule of Law* (2005). He is co-founder of the research group Biolaw and Symbolic Interaction (BioSI).

**Nicolle Zeegers** is an Assistant Professor in Political Science at the Faculty of Law, University of Groningen, the Netherlands. In 2005, she co-edited a book entitled *Social and Symbolic Effects of Legislation Under the Rule of Law* (Edwin Mellen Press). Her more recent publications are 'Convergence in European Nations' Legal Rules Concerning the Use of Human Embryos in Research?' in *European Journal of Health Law*. 21, 5, p. 454–472 and 'Devolved regulation as a two-stage rocket to public acceptance of experiments with human gametes and embryos' in *European Journal of Comparative Law and Governance*. 1, 1, p. 10–18.

# Chapter 1
# Introduction: Symbolic Dimensions of Biolaw

**Lonneke Poort, Britta van Beers, and Bart van Klink**

## 1.1 A Short History of Symbolic Legislation

In its effective history, the concept of 'symbolic legislation' has acquired various, not always compatible meanings. In socio-legal studies, it usually refers to instances of legislation that are ineffective and that serve other political and social goals than the goals officially stated. A classic example is the Norwegian Law on Housemaids, investigated by Vilhelm Aubert (1966). The legislation at hand was never meant to be effective, but was enacted in order to give recognition to the rights of housemaids on an immaterial or 'symbolic' level. It served to demonstrate that these rights were taken seriously, at least on paper. However, in practice nothing much changed in the position of housemaids. As Arnold (1938) and Edelman (1976) have argued, the legal system in general can be seen as a collection of symbols that fulfill a power maintaining and status quo preserving function. This could be called symbolic legislation in the negative sense, since it is associated with power simulation and deception of the people (Van Klink 1998: chapter 3; see also his contribution to this volume, Chap. 2). In contrast to instrumental legislation, symbolic legislation in this negative sense does not aim at enforcing the enacted rules of behaviour and, thereby, at reaching the goals officially stated. Different effects are associated with symbolic legislation, such as changing the status distribution or reconciling antagonistic groups in society, which have nothing to do with rule compliance.

During the 90s of the last century, however, a more positive notion of symbolic legislation emerged, in particular in Dutch legislative theory (see for instance

L. Poort (✉)
Department of Sociology, Theory and Methodology, Erasmus University Rotterdam, Erasmus School of Law, Burgemeester Oudlaan 50, 3062 PA Rotterdam, the Netherlands
e-mail: lonnekepoort@gmail.com

B. van Beers • B. van Klink
Department of Legal Theory and Legal History, Vrije Universiteit Amsterdam, De Boelelaan 1105, 1081 HV Amsterdam, the Netherlands

 
B. van Klink et al. (eds.), *Symbolic Legislation Theory and Developments in Biolaw*, Legisprudence Library 4, DOI 10.1007/978-3-319-33365-6_1

Witteveen 1991 and Van Klink 1998).[1] From this perspective, symbolic legislation was conceived as an alternative legislative approach that differs from the traditional top-down approach. The legislature no longer merely issues commands backed up with severe sanctions, as in instrumental legislation. Instead, lawmakers provide open and aspirational norms that are meant to change behavior not by means of threat but indirectly, through debate and social interaction. For example, in many countries, the enforcement and promotion of human rights are assigned to commissions that cannot take legally binding decisions, such as the Netherlands Institute for Human Rights which only has an advisory role in conflicts between two parties in matters of discrimination and other human rights violations (see Van Klink 1998: chapter 4). Other labels used for the same legislative approach or similar approaches are 'responsive regulation' (Ayres and Braithwaite 1995), 'communicative legislation' (Van Klink 1998: chapter 3) and 'interactive legislative approach' (Poort 2013 and Van der Burg and Brom 2000).

So, roughly speaking, two different concepts of symbolic legislation can be distinguished, a negative and a positive one. Symbolic legislation in the negative sense refers to non-effective legislation which is promulgated predominantly for political purposes (such as crisis management or power simulation), whereas symbolic legislation in the positive sense does aim at achieving its manifest goals, albeit in a communicative and interactive way. The debate that followed mainly focused on the later, positive sense of symbolic legislation which was developed under the banner of either a communicative or an interactive legislative approach (see, amongst others, Stamhuis 2006; Zeegers et al. 2005). Some of the questions raised were: In which respects does a communicative or interactive approach differ from the traditional instrumentalist approach? Or is it a more subtle and refined form of instrumentalism aimed at influencing and controlling people's minds? Which actors and organizations have to be involved in the implementation and interpretation of symbolic legislation? Is consensus a value worthwhile to pursue in the legislative process? Is symbolic value a general quality of the law or of specific instances of legislation only?

## 1.2 Symbolic Legislation in a New Context: Biolaw

With this edited volume we intend to give a new impulse to the academic debate on symbolic legislation by applying it to the field of biomedical law (hereafter: 'biolaw'). Since the 1990s, legal frameworks have been developed for the regulation of recent and emerging biomedical technologies, ranging from organ donation to

[1] In the previous decade, the symbolic dimension of law in this more positive sense was already discussed in sociological studies in Germany (see, for instance, Zielke 1980) and France (inspired by the Pierre Bourdieu's notion of 'symbolic power'; see, for instance, Commaille 1984). However, we are referring here primarily to legal theoretical studies which conceive of symbolic legislation as an alternative legislative strategy (that is, to symbolic legislation theory).

assisted reproductive technologies, from synthetic biology to direct-to-consumer genetic testing, and from stem cell research to genetic selection and modification. A brief sketch of this new area of law can illustrate how these developments have served to revive discussions on symbolic legislation.

On a national level, these technologies have given rise to different types of legislative activity. In most legal orders, broad parliamentary acts on biomedical issues have been issued, setting the stage for further legal-political decision making, such as acts on the use of embryos, organ donation or the use of human biological materials. Because of the complexities surrounding the legal status of the human body and derived materials, the ethical controversies surrounding these matters, and the novelty of the questions raised by these technologies, these laws have, in most cases, been preceded by years of parliamentary struggle and debate to reach a political agreement. Consequently, the use of delegated legislation, such as medical-professional guidelines for assisted reproductive technologies, or permits for certain forms of human experimentation, remains a popular alternative for the governance of this field.

On an international level, human rights discourse has been the prime source of inspiration for biolegal frameworks. International declarations, directives and conventions have been developed by, most prominently, the Council of Europe, the EU and UNESCO. Prime examples are the Convention on Human Rights and Biomedicine (Council of Europe 1997), the EU Directive on the legal protection of biotechnological inventions (Directive 98/44/EC) and the Universal Declaration on the Human Genome and Human Rights (UNESCO 1997). However, when it comes to the regulation of ethically and religiously sensitive matters in this field, these international legal documents have mostly proven to be quite ineffective. In most cases, either the wording of the provisions in question is open to conflicting interpretations, or the legal instrument in question is regarded as merely soft law.

The symbolic dimensions of biolaw can be recognized in this brief sketch in several ways. To begin with, one of the main reasons why regulation of biomedical developments has lead to such heated debates, is that these developments touch on deeply rooted, symbolic-cultural representations of the biological aspects of human life. More specifically, biomedical approaches to the human body are often characterized by instrumental, utilitarian or even economic views of the value of human body materials, thereby questioning existing, status-based approaches to the human body, which take the symbolic connectedness between persons and their bodies as their starting point. This raises the question whether and how the law should continue to reflect these symbolic values and meanings. Moreover, how can we decide what these symbolic values are, given the fact that we live in a pluralistic society?

Also in other ways the governance of medical biotechnology offers a striking example of the symbolic dimensions of law. As it is often impossible to reach consensus on these controversial questions, legislators have sought alternative ways to develop legal frameworks. The aforementioned communicative or interactive approach to legislation is therefore prominent within biolaw. For example, much of the regulation in this field consists of highly aspirational and abstract norms requiring further crystallization. Further interpretation of these norms has to be established within (e.g. expert committees and licensing) and outside (e.g. public debate and societal interaction) institutional frameworks. Moreover, as mentioned, legal-political decision making on biomedical developments is often not only informed by, but also delegated to medical-ethical commissions or semi-governmental bodies, which will take the viewpoints of various stakeholders into account.

Nevertheless, the symbolic aspirations and effects of bio-ethical legislation remain an often neglected and overlooked dimension of biolaw. This volume aims to fill that gap by bringing together perspectives from symbolic legislation theory on the one hand, and from biolaw and bioethics on the other hand. One of the volume's central claims is that biolegal approaches can profit from perceiving law from a symbolic perspective and, vice versa, that insights from biolegal and bioethical research can enrich current understandings of the meaning, value and functioning of symbolic legislation in its various manifestations.

## 1.3 Structure of the Book

The book consists of three parts: (1) Symbolic Legislation: The Symbolic Quality of Law; (2) Symbolic Approaches to Biolaw: Biolaw as a Symbolic Order; and (3) Legislative Strategies: Regulating Biomedical Developments from a Symbolic Perspective. The first part explores the theme of symbolic legislation on a predominantly theoretical level. It focuses, in particular, on the normative and empirical dimensions of this concept. The second part offers a theoretical and legal-philosophical reflection on the symbolic values and categories which underpin regulation of biomedical developments. The third part sheds light on symbolic legislation as a legislative strategy and discusses several examples of attempts to regulate biotechnological issues. In the conclusion, the various approaches to symbolic legislation in general and biolaw in particular, as presented in this volume, will be compared and the main outcomes of the present discussion will be discussed.

In the contributions to this volume the various, negative or positive, senses of symbolic legislation as distinguished above will recur. Some authors refer to the negative sense, by arguing that biolaw in specific cases does not live up to its aspirations and fails to promote a meaningful communication between the actors involved (see, for instance, the contributions of Sterckx & Cockbain and Lee & Stokes, in Chaps. 13 and 14 respectively). Lembcke (Chap. 6) discusses the implications for the state's authority when it issues legislation that is, by and large, not effective. Priban (Chap. 7) argues that legislation will never succeed in unifying people on the symbolic level of values. Other authors, from a more normative perspective, explore the positive sense of symbolic legislation, on a general theoretical level or particularly in the field of biolaw. For instance, Van Beers (Chap. 11), De Dijn (Chap. 9) and Pessers (Chap. 12) stress the constitutive role which biolaw has to play in the the symbolic mediation of the most important biological facts in life, such as birth, reproduction and death. Moreover, conditions are investigated under which a meaningful communication through legislation can take place (see the contributions of

Poort and Zeegers, in Chaps. 5 and 15 respectively). The relation and possible distinction between the communicative and the interactive approach to legislation are discussed by Poort (Chap. 5), Van der Burg (Chap. 3) and Van Klink (Chap. 2).

Below, the general themes of the three parts of the volume will be introduced and connected to the specific themes of the individual chapters.

## 1.4 Part I: Symbolic Legislation: The Symbolic Quality of Law

### *1.4.1 Theme*

In academic as well as public discourse, the concept of 'symbolic legislation' is charged with negative connotations. Symbolic legislation is generally understood as ineffective or 'toothless' law that serves other goals than the goals officially stated. It is promulgated, for instance, to pay lip service to some moral value or to simulate power in times of crisis. During the 90s of the last century, as indicated earlier, a more positive understanding of symbolic legislation was developed. The communicative or interactive turn in legislation and legislative theory has been discussed in several books, among which Van Klink and Witteveen (2000), Stamhuis (2006), and Zeegers et al. (2005). In this discussion, several points of criticism were put forward. To clarify the intellectual context from which especially this part of the volume arose, we will briefly discuss three points here (see further Van Klink 2014).

Firstly, the communicative approach was accused of being too much top-down oriented, because it placed the legislator in the center of the legislative process (Hertogh 2005 and Westerman 2005). It is based on a model of communication in which the legislator sends out to society a general message in the form of a law and, subsequently, expects the law to be applied faithfully to its spirit by legal officials in discussion with citizens. Proponents of the interactive legislative approach rejected this hierarchical view and conceived of legislation instead as an ongoing communication process in which various actors – not only the legislator, but also citizens and organizations in society – interact with each other on more or less the same level and make together the law (see, for instance, Van der Burg 2005 and Poort 2013, as well as their contributions to this volume). According to Van der Burg (2005: 257–258), the process of creating and applying law do not presuppose a regulative center but a network of actors:

> [T]aking legislation as a starting point for analysis easily leads to regarding the legislature as the central actor in normative analysis. In a consistently interactionist approach, the normative perspective of the actor should also be broadened. Society should be analyzed more in terms of a network of actors interacting with each other than of one central actor interacting with all other actors.

Secondly, it was argued that both the communicative and interactive approach are based on an idealistic, politically naïve notion of consensus: in political reality, legislation usually does not result from consensus but from compromises between the party involved (Stamhuis 2005; Poort 2013). In Stamhuis' view, the appeal to consensus is nothing but a "cover-up" for the political choices made by the legislator:

> Yet, it seems to be more realistic to assert that the legislator, while formulating a value or norm, may pretend to have found consensus, but in fact reflects the compromise that has been reached during the legislative process. During this process, the legislator has been prioritizing norms and making a selection between them. (Stamhuis 2005: 286)

Thirdly, some critics debunked the symbolic legislative strategy as instrumentalism in disguise (see, in particular, Hertogh 2000, 2005; Lindahl 2000; and Westerman 2005). Symbolic legislation is less overtly commanding than instrumental legislation based on clear, well-defined rules backed up with sanctions. However, it aims at controlling society in a more subtle manner by means of communication through general clauses. So, instead of "***command*** *and control*" the legislator now seeks to "***communicate*** *and control*" (Hertogh 2000: 54; italics and bold in the original text). According to Lindahl (2000: 179), there is no fundamental but (at best) only a gradual difference between an instrumental and a communicative or interactive approach when it comes to the application of the law: from a hermeneutic perspective, every law application – both in the case of clear rules and general clauses – requires interpretation and specification. By emphasizing the role of dialogue and persuasion in the legislative process, it hides from view the violence that legislation in whatever shape or form presupposes. As Lindahl (2000: 188) argues, dialogue and persuasion are only possible within a concrete legal order, where already normative boundaries have been drawn which cannot be justified fully on communicative grounds. Before a legal conversation on the law's meaning can start, there already has taken place an exclusion and inclusion of norms. Westerman (2005: 307; original italics) makes a similar point: "I think that the obligatory character of aims should not be hidden under the mask of the benevolent legislator. Legislators, even interactive ones, not only *propose* but also *impose* aims and ideals." Like instrumental legislation, symbolic legislation cannot dispense with the notion of command (and, therefore, with violence).

In the first theoretical part of the present volume, some new themes related to symbolic legislation which have not been discussed so far or have not been discussed in depth, will be taken up. It explores in particular the following questions. To begin with, how can the two concepts of symbolic legislation – a realist understanding and a normative, communicative or interactive interactionist understanding – be distinguished from each other? Subsequently, on what model of communication is the communicative or interactive approach based and how can it achieve legal closure? Is this approach compatible with democracy and the rule of law? Does it offer a satisfactory account of how norm compliance is achieved? Moreover, how does symbolic legislation affect political authority? And, finally, to what extent can legal symbolism contribute to the unity of society?

### *1.4.2 Contributions*

**Bart van Klink** addresses the question whether there is a real – in the sense of epistemologically real – difference between the two concepts of symbolic legislation: negative and positive symbolic legislation. He raises the question on what grounds the allegedly negative and positive concepts can be differentiated from each other. Is it possible that one instance of legislation can be classified as symbolic legislation both in the negative sense and in the positive sense? Are they two sides of the same coin or do they constitute mutually exclusive categories? Van Klink argues that the distinction between the two concepts cannot be made on scientific grounds only, but involves considerations of a political kind.

**Wibren van der Burg** presents the theory of interactive legislation in the context of a broader interactionist paradigm. Law-making becomes a cooperative effort on the part of various stakeholders, of which the state is one, but not necessarily the most important one. Van der Burg discusses two points of criticisms that are often expressed in context of this interactionist paradigm: interactive legislation is accused of both weakening democracy and contravening the rule of law, because it, for instance, does not provide sufficient legal certainty. He argues, on the contrary, that, interactionist law may, under specific conditions, reinforce democracy and the rule of law. For that purpose he examines and reconceptualizes the values that are at stake. When discussing legal certainty – a core value of the rule of law – we should, as he argues, distinguish between doctrinal or epistemic certainty and practical certainty. Interactive legislation is clearly detrimental to doctrinal certainty, but practical certainty may be improved rather than impaired by interactive legislation.

From a sociological point of view, **Rob Schwitters** questions some presumptions of the communicative approach to law. Building on Habermas, he maintains that in current complex societies the coercive power of the state and the formal procedural legitimacy of the law should be seen as building blocks of communicatively structured compliance. Schwitters argues that especially in circumstances in which the law contributes to overcoming problems of collective action, its effect will rely on the simultaneous impact of deterrent effects and persuasion-based effects. This shows that the coercive power of the state and the formal procedural legitimacy of law can help to foster a persuasion-based acceptance of the law.

**Lonneke Poort** approaches the debate between opponents and proponents of the communicative approach to law by addressing the tension between law's basic function of ending conflict and the interactive function of stimulating dynamics. Poort argues that this tension can be resolved by means of a two-track approach to the development of legal norms, based on an ethos of controversies. An ethos of controversies focuses primarily on structuring decision-making around the controversies that characterize complex issues in situations when aiming for consensus is premature. The two-track approach consists of a legal track, in which legal decisions can be made, and a moral track, in which moral debate may continue even after a decision has been made. It is argued that an interplay between these two tracks ensures that norm development may continue whenever the context requires, while legal conflicts can also be brought to an end. Therefore, as Poort, argues, the tension between ending conflict and stimulating dynamics need not be as big as is often assumed.

**Oliver W. Lembcke** examines the relation between symbolic legislation and authority. Does symbolic legislation undermine or strengthen authority? In the common view, symbolic legislation (in the negative sense) is conceived as ineffective law-making, which aims at protecting the current state of affairs. The relationship between authority and symbolic legislation becomes then rather simple: symbolic legislation is a sign of political crisis and, therefore, it undermines authority. Lembcke challenges the common view of symbolic legislation as a sign of political crisis on two grounds. First, he argues that the symbolic dimension of symbolic legislation is an integrative element of politics. Second, he claims that the common view is based on a simplistic notion of authority. It presupposes a concept of authority that depends on the performance of the political actors. In this view, political authority equals more or less output legitimacy. Lembcke introduces three alternative notions of authority, referred to as *being in authority*, *being an authority*, and *the authoritative*, in order to demonstrate that authority is less strictly tied to legitimacy than the common view suggests. He concludes that symbolic legislation may become instrumental, for instance in the case of conflict management. The lack of effectiveness attributed to symbolic legislation may result in disenchantment with politics and may also have a negative effect on political legitimacy, but it does not necessarily or automatically undermine political authority.

In his chapter, **Jiří Přibáň** is very critical about the claim that the legislature, by issuing symbolic legislation, can achieve social integration. He argues that legal symbolism is part of the process of functional differentiation. Therefore, no piece of symbolic legislation can provide access to the moral values and principles guaranteeing the normative unity of modern pluralistic society. It would be too ambitious to expect that the contemporary system of positive law can grant our access to a shared universe of symbols which would normatively reflect on the meaningful existence of society and its members. Instead of being treated as a foundational determination of legality's meaning and content, symbolic communication needs to be analyzed as specific communication internalized by the legal system which only contributes to further differentiation between the instrumental and symbolic rationality of law.

## 1.5 Part II: Symbolic Approaches to Biolaw: Biolaw as a Symbolic Order

### *1.5.1 Theme*

The second part of this volume examines legal regulation of emerging biomedical technologies from a symbolic perspective. The human body has been surrounded by symbolic-cultural values and rituals since the beginning of civilization. These symbolic values are also reflected in law. Secularization processes notwithstanding, the human body and derived materials still have a special legal status, distinguishing the body from ordinary legal objects. As a result, there are certain legal restrictions to what one is allowed to do with one's body or reproductive materials. More specifically, the gift approach to the human body (Mauss 2000; Titmuss 1997) remains an influential normative framework for the legal regulation of this field.

However, these legal-symbolic visions of the body are increasingly being challenged. Since the emergence of biomedical technologies, human body materials have acquired a new scientific, medical and even commercial value. One of the effects of these developments is a gradual commodification of the human body, a tendency which is reinforced by the much discussed rise of global markets in human tissues and reproductive materials (Dickenson 2007; Waldby and Mitchell 2007; Goodwin 2010).

This biomedical transformation of the human body has led to a fundamental interrogation and contestation of traditional legal representations and symbolizations of the human body. As human biological materials to an increasing extent circulate within contemporary bioeconomies as tradable commodities, the question has arisen whether this technological objectification should also lead to a legal objectification of the human body. Indeed, a burgeoning literature has come into existence which discusses the legal status of human biological materials, and the question to what extent these materials can be regarded as the object of property rights (e.g. Hoppe 2009; Goold et al. 2014).

Much of the current controversy in biolegal academia seems to go back to the fact that biomedical developments touch upon the foundational categories of our symbolic order which surface in our dealings with the body: the distinction between person and thing, organism and machine, and gift and commodity (Smits 2006; De Dijn 2003; Pessers 2005). This "dedifferentiation, through biotechnology, of deep-rooted categoric distinctions", in Habermas' words (2005), also necessitates renewed reflection upon the symbolic dimensions and effects of biomedical legislation.

From a symbolic perspective on biolaw, the following questions become of central importance. Do biomedical technologies show that we should be careful to uproot existing symbolic categories and values (Sandel 2009; Habermas 2005); that a new vocabulary and process of resymbolization is required (Hoeyer 2013); or rather that these symbolic representations have become redundant (Iacub 2002; Bortolotti and Harris 2006)? Moreover, one may ask whether law should fulfill a special role within this process. Does law have a special, mediating function in the symbolical and cultural representation of novel hybrid 'entities' (Jasanoff 2011; Latour 1991; Hoeyer 2013), such as human stem cells, tissue engineered products and egg cells? Can law even be regarded as a prime force in the formation of an *imaginaire social* to represent these biomedical artifacts of human origin (Supiot 2005)? Such an approach would shed light on the potential of symbolic legislation in its positive meaning for the regulation of biomedical technologies.

However, a strategy of pragmatic tolerance within regulation of bioethical matters seems to be on the rise. Due to a variety of factors, biolegal regulation is often not effective, and therefore referred to as symbolic legislation in the negative sense. National restrictions on the use of biomedical technologies, for instance, are increasingly evaded by traveling abroad, as evidenced by the rise of new forms of medical tourism, such as reproductive and transplant tourism (Van Beers 2015). Both legal officials and legal scholars have responded to these forms of tourism by adopting a pragmatic attitude, favoring regulation instead of prohibitions or restrictions (e.g. Pennings 2004). From this pragmatic perspective, in which only the practical effects of law are emphasized, the symbolic dimensions and effects of biolaw are hardly relevant.

### *1.5.2 Contributions*

Part II starts with a chapter by **Jonathan Herring**. Herring offers a critical examination of the prevailing legal view of the human body and human biological materials as objects of property rights. This predominant approach in legal academia is usually defended through pragmatic arguments, which ignore the symbolic value of the human body. Herring's main argument is that the property approach rests on a false understanding of what it means to have a body, and that this false image subsequently reinforces certain misplaced attitudes towards the human body. For instance, property approaches disregard the fact that our bodies are essentially interconnected, communal, leaky, vulnerable and in constant flux. According to Herring, the main problem lies with the individualized view of the self which underlies the property approach. Instead, he proposes to adopted a more relational understanding of both the self and the human body.

In the subsequent chapter, **Herman De Dijn** traces the current lack of understanding of the special status of the human body, as also analyzed and criticized by Herring, back to the emergence of a revisionist type of moral reasoning, which can be recognized in contemporary bioethics and biolaw. This revisionist approach aims to solve complex bioethical and biolegal problems by a combination of scientific facts and purely rational normative principles. From that perspective, taboos, traditions, symbolic distinctions and references to lived experience no longer matter. Instead, revisionist ethics heavily relies on instrumental and pragmatist ways of thinking, a tendency which is exacerbated by the influences of market thinking. According to De Dijn, this revisionist attitude has led to an overall inability to critically reflect on the most important normative dimensions of technological developments. As an illustration of this tendency he offers an analysis of recent debates on egg cell markets.

Where De Dijn focuses on markets in egg cells, **Klaus Hoeyer**'s contribution discusses the commercialization of human biological materials in general. Although public and scholarly debates on the regulation of markets in human biological materials are heavily polarized, most participants seem to agree on the following: international laws which pose restrictions to the emerging 'bioeconomy' are merely symbolic, that is, legal retorics without real effects. Indeed, the European bioeconomy has thrived inspite of the numerous provisions in European law which are based on the principle that the human body and its parts shall not give rise to financial gain. Hoeyer's contribution, however, argues that this *communis opinio* is false: the symbolism involved in international laws which curb markets in human biological materials has real effects, even if these effects are probably different from what one would expect. More specifically, even if existing laws which regulate the bioeconomy are characterized by contradictory governmental ambitions and claims, it is exactly this friction which has performative effects.

**Britta van Beers** addresses the positive symbolic effects of biomedical laws from a different angle. Her main argument is that legal discourse fulfills an essential role in the cultural-symbolic process of coming to terms with the hybrid entities which are brought into existence by biomedical technologies, such as human-animal cybrids, frozen embryos and artificially grown human tissues. Admittedly, other systems of value, such as economic, religious or medical discourse, also contribute to this process of symbolization. However, when the symbolic orders of these value systems collide, as it generally the case in biomedical issues, it is ultimately the legal system which has to balance and mediate between these competing values and symbolisations. As a result, legal discourse has gained a certain autonomy in the symbolization process of biomedical hybrids.

**Dorien Pessers** offers reflection on the special, symbolic function of law by analyzing the symbolic meaning of legal subjectivity within the regulation of biomedical developments. In his essay *Rules for the Human Zoo*, Peter Sloterdijk argues that traditional anthropotechnologies, such as reading and education, are now gradually being replaced by technological anthropotechnologies, such as medical biotechnology. Pessers, in her turn, argues that the law can also be regarded as an 'anthropotechnology'. According to her, the process of legal subjectification is in essence the symbolic insertion of new-borns into the community of legal subjects. This second birth, as a subject in the symbolic, transcendent order of law, contributes in fundamental ways to the humanization of human beings and gives rise to the *homme rêvé* of human dignity, which is at the root of law's symbolic order. Biomedical technologies, however, seem to promise a shortcut in this process of humanization, through the prospect of realizing this *homme rêvé* by technological means. Accordingly, post-humanist philosophers propose to strip law of its transendent dimensions. As Pessers argues, this posthumanist shortcut is not without risk, as without its transcendent dimensions, the law may loose its legitimacy altogether.

## 1.6 Part III: Legislative Strategies: Regulating Biomedical Developments from a Symbolic Perspective

### *1.6.1 Theme*

The third part discusses the characteristic elements of the legislative process drafting symbolic legislation and the legislative practice in which general clauses are interpreted. It focuses on the legislative processes of the regulation of bioethical issues. The legislative process plays an important role in implementing symbolic legislation and developing new frameworks and vocabulary to express the symbolic value of the legal norms at hand. In the communicative or interactive approach, drafting symbolic legislation requires a specific kind of legislative process in which communication is emphasized. In this part, two central features of the interactive approach are critically analyzed: the call for broad participation in the legislative process and for permanent norm development.

In order to give meaning to the general clauses contained in symbolic legislation, the legislature has to engage in processes of communication with the norm addressees. Law-making becomes a horizontal process instead of a top-down approach. However, how to structure the communication in the legislative process in such way that 'meaningful' symbolic legislation can be designed? Who needs to be involved in the communication process: experts and/or citizens? What is the central driving force of the communication: pluralistic expertise or democratic decision-making?[2] Who represents the public? Does input from experts only involve scientific expertise? How can we take into account the normative questions that are raised by regulatory issues on bioethical issues?

Moreover, because of the communicative function of symbolic legislation, the legislative process does not end with the promulgation of the law; it becomes an ongoing process. Legal norms have to be responsive to the rapid developments in life sciences and in society. Therefore, they need to be interpreted in critical exchange with these new developments. Also a broad involvement in the development of the norms has to be promoted. It is important that all those involved have the opportunity to participate in the creation and application of the law. Otherwise, symbolic legislation risks to remain a dead letter. But how can we keep legal norms open for discussion?

### *1.6.2 Contributions*

The legislative process plays an important role in implementing symbolic legislation and developing new frameworks and vocabulary to express the symbolic value of the legal norms at hand. **Sigrid Sterckx & Julian Cockbain** review recent

[2] See Jasanoff (1998).

developments in patent law in the US and Europe in relation to the patenting of 'isolated' human body materials, noting the tension in patent law between its general aspiration of promoting the (bio)economy while at the same time ring-fencing products of nature from effective monopolization by patentees. A central feature of the law is that patents should be granted for 'inventions' but not for 'discoveries', two symbolic terms that lack clear definitions. In this context, Sterckx & Cockbain conclude that hard law has been used in Europe to define a boundary between these categories that is industry-favorable – mere isolation transforms a discovery into an invention.

In their contribution, **Robert Lee & Elen Stokes** pay attention to the communication process of symbolic legislation and how the law acquires its meaning. They examine the symbolic qualities of the EU regulation of nanotechnologies. In particular, they ask whether and to what extent the categorization of the different effects of symbolic legislation – functional and communicative – is reflected in the EU's regulatory approach to this field. Lee & Stokes distinguish between the regulatory responses of the European Commission and the European Parliament to nanotechnology, and illustrates how those responses may have different symbolic effects. They conclude that the balance of existing regulatory arrangements is still tilted in favor of symbolism in its narrower, functional sense.

**Nicolle Zeegers** explores the differences between European member States' and interest groups' positions concerning the moral status of the early human embryo. The European Citizen's initiative *One of Us* has recently pleaded for a total ban on human stam cell research. On the basis of this plea, Zeegers investigates whether the EU norms for funding research with human embryos have been established in a democratically legitimate way by elaborating on pluralist theory incorporating the idea of deliberate democracy. According to her, the EU decision-making process in this are already followed to some extent the interaction that characterizes the interactive approach to legislation. However, she challenges the *dynamics* that such decision making offers by developing an argument in which the basic condition of an ethos of controversies (see Poort, Chap. 5) are translated in tools to determine whether the achievements are democratically legitimate.

**Lonneke Poort & Bernice Bovenkerk** discuss the role of experts in the legislative process. They question whether expert involvement only refers to scientific expertise and how normative questions raised by the regulatory issues on bioethical matters can be taken into account. Poort & Bovenkerk focus on the symbolic role that ethics committees play. Disagreement about complex policy issues can often be traced back to fundamental value differences. Governments tend to avoid political conflicts based on these value differences. In order to avoid political conflicts, they can resort to either seeking advice or even deferring decision-making to expert committees, in particular ethics committees. Poort & Bovenkerk argue that governments and the public foster wrong expectations regarding the role and mandate of ethics committees. Normative expertise is essentially something different than scientific expertise. Therefore, it is necessary to explicate the roles the various players have and to define what is to be expected from them. They conclude that ethics committees can fulfill a better communicative role when they act as moderator of the debate.

## 1.7 Final Note

We hope that this volume, through its combination of symbolic legislation theory and reflection on biomedical regulation, is able to give a new impulse to the academic debate on the symbolic dimensions of law. We are grateful that the contributors to this book, coming from various academic disciplines, were prepared to use a different theoretical lens than they were used to, or to broaden and combine various theoretical lenses to come to a new understanding of the relation between symbolic legislation and biolaw. As editors, we are very happy with the richness of the various contributions and the attempts made to confront or combine biolaw and symbolic legislation with each other. Hopefully, the present volume will also encourage the reader to reflect on possible relations between these fascinating legal phenomena.

## References

Arnold, Thurman W. 1938. *The symbols of government*. New Haven: Yale University Press.

Aubert, Vilhelm. 1966. Some social functions of law. *Acta Sociologica* 10(1/2): 98–120.

Ayres, Ian, and John Braithwaite. 1995. *Responsive regulation. Transcending the deregulation debate*. Oxford: Oxford University Press.

Bortolotti, Lisa, and John Harris. 2006. Embryos and eagles. Symbolic value in research and reproduction. *Cambridge Quarterly of Healthcare Ethics* 15: 22–34.

Brunnee, Julia, and Stephan Toope. 2010. *Legitimacy and legality in international law. An international account*. Cambridge: Cambridge University Press.

Commaille, Jacques. 1984. Le divorce en Europe occidentale. La loi et le nombre. Présentation d'un ouvrage. *Population* 39(2): 357–361.

Council of Europe. 1997. Convention on human rights and biomedicine. CETS No. 164.

Dickenson, Donna. 2007. *Property in the human body, feminist perspectives*. Cambridge: Cambridge University Press.

De Dijn, Herman. 2003. *Taboes, monsters en loterijen. Ethiek in de laat-moderne tijd*. Kapellen/Kampen: Pelckmans en Klement.

Edelman, Murray. 1976. *The symbolic uses of politics*. Urbana: University of Illinois Press.

Goodwin, Michele Bratcher (ed.). 2010. *Baby markets. Money and the new politics of creating families*. Cambridge: Cambridge University Press.

Goold, I., K. Greasley, J. Herring, and L. Skene (eds.). 2014. *Persons parts and property. How should we regulate human tissue in the 21st century?* Oxford: Hart.

Habermas, Jürgen. 2005. *Die Zukunft der menschlichen Natur. Auf dem Weg zu einer liberalen Eugenik?*. Frankfurt a.M.: Suhrkamp.

Hertogh, Marc. 2000. De wondere wereld van de wetgever. Feit en fictie van communicatieve wetgeving'. In *De overtuigende wetgever*, ed. Bart van Klink and Willem Witteveen, 45–60. Zwolle: W.E.J. Tjeenk Willink.

Hertogh, Marc. 2005. The social making and social working of legislation. In *Social and symbolic effects of legislation under the rule of law*, ed. Nicolle Zeegers, Willem Witteveen, and Bart van Klink, 163–181. Lewiston: The Edwin Mellen Press.

Hoeyer, Klaus. 2013. *Exchanging human bodily material. Rethinking bodies and markets*. New York: Springer.

Hoppe, Nils. 2009. *Bioequity – property and the human body*. Farnham: Ashgate.

Iacub, Marcela. 2002. *Le crime était presque sexuel*. Paris: Flammarion.

Jasanoff, Sheila (ed.). 2011. *Reframing rights. Bioconstitutionalism in the genetic age*. Cambridge: MIT Press.

Latour, Bruno. 1991. *Nous n'avons jamais été modernes. Essai d'anthropologie symétrique*. Paris: La Découverte.
Lindahl, Hans. 2000. Communicatieve wetgeving en volkssoevereiniteit. In *De overtuigende wetgever*, ed. Bart van Klink and Willem Witteveen, 173–188. Zwolle: W.E.J. Tjeenk Willink.
Mauss, Marcel. 2000. *The gift. The form and reason for exchange in archaic societies*. New York: W.W. Norton & Company.
Pennings, Guido. 2004. Legal harmonization and reproductive tourism in Europe. *Human Reproduction* 19(12): 2689–2694.
Pessers, Dorien. 2005. *Menselijke waardigheid en het persoonsbegrip in het recht*. Utrecht: Lemma.
Poort, Lonneke. 2013. *Consensus & controversies in animal biotechnology. An interactive legislative approach to animal biotechnology in Denmark, Switzerland, and the Netherlands*. Den Haag: Eleven International Publishing.
Sandel, Michael. 2009. *The case against perfection. Ethics in the age of genetic engineering*. Cambridge: Belknap Press of Harvard University Press.
Smits, Martijntje. 2006. Taming monsters. The cultural domestication of new technology. *Technology in Society* 28(4): 489–504.
Stamhuis, Jellienke. 2005. Communicative law. A quest for consensus. In *Social and symbolic effects of legislation under the rule of law*, ed. Nicolle Zeegers, Willem Witteveen, and Bart van Klink, 277–297. Lewiston: The Edwin Mellen Press.
Stamhuis, Jellienke. 2006. Conflicting interests in corporate regulation. An exploration of the limits of the interactionist approach to legislation in employee participation and corporate governance (diss. Groningen). Groningen: S.N.
Supiot, Alain. 2005. *Homo Juridicus. Essai sur la fonction anthropologique du droit*. Paris: Seuil 2005.
Van Beers, Britta. 2015. Is Europe 'giving in to baby markets'? Reproductive tourism in Europe and the gradual erosion of existing legal limits to reproductive markets. *Medical Law Review* 23: 103–134.
Titmuss, Richard. 1997. *The gift relationship. From human blood to social policy*. New York: The New Press.
UNESCO. 1997. *Universal declaration on the human genome and human rights*. Paris: UNESCO.
Van der Burg, Wibren. 2005. The irony of symbolic crusades. The debate on opening up civil marriage to same-sex couples. In *Social and symbolic effects of legislation under the rule of law*, ed. Nicolle Zeegers, Willem Witteveen, and Bart van Klink, 245–275. Lewiston: The Edwin Mellen Press.
Van der Burg, Wibren, and Frans W.A. Brom. 2000. Legislation on ethical issues. Towards an interactive paradigm. *Ethical Theory and Moral Practise* 3: 57–75.
Van Klink, Bart. 1998. *De wet als symbool. Over wettelijke communicatie en de Wet gelijke behandeling van mannen en vrouwen bij de arbeid (diss. Tilburg)*. Deventer: W.E.J. Tjeenk Willink.
Van Klink, Bart. 2014. Symboolwetgeving. Opkomst, ondergang en wederopstanding van een begrip. *RegelMaat* (*Journal for Legislative Studies*) nr. 29: 5–26.
Van Klink, and Willem Witteveen (eds.). 2000. *De overtuigende wetgever*. Zwolle: W.E.J. Tjeenk Willink.
Waldby, C., and R. Mitchell. 2007. *Tissue economies. Blood, organs, and cell lines in late capitalism*. Durham/London: Duke University Press.
Westerman, Pauline. 2005. Some objections against an aspirational system of law. In *Social and symbolic effects of legislation under the rule of law*, ed. Nicolle Zeegers, Willem Witteveen, and Bart van Klink, 299–315. Lewiston: The Edwin Mellen Press.
Witteveen, Willem. 1991. De jacht op de wet. In *Wat maakt de wet symbolisch?* ed. W.J. Witteveen, P. van Seters, and G. van Roermund, 115–136. Zwolle: W.E.J. Tjeenk Willink.
Zeegers, Nicolle, Willem Witteveen, and Bart van Klink (eds.). 2005. *Social and symbolic effects of legislation under the rule of law*. Lewiston: The Edwin Mellen Press.
Zielke, Andreas. 1980. *Die symbolische Natur des Rechts. Analyse der Rechtssoziologie Niklas Luhmanns*. Berlin: Duncker & Humblot.

# Part I
# Symbolic Legislation: The Symbolic Quality of Law

# Chapter 2
# Symbolic Legislation: An Essentially Political Concept

**Bart van Klink**

## 2.1 Two Concepts of Symbolic Legislation

Symbolic legislation commonly has a bad name. In critical sociological studies, it refers to instances of legislation that are to a large extent ineffective and that serve other political and social goals than the goals officially proclaimed. A well-known example is the Norwegian 1948 Housemaid Law, studied by Vilhelm Aubert (1966). In Aubert's view, the legislation at hand was never meant to be effective, but was enacted in order to give recognition to the rights of housemaids on an immaterial or 'symbolic' level. It served to demonstrate that these rights were taken seriously, at least on paper. However, in practice nothing much changed in the position of housemaids. European and international environmental regulation and policy have often been analysed in terms of 'symbolic politics', because of the weak enforcement mechanisms it offers (see, for instance, Matten 2003 and Cass 2009). Recently, the Dutch Government has proposed to penalise foreigners who have no legal permit of residence in the Netherlands. According to the criminal law scholar Theo de Roos (2013, 9), this proposal is an example of symbolic legislation 'in the worst sense of the word'. He considers the penalisation of unauthorised residence to be merely symbolic, because it will be very difficult to enforce it and, moreover, it will probably not deter immigrants from coming to the country. As Arnold (1938) and Edelman (1976) have argued, the legal system in general can be seen as a collection of symbols that fulfil a power maintaining and status quo preserving function.

Since the nineties of the last century another concept of symbolic legislation is developed, in particular in Dutch legislation theory (see, e.g., Witteveen 1991, 2005; Van Klink 1998; Witteveen and Van Klink 1999; Van der Burg and Brom 2000; and Poort 2013). In this more recent and more positive understanding, symbolic

B. van Klink (✉)
Department of Legal Theory and Legal History, Vrije Universiteit Amsterdam,
De Boelelaan 1105, 1081 HV Amsterdam, the Netherlands
e-mail: b.van.klink@vu.nl



B. van Klink et al. (eds.), *Symbolic Legislation Theory and Developments in Biolaw*, Legisprudence Library 4, DOI 10.1007/978-3-319-33365-6_2

legislation is an alternative legislative technique that differs from the traditional top-down approach. The legislature no longer issues commands backed up with severe sanctions, as in an instrumentalist approach, but provides open and aspirational norms that are meant to change behaviour not by means of threat but indirectly, through debate and social interaction. General clauses, such as 'human dignity', are favoured because they are supposed to make the law more flexible. They are supported by relatively soft enforcement mechanisms which have to stimulate discussion and raise awareness in society. Other labels used for the same legislative approach or similar approaches are 'responsive regulation' (Ayres and Braithwaite 1995), 'communicative legislation' or 'communicative approach' (Van Klink 1998, chapter 3) and 'interactive legislative approach' (Van der Burg and Brom 2000; Poort 2013; and Van der Burg 2014).

The question that I will address in this chapter, is whether there is a real – in the sense of epistemologically real – difference between the two concepts of symbolic legislation. On what grounds can the allegedly negative and positive concepts be differentiated from each other? Is it possible that one instance of legislation can be classified as symbolic legislation both in the negative sense and in the positive sense? Are they two sides of the same coin or do they constitute mutually exclusive categories? As I will argue below, the distinction between the two concepts cannot be made on scientific grounds only, but involves considerations of a political kind. (What 'political' means in this context will be specified below.) So, in the final analysis, I consider symbolic legislation to be a political concept. It is a concept that is used to describe and evaluate political and legal practise, that is, the process of legislation as well as its products. However, the concept can only do its descriptive and evaluative work on the basis of certain political presumptions and ideological choices. As a consequence, symbolic legislation theory does not stand outside legal and political practise and its power struggles, but is an integrative part of this practise. It is my aim to reveal the often hidden political basis or bias underlying the concept of symbolic legislation and the differentiation between the two conceptions thereof.

What I offer here, is an exercise in political methodology based on a *transcendental*, not an empirical or a normative, argumentation. So my question is not: on what empirical grounds can it be established that symbolic legislation 'really' exists nor what is, normatively speaking, the best approach to symbolic legislation, but I want to know primarily what has to be presupposed in political terms, before an empirical or normative theory of symbolic legislation can be developed in the first place. To begin with, I will describe and compare the two different concepts or conceptions of symbolic legislation. In Sect. 2.2 I will analyse the current 'negative' concept, according to which symbolic legislation is a mere instrument of power used for dubious political purposes. In Sect. 2.3 I will present an alternative, 'positive' understanding of symbolic legislation, which I have developed in my earlier writings but which I have criticised later.[1] In Sect. 2.4 I intend to demonstrate

[1] Sections 2.2 and 2.3 are based on my PhD thesis (Van Klink 1998, chapter 2 and 3). On some points I have revised and updated the text (following Van Klink 2014). Moreover, I have added some examples from other contributions to this volume.

subsequently that the two concepts or conceptions of symbolic legislation cannot be distinguished from each other on purely scientific grounds. Finally, in Sect. 2.5 I will evaluate briefly the two opposing, critical sociological and communicative, approaches to symbolic legislation.

## 2.2 Symbolic Legislation 'in the Worst Sense of the Word'

In legal sociology, a specific instance of legislation is usually called 'symbolic', if a law is promulgated primarily to function as a symbol. Symbols are, in semiotic terms, a special kind of connotative signs, that is signs which, beside their literal or conventional meaning, convey another, secondary meaning (or connotation, see Eco 1984, 131–163). An established unit of expression (or *signifier*) and content (or *signified*) which constitutes an 'ordinary' sign, functions as the expression for a new content. One can think of the sign 'sun' which not only has the primary meaning of a celestial body, but which also transfers connotations such as fertility, warmth and recreation. Like other connotative signs, a symbol contains a layered semantic structure. However, the secondary meaning that it conveys is of a more general and indeterminate nature, compared to other, more conventional connotations. There is no code available that may determine or guide the interpretation of the symbol. As Nöth (1999, 119) indicates, symbols represent an 'immaterial content of importance to human life.' In many cases, values of a 'higher', spiritual order are at stake. For example, after his release, Nelson Mandela became a symbol for ideals such as justice, equality and forgiveness. At the same time, the literal meaning does not fully disappear, though it is less dominant: Mandela was also the person who resisted the Apartheid regime, was imprisoned on Robben Island for many years, became president after that and died in 2013 at the blessed age of 95. In due course, these plain facts of his life are moved to the background in favour of the higher values that the former politician symbolises for many people all over the world. A well-known and somewhat worn-out legal symbol is that of Lady Justice, who represents an impartial and fair trial.

When an instance of legislation is conceived as symbolic legislation, the law at hand signifies something else and something more than appears at first sight. By the sheer promulgation of the law, the legislature aims at construing a sign whose meaning transcends the enacted rules (Kindermann 1989, 265). Values are added to it, so that the law acquires 'added value' on a semantic level for some group(s) in society or the whole of society. In the conventional understanding, a law is collection of substantive provisions that prescribe, forbid or permit certain types of behaviour. These substantive provisions are often supported by provisions that have to secure that the rules of behaviour are complied with, for instance by putting a sanction on non-compliance. In case of symbolic legislation this primary or literal layer of meaning is supplemented by a second layer of meaning. Thus, symbolic legislation is characterised, as the symbol as such, by a layered structure of meaning: on the primary or literal layer of meaning, we find the conceptual content of the substantive provisions (rules of behaviour) and the provisions to secure law compliance (that is, rules to enforce these rules of behaviour), whereas the secondary or symbolic layer contains immaterial values that are attached to this conceptual content. The secondary meaning that is attached to symbolic legislation, replaces or overshadows to a large extent the 'surface' or apparent meaning of the law. In order words, the law is not meant to be complied with, but its main purpose is to give expression to values in the political sphere.[2] For that reason the legislature deliberately fails to provide for sufficient means to enforce the law. Moreover, in many cases symbolic legislation consists of vague norms which are unclear and open to multiple interpretations, of even contradictory norms. The meaning that is given to the immaterial values expressed through symbolic legislation can be different for different groups in society and may change over time. As Aalders (1984) has shown, initially the environmentalist movement considered the Dutch Public Nuisance Act ('Hinderwet') to be a moral victory over the industry. Eventually, when it became apparent that the Act would remain largely ineffective, it became a symbol of the government's betrayal.

In order to determine whether a particular instance of legislation is symbolic, several criteria are used, taken either from the textual qualities of the law at hand (semantic criteria) or the context in which the legislative process takes place (pragmatic criteria). Semantic criteria are: in a symbolic law, the substantive provisions are not backed up with provisions to enforce them and this discrepancy cannot be justified on rational grounds (*criterion of discrepancy*); the text of the law is incomprehensible for the citizens who have to comply with it as well as for the legal and political actors who have to apply it (*criterion of obscurity*); and the rules of which the law consist can be interpreted in various ways (*criterion of vagueness*). Pragmatic criteria are: in the legislative process two or more groups with conflicting or incompatible interests are fighting each other (criterion of conflict of interest); one of the groups involved considers the enactment of the law as a moral victory over the other group (or groups) and a confirmation of its values (*criterion of politisation*); and society is in a state of emergency that calls for immediate governmental action (*crisis criterion*). If two or more of these criteria are met, it is likely that the law in question serves symbolic rather than instrumental goals.

In contrast to traditional instrumental legislation, symbolic legislation does not aim at enforcing the enacted rules of behaviour (direct effect) — especially by means of 'hard and fast' rules backed up with sanctions — and at reaching more fundamental goals (indirect effect). Different effects are associated with symbolic legislation, such as changing the status distribution or reconciling antagonistic groups in society, which have nothing to do with this kind of rule compliance and which therefore are called 'independent effects'.[3] Because of their political character, I have coined

[2] In this context, 'political' refers to the use or abuse of power by state officials for their own sake or their clientele's at the cost of the common good. In Sect. 2.4, I develop a more general concept of the political (of which 'dirty politics' is just one particular instance).

[3] The distinction between direct, indirect and independent effects I have borrowed from Griffiths (1978, 8–12).

these effects, which are mainly of an immaterial nature, 'negative-symbolic effects' (Van Klink 1998, 47). Generally speaking, symbolic legislation can be promulgated in order to attain three kinds of negative-symbolic effects: first, the confirmation of the value system defended by a particular group; second, the demonstration or simulation of power on the part of the government; and, third, the resolution of a social conflict between two (or more) groups or political parties. Three types of symbolic legislation correspond to these effects: status, illusionary and compromise legislation respectively (Kindermann 1988, 230–239). Status legislation, to begin with, aims at improving the prestige of a social group. According to Gusfield (1976), the American prohibition legislation from the beginning of the twentieth century, constituted an official recognition of the ascetic, Protestant morality at the expense of the more hedonistic lifestyle of catholic Irish immigrants. Status legislation has primarily an expressive function: it confirms values that are essential for those involved in the legislative process. This expression of values does not take place for its own sake,[4] but in order to change the status distribution in society. Secondly, by promulgating an illusionary law, the government tries to create the impression that a crisis situation is under control or a (possible) threat to society is averted. The Dutch bill that penalises unauthorised residence in the Netherlands, as discussed in Sect. 2.1, can be characterised as illusionary legislation, because it probably will not solve the problem of the illegal immigration but it sends out a signal to society that the government does not approve of it. In Chap. 14 of this volume, Lee & Stokes argue: 'Legislating to regulate nanomaterials suggests symbolically a capacity to control not just these materials but the prospects to which they will give rise. This may well be illusory, raising the question of what it is we hope for in legislation, especially when that legislation represents a political mêlée in which power is at issue and where regulation may be seen to facilitate dominant political choices in favour of the technology.' In other words, they consider the EU regulation of nanotechnologies primarily to be a simulation of power. An illusionary law can also be used in order to attract votes during an election campaign: just before the elections, the government wants to show that is capable to act and to achieve concrete results. Ad hoc laws that are enacted for the purpose of one particular case only offer another example (Jasiak 2010). Thirdly, in the case of compromise legislation antagonistic groups or parties are reconciled by means of a would-be compromise. In Aubert's classical study (Aubert 1966), the Norwegian 1948 Housemaid Law is presented as a compromise in this sense between two opposing parties in Parliament: a progressive party that wanted to improve the legal position of housemaids on the one hand and a more conservative party that defended the interests of housewives on the other hand. In the same vein, Aalders (1984) analyses the Dutch Public Nuisance Act as a 'peace pipe' that had to pacify members of the environmentalist movement and the industry.

As these examples show, symbolic legislation may very well serve instrumental goals. Therefore, contrary to popular belief, symbolic legislation is not the antithesis to instrumental legislation and does not simply equal non-effective legislation.

[4] As in the case of expressive legislation, see Sect. 2.3.

Symbolic laws do affect social reality, though in different ways than 'ordinary' instrumental laws and in order to achieve different goals, such as the pacification of antagonistic groups, the simulation of power in emergency situations, or the (re) distribution of status in society.

## 2.3 Symbolic Legislation as a 'Communicative Framework'

In the nineties of the last century another, positive conception of symbolic legislation was developed, in particular in Dutch legislation theory.[5] It was also meant to offer an alternative to the instrumental regulatory style, though not by way of deceit and power simulation, as in the negative conception, but through communication and interaction. According to this communicative or interactive approach, a law can be called 'symbolic' in the positive sense, if a law has acquired an extraordinary meaning within the legal and political community. The law is not merely a set of rules, but it is also a symbol for something higher, more valuable. It gives expression to values which are fundamental for the community, for example human dignity, equality or environmental protection. Moreover, a law is called 'symbolic' in the positive conception because of the general clauses it contains. Symbolic legislation offers a 'communicative framework'[6] consisting of general viewpoints by means of which people can settle their differences. General clauses function as symbols that are interpreted again and again in the light of new circumstances or changed legal and moral opinions. Since the general clauses which makes up symbolic laws consists embody important moral aspirations, I have called them 'aspirational norms' (Van Klink 1998, 109). Aspirational norms have no fixed meaning and have to be developed in an ongoing interaction between various legal, political and social actors. In symbolic legislation, two different messages are transmitted at the same time: the open, aspirational norm itself (the primary norm) as well as the assignment directed to the interpretative community to apply this norm and make it more concrete (the secondary norm).[7]

In order to identify symbolic legislation in the positive sense several, semantic and pragmatic, criteria have be met. Semantic criteria include the *criterion of openness*, which indicates that the aspirative norms enacted in symbolic legislation offer legal actors (the judge in particular) much, but not complete discretion in interpreting the law; the *criterion of positioning*, according to which the special value of the law can be deduced from the central position the law occupies within the existing hierarchy of norms; and the *criterion of discrepancy*, which holds that in a symbolic law there is a gap between the substantive provisions and the provisions to enforce them, which could not have been easily and reasonably avoided. The rationale

[5] This research project started with the edited volume Witteveen, Van Seters and Van Roermund (1991).

[6] I have borrowed this notion from Minow (1990, 294).

[7] See Allott (1980, 33). Below, the notion of interpretative community will be explained.

behind this approach is that in certain matters, persuasion is a more effective instrument than punishment. The pragmatic criteria are the *communication criterion*: a symbolic law plays an important part in the public debate on the matter regulated by the law; *the criterion of inclusiveness*: the ends the law pursues and the means chosen are widely supported by society; and finally the criterion of *symbolic working*: the law succeeds gradually in achieving the goals intended through communication and interaction. If all of these criteria are met to a greater or a lesser extent, the legislation at hand requires an important symbolic value.

Symbolic legislation in the positive sense fulfils various useful functions in society. One of the most important functions is the epistemic function: a symbolic law offers a vocabulary that affects the way in which legal and political actors perceive reality. Reality is accessed through the concepts and distinctions provided by the law. The anthropological function that Van Beers mentions in her contribution (Chap. 11) can be seen as an application of the epistemic function to biolaw: by introducing new categories and distinctions and redefining existing ones, biolaw provides for 'the symbolic mediation of the bare, biological facts of life (*zoè*) into the meaningful events of the lives that we lead (*bios*).' According to De Dijn (Chap. 9), '"behind" the law there were and are at least some deep ethical conceptions related to "thick" notions such as responsibility, human dignity, etcetera, themselves incomprehensible without their link to a human life world with its system of symbolic categories and distinctions (e.g., between humans and animals, life and death, family and non-family, etc.).' A symbolic law has also a rhetorical function, which consists of offering a source of arguments. In debates concerning law, these arguments or *topoi* can be used as trumps. Sterckx & Cockbain argue in their contribution (Chap. 13) that, through the exchanges between political actors and the courts, European patent law has influenced the public debate. Moreover, a symbolic law constitutes an interpretive community,[8] that is a group of legal and political actors as well as citizens who are involved in the application of the law. This is the so-called constitutive function. The communicative framework that the law offers facilitates mutual understanding among members of the interpretive community. According to Sterckx & Cockbain, European patent law has constituted an interpretative community consisting of, among others, politicians, officials, judges, patentees, patent applicants, opponents and alleged infringers, industry and patent attorney associations, NGO's, academics and also citizens. However, in their view, this community has not succeeded yet in defining key concepts in patent law, such as 'invention'.

By giving voice to essential values of a community, symbolic legislation fulfils an expressive function. In his contribution, Priban (Chap. 7) states that the symbolic dimension of legislation, in particular in the context of constitution-making, primarily consists of the expression of social identity: 'Legal symbolism is best understood as the legal system's specific reflection of social expectations of communal togetherness, goodness and justice.' However, he does not believe in 'law's

[8] Fish (1980, 171) uses the notion of 'interpretative community' in the context of literature studies and literary criticism. It is applied here to the legal and political sphere.

capacity to promote social cohesion by recognising moral values.' Several authors in this volume refer to the expressive function of biolaw. For instance, Herring (Chap. 8) argues that biolaw should incorporate a relational concept of the human body: 'We need a legal model that appreciates and promotes this more communal, mutable, interdependent nature of bodies.' In his view, biolaw does not merely serve instrumental goals, but has an important symbolic value as well: 'The legal classification of bodily material is not simply of pragmatic assessment, but says something profound about our understanding of bodies and ourselves.' According to Van Beers (Chap. 11), human dignity, 'as one of the central values of biolaw', 'can be understood as a legal-symbolic representation of what it means to be human.' This does not mean that this kind of legislation 'only' represents an expression of values; it also aims at affecting social reality.[9] However, the means by which it tries to steers the behaviour of citizens differ fundamentally from the approach chosen in the case of instrumental legislation. Instrumental legislation is based on an autonomous, authoritarian regulatory style: the ruler gives commands to the ruled, which ought to be obeyed unconditionally. The substantive provisions are written in a concrete and unambiguous language and are backed up, in case of non-compliance, by severe sanctions. By contrast, the communicating legislature chooses a less hierarchical and more interactive approach. It introduces abstract and multi-interpretive norms whose content has to be further determined by the members of the interpretive community. Moreover, structures for deliberation are created which stimulate the communication between the executive, the judiciary and the addressees of the law. For instance, in the enforcement of the Dutch Equal Treatment Act a central role is assigned to the Board for Human Rights and Equal Treatment which only has an advisory role in conflicts between two parties in matters of discrimination (see Van Klink 1998, chapter 4). According to Zeegers (see this volume, Chap. 15), legislation concerning the use of human embryos in research, implemented in the various European countries, can be seen as symbolic, both in the sense that it recognises the special legal status of human embryos (expressive function) and that it adopts a deliberative and interactive legislative approach.

Three types of symbolic legislation can be distinguished: constitutive legislation, educational legislation and expressive legislation. By means of constitutive legislation the legislature codifies fundamental legal, moral and political values that are shared widely in society. In many cases, written constitutions consist of constitutive legislation of this kind. For instance, in the first article of the German constitution the value of human dignity is codified in order to confirm the sanctity of human life against the atrocities of the Nazi regime (see Lembcke 2013). Furthermore, in constitutive legislation a division of roles is established among the legal and politi-

[9] In his contribution, Hoeyer (Chap. 10) also points to the instrumental value of symbolic legislation: '(…) we need to appreciate the constitutive social effects of legal symbols, even when the laws as such do not seem to work as intended.' See also Lembcke's contribution (Chap. 6) on this point.

cal actors who have to implement and realise these values.[10] Educational legislation is directed at changing mentality.[11] Contrary to constitutive legislation, it has a modifying character: it is introduced in order to de-automatise current patterns of thinking and evaluating and corresponding patterns of behaviour in a non-instrumental way. The effects pursued, which lie in the realms of cognition, language use and attitude, can be called positive-symbolic effects (Van Klink 1998, 47). In comparison to constitutive legislation, education laws are more controversial in society although they are expressive of widely shared ideals. Gender equality legislation offers well-known example: many people will support equality of men and women on a general level, but opinions differ fundamentally regarding the means to be chosen to achieve this – through preferential treatment, gender quota or less intrusive measures. The main purpose of expressive legislation is to give recognition to fundamental values. While this official recognition constitutes a value in itself, it does aims at influencing behaviour. In his study of Dutch embryo legislation, Wibren van der Burg (1996, 80) states: 'Legislation can be (…) an expression of the values that are essential to this society. Through legislation, it communicates those values to its members, expecting that they will take these values as guidelines.'

On a superficial level, symbolic legislation in the positive sense may seem to share some characteristics with its symbolic legislation in the negative sense, with respect to, for instance, the phrasing of the norms and the political circumstances in which they are arise.[12] However, they differ in many other respects substantially, in particular regarding the connotative meanings which are transmitted (dubious political aims versus ideals of a high moral standard), the kind of norms used (vague versus open norms) and the effects pursued (negative-symbolic effects such as electoral success versus positive-symbolic effects such as attitude change). Because of these fundamental differences and to avoid misunderstandings, I have proposed to rename 'symbolic legislation in a positive sense' *communicative legislation*, whereas 'symbolic legislation in a negative sense' may simply remain *symbolic legislation* (Van Klink 1998, 90). Van der Burg (2000, 2005), Van der Burg and Brom (2000) and Poort (2013) prefer to speak of an 'interactive legislative approach,' which promotes on on-going interaction between officials and citizens on all levels of norm creation and application.[13]

A communicative or interactive approach to legislation may possibly conflict with the Rule of Law. General clauses are controversial, because laws have to be

[10] For a more detailed analysis of constitutive legislation, see Witteveen and Van Klink (1999, 130–133).

[11] Cotterrell (1984, 57) uses the notion of 'educational legislation', whereas in Dutch legislation theory it used to be more common to speak of 'mentality legislation'.

[12] In hindsight, I consider these resemblances to be less superficial than I did while writing my PhD thesis. Now, I see them as indicative of the political nature of the distinction between the two concepts of symbolic legislation (see Sect. 2.4).

[13] For a general introduction to the interactive legislative approach, see Poort (2013, Chaps. 1 and 3) and Van der Burg (2014). See also Van der Burg's contribution to this volume (Chap. 3).

precise and clear.[14] Otherwise, the citizens cannot orient themselves to the rules. Moreover, too much power is transferred from the legislature to the administration and the courts. However, in certain circumstances it is inevitable or even desirable to leave the texture of the law open. This is especially the case when, to begin with, the legislature has insufficient technical knowledge to formulate clear and distinct rules (as in environmental or IT legislation); subsequently, the matter in hand is too complex conceptually to be regulated in detail (see the standard reasonableness and fairness in civil law); and, finally, the matter is ethically sensitive and controversial (as is the case with legislation on euthanasia or embryos). If the problems of a technical, conceptual and/or ideological nature are solved gradually after the promulgation of the law, it is possible to make the rules more concrete. Subsequently, it may be argued that in some cases it is desirable to leave the texture of the law open in order to preserve the responsive character of the law. The legislator does not place himself above society, but rather prefers a dialogue. General clauses enable legal and political actors to react flexibly, which does not mean uncritically, to changing opinions concerning law and justice in society. Finally, openness does not equal vagueness. Whereas general clauses are usually not very helpful, open norms — possibly in combination with more concrete rules — do guide, to a certain extent, the interpretive activities of the executive and the judiciary. It is then necessary that members of the interpretive community take seriously their collective responsibility to elaborate the given norms 'in the spirit of the law.'[15]

## 2.4 The Conflict of Interpretations

Obviously, 'negative' and 'positive' are no neutral but evaluative terms. They express the normative stance the investigator takes towards the object investigated. The concepts of symbolic legislation in the negative sense (or, as De Roos put it, 'in the worst sense of the word') and symbolic legislation in the positive sense (conceived as a 'communicative framework') seem to correspond to good and bad legislation respectively. However, things are not that simple. As some scholars have argued, symbolic legislation may serve useful political and social functions. When a society is in serious trouble and the feelings are running high, a would-be compromise or a merely verbal expression of values is sometimes the best or even the only possible way out (see also Lembcke in Chap. 6). This is the so-called safety valve function of symbolic legislation. Conversely, symbolic legislation in the positive

[14] According to Fuller (1969, 63), clarity of the law is one of the requirements of good law making. In his view, this does not rule out the possibility of using standards such as 'good faith' and 'due care' in the text of the law: 'Sometimes the best way to achieve clarity is to take advantage of, and to incorporate into the law, common sense standards of judgment that have grown up in the ordinary life outside legislative halls.' However, this does not mean that the legislator can 'always safely delegate his task to the courts or to special administrative tribunals' (Fuller 1969, 64).

[15] On the compatibility between the interactive approach and the requirements of the Rule of Law, see further Van der Burg's contribution (Chap. 3).

sense is never presented as a panacea for every social problem. A communicative or interactive approach is advocated only under specific conditions and for specific cases. As indicated above, I have argued that problems of a technical, conceptual or ideological nature may justify the use of general clauses (Van Klink 1998, see Sect. 2.3). Van der Burg and Brom (2000) and Poort (2013) have designed their interactive legislative approach exclusively for ethically sensitive matters, such as the embryo selection, euthanasia and animal biotechnology. According to some critics (among whom Griffiths 2005), communicative legislation equals bad legislation because it contains too vague norms and does not offer any serious means of enforcement. That does not change the fact though that every characterisation of symbolic legislation is based on, or at least inevitably contains, evaluative considerations; only *the kind of* evaluation may change – from negative to positive, or vice versa –, depending on who evaluates and/or what is evaluated. What exactly is the nature of this evaluation? And how does it make a qualification or disqualification in terms of symbolic legislation possible as well as a differentiation in two concepts or conceptions of symbolic legislation?

In my view, 'symbolic legislation' is an essentially political concept. By implication, I consider the distinction between the two concepts or conceptions of symbolic legislation to be political, as well as the communicative and interactive approaches which have been developed out of some notion of symbolic legislation and/or symbolic effects of legislation. Every symbolic legislation theory, implicitly or explicitly, presupposes some view on where regulative power should be located, how it should be distributed and executed, that is, how legislation should be fabricated and implemented, who has to be involved in the legislative process and in what way, which norms or values deserve legal recognition, and so on. For the purposes of the present analysis, I use the notion of the political in three different, but closely related senses, referring to: (i) a comprehensive normative view on the nature, aim and scope of regulatory power,[16] the division of power within the state (or some other social organisation), the relation between state and society, the values to be protected through regulation and so on (that is, a political philosophy or, in Marxist terms, an ideology); (ii) a political (not necessarily party-political) programme that translates the general and fundamental assumptions and aspirations into a set of specific prescriptions for the execution and division of regulatory power within the social organisation at hand and that identifies the central goals to be achieved; and (iii) a scenario that, on an even more concrete level, assigns to every actor or organisation a specific role in the regulatory process and selects the regulatory instruments (legislation or some other means) in order to achieve the political goals, laid down in the political programme. For instance, the political philosophy of liberalism has, besides freedom, equality as one of its core values (ideology); building on this value, equal treatment of men and woman can be identified as a concrete political goal which the state is expected to guarantee (programme) and which can be

[16] That is, the power in both senses of capacity and competency to create or maintain order within a certain social domain through the use of law or other means.

achieved by a means of, e.g. setting legal quota on the amount of women in higher positions or through policy instruments like subsidies (scenario).

### 2.4.1 *Symbolic Legislation in the Positive Sense: The Communicative and Interactive Approach*

Generally speaking, exponents of the communicative and interactive approach intend to make the process of law making and implementation more democratic and responsive. In opposition to supposedly traditional and instrumentalist notions of top-down steering in which the ruler issues unilaterally commands to the ruled, they conceptualise regulative poweras an interactive, two-way or bottom-up process, in order words, as a matter of interpretation, negotiation and communication. Opinions differ on the exact role and function of the state in its capacity of centralised legislative power. Some scholars consider the legislator to be just one of the many legal and political actors involved in the legislative process, and not necessarily the most important one (cf. Van der Burg 2005 and Poort 2013), whereas other scholars still assign to the legislative power a key role in initiating and co-ordinating the legislative process and eventually in determining and implementing the legal norms (cf. Witteveen and Van Klink 1999).[17] In both cases, the legislature appears to be a benevolent, non-authoritarian instance that responsibly and responsively, in a co-production with its citizens, makes laws and takes care of their implementation. Law and morality are seen as intertwined: though not identical they are necessarily connected, feeding and reinforcing each other like Siamese twins.[18] Ideally, the legal norm development parallels the moral norm development, so that law enforcement may no longer be needed: 'where moral norm development and legal norm development go hand in hand, actors working with legal rules in the field are likely to be in conformity with the new legal rules before these rules are enforced at all' (Poort 2013, 12). The values which are studied and, implicitly or explicitly, supported within a communicative and interactive approach are mostly of a social liberal kind, for instance equality of men and women, equal treatment of gay people, good labour conditions, and animal protection (see Van Klink 1998; Van der Burg 2005; Azimi 2007; and Poort 2013 respectively). However, since these progressive values have to be realised in a communicative and interactive way, one can never be sure that the law at hand is implemented in a good manner, that is, in accordance with the social liberal ideology. So there is a possible tension (which is not further theorised) between the adherence to certain substantive political values on the one hand and the formal ideal of democracy on other hand, which is taken to prescribe that the realisation of these values have to be outsourced, from the legislature to the courts or to society in general. Within the communicative and interactive approach

[17] This point is also discussed in the introduction (Chap. 1, Sect. 2.2).

[18] This metaphor is introduced in Van der Burg and Ippel (1994). On the relation between law and morality from an interactive perspective, see also Van der Burg (2003).

not much attention is paid to the dimension of violence which is a necessary part of every legislative act (not everyone can participate in the legislative process, not every voice can be heard and recognised).[19] The attempt to depolitise political power has, of course, important political effects: it gives scientific approval to a specific way of lawmaking and to the laws that are made that way and, thereby, it confirms and reinforces the given legal and political order. The existing power structure is studied from an internal perspective and constructive suggestions are made for improving its functioning, building on mainstream social liberal values.

### *2.4.2 Symbolic Legislation in the Negative Sense: The Critical Sociological Approach*

Legal sociologists who have studied symbolic legislation in the negative sense (such as Arnold 1938; Edelman 1976 and Gusfield 1976) or have criticised the positive conception of symbolic legislation (Griffiths 2005; Stamhuis 2005 and others) take, on the other hand, in general an external and very critical perspective towards the state and the status quo. In their view, the legislature is not so much a communicator or facilitator but a manipulator of symbols. By officially recognising certain legal norms, it wants to transmit the message that the state is still in control or that it takes seriously a certain value, morality or group in society (such as the protestant morality in case of the temperance legislation in the US). However, because the state does not provide for any serious means of enforcement, the overt message is distrusted and reinterpreted as a sign of unwillingness or inability to change anything substantial in the current situation. Law and morality may be connected (though not necessarily), but this connection is perceived from a political perspective, that is as one that is established for political purposes only or predominantly. Political action is unmasked as a theatre play to which citizens are doomed to remain spectators. This semiotic strategy of debunking has a delegitimising effect on state power: the legislative process and its products are exposed as mere signifiers of 'dirty politics.' As a consequence, the given legal and political order is not confirmed and reinforced, as in the communicative and interactive approach, but is fundamentally called into question. The exact political agenda of these critical sociological studies is not always clear, but it may be related to various ideological positions, ranging from a radical progressive position that is disappointed by the failure to achieve social change (as in Aubert 1966) and the lack of any real participation by the people in the political process (for instance, Edelman 1976, 1988) to a more conservative or even anti-political (but therefore no less political) position that distrusts political power *tout court* and dismisses any appeal to social change as make-believe (as in Arnold 1938).

[19] See also the introduction (Chap. 1, Sect. 2.2).

In order to differentiate between the two concepts or conceptions of symbolic legislation, it does not suffice to refer merely to scientific and 'objective' data. Significantly, there are striking similarities between the criteria used in identifying symbolic legislation in the two different, negative and positive, senses. In both cases, legislation has to be made under socially and politically very difficult circumstances, in particular when the ethical convictions in society differ fundamentally and a widely held compromise seems unattainable. Subsequently, the clauses used – whether they are characterised as 'open' or 'vague' norms – are of very general nature and in desperate need of interpretation. Finally, there is a discrepancy or mismatch between the substantive provisions in the law and the provisions to enforce them. In both cases, value expression is a value in itself and seems to be more important than the law's actual enforcement, at least for the time being. These 'bare' facts only – difficult circumstances, general clauses, and weak law enforcement – are not enough to carry out an analysis in terms of symbolic legislation, either in the negative or in the positive sense. What is needed in addition is an appreciation or evaluation of these facts from a specific political or ideological point of view. If one takes the external, critical sociological position of distrust and suspicion, one sees dubious political motives at work everywhere and an unwillingness or inability to change the existing state of affairs in any fundamental way. If one is prepared, on the contrary, to put on one's pair of rose-coloured glasses and approach the phenomena internally from an interactionist or communicative perspective, one might see interesting opportunities for debate and interaction and chances for moral and legal norm development. Suddenly, a value expression through legislation no longer serves political purposes but has an 'inherent value', or at least it is an important step in the right direction. The communicative and interactive approach is idealistic in the double sense of focusing on ideals (the world wished for instead of the world as it is) as well as on effects at the immaterial level of thought and speech, whereas critical sociological studies of symbolic legislation (and symbolic politics in general) are more 'down to earth' and demand 'real', immediate and material results. In very rough terms, one could say that the two approaches can be distinguished by means of the opposition between *Idealpolitik* and *Realpolitik*.

## 2.5 The Politics of Symbolic Legislation Theory

It is not my intention now to take sides in this 'conflict of interpretations',[20] though I am ready to admit that I still have more sympathy for the communicative and interactive approach, despite its political naivety and its methodological and conceptual flaws. Critical sociological studies of symbolic legislation are very entertaining reading material, if you enjoy myths debunked (as I do). These studies have drawn attention to effects of legislation that are often neglected in traditional impact studies. Furthermore, they may serve a useful political purpose in criticising

[20] The phrase is taken from Ricœur's famous book title (Ricœur 1974).

political power and demanding more tangible results. However, critical sociological studies in this branch can be very sweeping and tend to overstate their case. They reduce politics to a theatre play.[21] Although there are undeniably theatrical aspects to politics, there is more to the political than mere theatre. If one is prepared to take an internal perspective to the legislative process – which I consider to be a pre-scientific, political choice –, one can see that, at least in well-functioning states, legislation mostly serves various functions and may have various effects, both of an instrumental and a symbolic kind. Law is never entirely symbolic or entirely instrumental, but a dynamic and unstable mixture of both.[22] Critical sociological studies focus exclusively or predominantly on negative-symbolic effects of legislation (power simulation, changes in status distribution, pacification of antagonistic groups, and so on), thereby ignoring or downplaying possible positive-symbolic effects on speech and thought that are essential for the working of the law (see Schwitters 2005). What is lacking, in short, is a feeling for the complex hermeneutic processes that necessarily accompany the fabrication and implementation of the law. Furthermore, in most cases no serious, practical or normative, alternative to the legislation dismissed as symbolic is offered. To be critical is one thing, but to make constructive suggestions for improving the current situation is something else.

On the other hand, the communicative and interactive approach may be accused of confusing factual and normative statements, of misrepresenting the inevitably hierarchical relation between state and society, and of ignoring elements of power and violence in the law. I agree with all that.[23] It definitively lacks the playfulness and the critical attitude that characterises many of the critical sociological studies mentioned above. At the same time, I still believe it is worthwhile to think about legislation in a normative way and to look for possibilities to enhance the democratic quality of the law and its responsiveness, while recognising the necessary limitations thereof – not everyone can participate in the legislative process, not every viewpoint can acquire official recognition. Moreover, I think that the law's contribution to the symbolic order – its influence on thought, speech and attitude formation – deserves more attention than it usually gets.

What I wanted to show here primarily, is that the scientific debate on symbolic legislation is no neutral affair, but a scholar has to take a normative stance in political and ideological matters concerning the role and function of the state, the relation between state and society, the values to be supported and the way to support them, before s/he can apply the conceptual and methodological tools, offered by the two opposing approaches, for describing and evaluating symbolic legislation or, more

---

[21] Edelman (1988) frequently speaks of politics as a 'spectacle' in which citizens are assigned the role of mere spectators.

[22] Jellinek already pointed out that law not only serves an instrumental function by establishing order, but it also possesses a symbolic dimension in that it gives expression to the self-understanding of a given society. In its symbolic dimension law has, according to Jellinek, an orientational function (see Lembcke and Van Klink forthcoming). Van der Burg (2005) also argues that legislation in general serves both functions.

[23] In Van Klink (2005, 2014) I provide a more extended evaluation of the communicative and interactive approach.

generally, symbolic effects of legislation. I would welcome it if both approaches would be more explicit about the political implications of their theories of symbolic legislation.

## References


Aalders, M.V.C. 1984. *Industrie, milieu en wetgeving. De Hinderwet tussen symboliek en wetgeving*. Amsterdam: Kobra.

Allott, Antony. 1980. *The limits of law*. London: Butterworths.

Arnold, Thurman W. 1938. *The symbols of government*. New Haven: Yale University Press.

Aubert, Vilhelm. 1966. Some social functions of law. *Acta Sociologica* 1–2: 98–120.

Ayres, Ian, and John Braithwaite. 1995. *Responsive regulation. Transcending the deregulation debate*. Oxford: Oxford University Press.

Azimi, A. 2007. *Open norm als maatwerk? De communicatieve benadering toegepast op de Arbeidsomstandighedenwet 1998* (diss. Tilburg). Nijmegen: Wolf Legal Publishers.

Cass, Loren R. 2009. The symbolism of environmental foreign policy. Power, interest, and ideas. In *Environmental change and foreign policy. Theory and practice*, ed. Paul G. Harris, 41–56. Oxon: Routledge.

Cotterrell, Roger. 1984. The Sociology of Law. An Introduction. London: Butterworths.

De Roos, Theo. 2013. *Grimmig spiegelbeeld. Ontwikkelingen in de strafwetgeving in de laatste decennia*. Tilburg: Tilburg University Press.

Eco, Umberto. 1984. *Semiotics and the philosophy of language*. London: MacMillan Press.

Edelman, Murray. 1976. *The symbolic uses of politics*. Urbana: University of Illinois Press.

Edelman, Murray. 1988. *Constructing the political spectacle*. Chicago: University of Chicago.

Fish, Stanley. 1980. *Is there a text in this class? The authority of interpretative communities*. Cambridge: Harvard University Press.

Fuller, Lon L. 1969. *The morality of law* (revised edition). New Haven: Yale University.

Griffiths, John. 1978. *Is law important?* (inaugural lecture, Groningen). Deventer: Kluwer.

Griffiths, John. 2005. Do laws have symbolic effects? In *Social and symbolic effects of legislation under the rule of law*, ed. Nicolle Zeegers, Willem Witteveen, and Bart van Klink, 147–161. Lewiston: The Edwin Mellen Press.

Gusfield, Joseph R. 1976. *Symbolic crusade. Status politics and the American temperance movement*. Urbana: University of Illinois Press.

Jasiak, Anna. 2010. *Constitutional constraints on ad hoc legislation. A comparative study of the United States, Germany and the Netherlands (diss. Tilburg)*. Nijmegen: Wolf Legal Publishers.

Kindermann, Harald. 1988. Symbolische Gesetzgebung. In *Gesetzgebungstheorie und Rechtspolitik*, ed. Dieter Grimm and Werner Maihofer, 222–245. Opladen: Westdeutscher Verlag.

Kindermann, Harald. 1989. Alibigesetzgebung als symbolische Gesetzgebung. In *Symbole der Politik, Politik der Symbole*, ed. Rüdiger Voigt, 257–273. Opladen: Lese Verlag & Budrich.

Lembcke, Oliver W. 2013. Menschenwürde als Verfassungsprinzip. In *Wörterbuch der Würde*, ed. Rolf Gröschner, Kapust Ante, and Oliver W. Lembcke, 344–346. Stuttgart: UTB.

Lembcke, Oliver W., and Bart van Klink. forthcoming. Das mag in der Praxis richting sein, taugt aber nicht für die Theorie. Zum Verhältnis zwischen Faktizität und Geltung bei Jellinek und Kelsen. In *Faktizität und Normativität. Georg Jellineks freiheitliche Verfassungslehre*, eds. Brugger, Winfried, Rolf Gröschner, and Oliver W. Lembcke. Tübingen: Mohr Siebeck.

Matten, Dirk. 2003. Symbolic politics in environmental regulation. Corporate strategic responses. *Business Strategy and the Environment* 12: 215–226.

Minow, Martha. 1990. *Making all the difference. Inclusion, exclusion, and American law*. Ithaca: Cornell University Press.

Nöth, Winfried. 1999. *Handbook of semiotics*. Bloomington: Indiana University Press.
Poort, Lonneke. 2013. *Consensus & controversies in animal biotechnology. An interactive legislative approach to animal biotechnology in Denmark, Switzerland, and the Netherlands*. Den Haag: Eleven International Publishing.
Ricœur, Paul. 1974. *The conflict of interpretations. Essays in hermeneutics*, ed. Ihde, Don. trans. Domingo, Willis, et al. Evanston: Northwestern University Press.
Schwitters, Rob. 2005. Symbols work. In *Social and symbolic effects of legislation under the rule of law*, ed. Nicolle Zeegers, Willem Witteveen, and Bart van Klink, 89–111. Lewiston: The Edwin Mellen Press.
Stamhuis, Jellienke. 2005. Communicative law. A quest for consensus. In *Social and symbolic effects of legislation under the rule of law*, ed. Nicolle Zeegers, Willem Witteveen, and Bart van Klink, 277–297. Lewiston: The Edwin Mellen Press.
Van der Burg, Wibren. 1996. Legislation on human embryos. From status theories to value theories. *Archiv für Rechts- und Sozialphilosophie* 82(1): 73–87.
Van der Burg, Wibren. 2000. Het primaat van de praktijk. In *De overtuigende wetgever*, ed. Bart van Klink and Willem Witteveen, 105–122. Zwolle: W.E.J. Tjeenk Willink.
Van der Burg, Wibren. 2003. An interactionist view on the relation between law and morality. In *The importance of ideals. Debating their relevance in law, morality, and politics*, ed. Wibren van der Burg and Sanne Taekema, 197–218. Bruxelles: Peter Lang.
Van der Burg, Wibren. 2005. The irony of symbolic crusades. The debate on opening up civil marriage to same-sex couples. In *Social and symbolic effects of legislation under the rule of law*, ed. Nicolle Zeegers, Willem Witteveen, and Bart van Klink, 245–275. Lewiston: The Edwin Mellen Press.
Van der Burg, Wibren. 2014. *The dynamics of law and morality. A pluralist account of legal interactionism*. Farnham: Ashgate.
Van der Burg, Wibren, and Frans W.A. Brom. 2000. Legislation on ethical issues. Towards an interactive paradigm. *Ethical Theory and Moral Practise* 3(1): 57–75.
Van der Burg, Wibren, and Pieter Ippel (eds.). 1994. *De Siamese tweeling. Recht en moraal in de biomedische praktijk*. Assen: Van Gorcum.
Van Klink, Bart. 1998. *De wet als symbool. Over wettelijke communicatie en de Wet gelijke behandeling van mannen en vrouwen bij de arbeid* (diss. Tilburg). Deventer: W.E.J. Tjeenk Willink.
Van Klink, Bart. 2005. An effective-historical view on the symbolic working of law. In *Social and symbolic effects of legislation under the rule of law*, ed. Nicolle Zeegers, Willem Witteveen, and Bart van Klink, 113–145. Lewiston: The Edwin Mellen Press.
Van Klink, Bart. 2014. Symboolwetgeving. Opkomst, ondergang en wederopstanding van een begrip. *RegelMaat (Journal for Legislative Studies)* 1: 5–26.
Witteveen, W.J. 1991. De jacht op de wet. In *Wat maakt de wet symbolisch?* ed. W.J. Witteveen, P. van Seters, and G. van Roermund, 115–136. Zwolle: W.E.J. Tjeenk Willink.
Witteveen, Willem J. 2005. Turning to communication in the study of legislation. In *Social and symbolic effects of legislation under the rule of law*, ed. Nicolle Zeegers, Willem Witteveen, and Bart van Klink, 17–44. Lewiston: The Edwin Mellen Press.
Witteveen, W.J., and B.M.J. van Klink. 1999. Why is soft law really law? *RegelMaat (Journal for Legislative Studies)* 3: 126–140.
Witteveen, W.J., P. van Seters, and G. van Roermund (eds.). 1991. *Wat maakt de wet symbolisch?* Zwolle: W.E.J. Tjeenk Willink.
Zeegers, Nicolle, Willem Witteveen, and Bart van Klink (eds.). 2005. *Social and symbolic effects of legislation under the rule of law*. Lewiston: The Edwin Mellen Press.

# Chapter 3
# The Emerging Interactionist Paradigm and the Ideals of Democracy and Rule of Law

**Wibren van der Burg**

## 3.1 Introduction

The emergence of interactive legislation and its close cousin communicative legislation is not an isolated phenomenon.[1] It fits within a broader gradual shift towards an interactionist paradigm (Van der Burg 2014, 148; see also Witteveen 2007 and Chap. 1 in this volume). In this paradigm, interactional law becomes more important. The interventionist state model of top-down legislation by the state claiming exclusive and ultimate authority, using legislation for instrumental purposes, is becoming less dominant. Instead, law-making becomes a cooperative effort on the part of various stakeholders, of which the state is one, but not necessarily the most important.

This shift is a gradual and partial one. The interactionist paradigm certainly has not replaced the traditional paradigm of top-down instrumental legislation in the interventionist state; the two paradigms co-exist and intertwine. The shift may be more strongly visible in law with regard to ethically controversial fields such as biotechnology than in criminal law. Nevertheless, even in a field dominated by highly instrumentalist use of detailed black-letter law like tax law, we may encounter more horizontal and interactionist phenomena such as horizontal monitoring in the Netherlands and responsive regulation in Australia (Huiskers-Stoop 2012, 13 and 2015; Braithwaite 2003).

---

A draft version of this chapter was discussed at the authors' conference in Amsterdam. I would like to thank the conference participants, and especially Meike Bokhorst and Lonneke Poort, as well as John Braithwaite and Esther Huiskers-Stoop, for their helpful comments.

[1] In accordance with Van Klink (2014), I prefer the phrase 'communicative legislation' for the latter type rather than 'symbolic legislation'.

W. van der Burg (✉)
Department Sociology, Theory and Methodology, Erasmus School of Law, EUR,
Burgemeester Oudlaan 50, 3062 PA Rotterdam, the Netherlands
e-mail: vanderburg@law.eur.nl



B. van Klink et al. (eds.), *Symbolic Legislation Theory and Developments in Biolaw*, Legisprudence Library 4, DOI 10.1007/978-3-319-33365-6_3

In this chapter, I will start by presenting the theory of interactive legislation in the context of this broader interactionist paradigm. I will show how different types of newly emerging legal phenomena can be seen as part of it. In addition to interactive legislation, we can mention responsive regulation (Ayres and Braithwaite 1992), communicative legislation (Witteveen and Van Klink 1999), negotiated rule-making (Coglianese 2003), co-regulation (Senden 2005), legally conditioned self-regulation (see Witteveen 2005), and regulatory capitalism (Braithwaite 2008). We can also include the theory of interactional international law as developed by Jutta Brunnée and Stephen Toope (2010). These phenomena all fit within the shift towards an interactionist paradigm, and can easily be fitted within the framework presented here, even if minor and major differences also exist between the various theories in this paradigm.

My purpose in the first part of the chapter is to identify in which respects these types of law differ from traditional instrumentalist legislation. I will compare these two paradigms using an analytic framework of four different dimensions. This comparison will be helpful not only to understand the broader interactionist paradigm but also the variations between the various types of legal phenomena belonging to it. Against this background, I will elaborate upon one specific subtype: namely, interactive legislation.

The second part of the chapter addresses normative concerns. If legislation by a democratically elected parliament is no longer the primary source of law, does this endanger the ideal of democracy? Does the decreased importance of formal legislation not also weaken the rule of law? I will argue that, on the contrary, interactionist law may reinforce both democracy and the rule of law – but only under specific conditions.

An important caveat is that I restrict myself to philosophical analysis in this chapter. Many empirical claims have been made by the proponents of the various positions discussed as well as numerous empirical criticisms.[2] Insofar as I discuss these claims and criticisms, I do so as a legal philosopher, trying to understand what these criticisms can tell us about how to understand and critically reconstruct the substantive contents of the respective positions, rather than inform us about their empirical validity.

One remark about the distinction between symbolic legislation and interactive legislation. Interactive legislation and communicative – or symbolic – legislation are distinct theories, although they belong to the same interactionist paradigm and there is much more overlap than difference between their central tenets. They overlap, for example, in emphasizing the expressive and communicative functions of legislation, the combination of which may be called the symbolic function (see Chap. 1 in this volume).[3] Some other shared tenets are the central role of ideals,

[2] See, for example, various contributions in Zeegers et al. (2005); Stamhuis (2006); Poort (2013); and Huiskers-Stoop (2015). For an overview and a reply, see Van Klink (2014).

[3] However, I believe we should avoid talking about 'symbolic legislation' as a distinct type of legislation. In my view, every statute may have expressive and communicative functions. The extent to which statutes have these functions may vary, but there is a continuum rather than a clear demar-

communication, and horizontal elements in law. Moreover, there has clearly been a strong mutual influence, especially because after 1994 most of the core authors collaborated closely in Tilburg.[4] In my view, the difference is primarily one of perspectives. The theory of communicative legislation has been developed initially as an alternative theory of legislation and, consequently, the primary perspective was that of the legislator authoritatively creating a statute. The focus was on what happened after the enactment. The theory of interactive legislation has been developed in the context of emergent biolaw in fields where the legislator did not yet play a central role; the perspective was that of citizens and self-regulating practices. So the perspective on legal norm development was primarily horizontal; it included both the interactive process leading up to a statute and the process after the enactment – but embedding both processes in a broader process of moral and legal norm development.[5] Most of the differences between the two theories can be associated with these differences in perspectives. For example, the constitutive function defended by Witteveen and Van Klink (1999; see also Chap. 2 in this volume), is rejected by interactive theorists, because in their view the community of discourse is not constituted by the statute, but usually exists already long before the enactment; at most, we could say that this community is modified by the statute. A debate about the precise differences seems rather futile to me as, partly because of those different perspectives, various participants in the debate would draw the lines differently. Therefore, I will simply present the theory of interactive legislation as developed by Frans Brom and myself – however, with certain important modifications in light of Poort's internal criticism of the strong orientation towards consensus in our initial presentations.

---

cation here. In fact, both Witteveen and Van Klink have also often carefully avoided such a reification of symbolic functions or symbolic elements. See Van der Burg (2005, 256) for an elaboration of this point and further references.

[4] For a more elaborate discussion of how the two lines of research have developed, see Van der Burg (2005, 245 ff.)

[5] There has been a long debate between the main proponents of communicative legislation, Willem Witteveen and Bart van Klink, and various critics, including myself. Their 1999 article on soft law was translated into Dutch, and presented as the opening article in Van Klink & Witteveen 2000. A number of contributors – including some who are sympathetic to an interactionist approach – then criticised it in that same volume; see the contributions by Marc Hertogh, Hans Lindahl, Bert van den Brink, and myself. A further round of critical discussion may be found in various contributions to Zeegers et al. (2005) and in Stamhuis (2006) and Poort (2013). These critics have argued that, at least in the original article from 1999, communicative legislation is still characterised by strong instrumentalist and top-down tendencies. Moreover, Lonneke Poort and I argue that there are important differences between communicative and interactive legislation precisely on those points of critique, and that the latter is less vulnerable to most of the criticisms. Willem Witteveen has emphatically rejected these criticisms. Bart van Klink has argued that there are only minor differences with interactive legislation (see, for instance, Van Klink 2014). Neither party in the debate has been able to convince the other, so I will leave the debate aside here. See Chap. 1 in this volume for a discussion of some of the main critiques.

## 3.2 The Emerging Interactionist Paradigm

In the modern interventionist state, legislation – including delegated legislation – has a central role. We may identify a number of characteristics of legislation in the interventionist state, in order to compare them with how law-making in the interactionist paradigm and its various sub-types is perceived. First, legislation is top-down. Legislator and citizens are in a vertical relationship in which the legislator has the authority to command the citizens. Second, legislation is an instrument in the hands of the political powers, to advance certain policy goals. Third, the state is the main initiator of legal change by creating new statutory rules. Fourth and finally, legislation is considered to be the main source of the law: the law changes as a result of the enactment.

In summary, law in the interventionist paradigm has four distinct characteristics:

1. Top-down perspective: Law-making is embedded in a vertical relationship;
2. Instrumentalism: Law is an instrument in the hands of the political authorities;
3. State-centricity: The state is the primary initiator of legal change;
4. Legislative supremacy: Legislation is the primary source of law.

Of course, this is an ideal typical sketch. Even in the heyday of the interventionist state, it did not completely fit reality. The state does not and did not control society in the way envisioned by the model, as numerous sociological studies have shown.[6] Even so, it is a model that has dominated legal and political discourse, and is still very much the general common sense. Constructing this model helps to identify in which respects law in the emerging interactionist paradigm differs from that in the traditional paradigm, and also in which respects the various subtypes of the interactionist paradigm differ.

The interactionist paradigm is radically different in all four dimensions. An ideal typical sketch includes the following:

1. Horizontal perspective: Law-making is embedded in a horizontal relationship;
2. Broad dialectical means-ends relationship: Law is both a means and an end for political authorities and other actors alike;
3. Societal orientation: Interaction within society is the motor of legal change[7];
4. Legislation is only one of the sources of law, alongside others such as interactional law and contract.

None of these dimensions is a question of black or white, as they allow for gradual answers. Most subtypes of law that I have mentioned above as belonging to the interactionist paradigm are hybrid in that they combine elements of the interventionist

[6] See, for example, Moore (1973); Griffiths (1986).

[7] Cf. the famous quote of Eugen Ehrlich in the 'Foreword' to Ehrlich (2002): 'the center of gravity of legal development lies not in legislation, nor in juristic science, nor in judicial decision, but in society itself'.

and the interactionist paradigms. The differences between the various subtypes of the interactionist paradigm can be understood mainly as different mixtures of these elements.[8]

We may identify a pure ideal-typical form of the interactional paradigm, exemplified by Brunnée and Toope's theory of interactional international law. For Brunnée and Toope (2010), law-making is embedded in horizontal relationships. It is not an instrument in the hands of the state but emerges from the cooperation between many actors, including states that may have instrumentalist strategies. The emergence of these norms is not necessarily initiated by states, but can be initiated by each of these actors, and the basic source of law is interactional, horizontal law.

However, most interactionist theories are somewhere on the spectrum between the purely horizontal and purely vertical perspectives. Interactive legislation is quite close to interactional international law. Responsive regulation and communicative legislation – at least in their initial formulations (Ayres and Braithwaite 1992; Witteveen and Van Klink 1999) – are in some respects still close to the traditional interventionist paradigm. In my view, responsive regulation and communicative legislation started initially from traditional top-down, instrumentalist, and state-centric perspectives, but adapted and enriched these perspectives with a more sophisticated instrumentalism and a partly horizontal orientation. They emphasised dialogue, and regarded co-regulation and self-regulation as auxiliary sources of law. These initial formulations still had a hybrid character, as they contained elements of both paradigms. However, in their later work, both Braithwaite (2008) and Witteveen (2005, 2007) elaborated their ideas in a more consistently interactionist way.

Apart from variations on each of these four dimensions, various other differences may be important for fully understanding the theories and their mutual differences. A first distinction is that some theories are strongly prescriptive (Ayres and Braithwaite 1992) and others claim to be primarily descriptive (Van Klink 1998), whereas the theory of interactive legislation contains both descriptive and normative theses (Van der Burg and Brom 2000; Van der Burg 2005). A second difference is the field in which they have been developed. Is it one in which there is hardly any positive law – like emerging health law in its early stages – or in which there is much black-letter law – such as tax law – in which the debate focuses on whether there is overregulation? A third difference concerns whether the theory focuses primarily on conflicting interests – as in tax law and economic regulation – or on conflicting normative views – as in bio-ethical issues. For example, the theory of interactive regulation has only twice been applied to fields where conflicting economic interests dominate the scene. The first case study in which it was tested in such fields (Stamhuis 2006, discussing legislation on employee participation and

---

[8]Whereas Bart van Klink in his contribution to this volume groups the various types together, I suggest it is important to be perceptive to the differences between them. In my view, the initial formulations of communicative or symbolic legislation and of responsive regulation are much more vulnerable to various criticisms brought forward in the Dutch debate (see note. 5 above, and see also Van Klink's own critique in this volume) than are the more consistently interactionist versions such as interactive legislation and interactional international law.

corporate governance) seemed to suggest that it is not fully applicable in these contexts. However, the recent study by Huiskers-Stoop (2015) concludes on the basis of extensive empirical research that horizontal monitoring by tax authorities – which can be understood as a partial realization of interactive legislation – works better than traditional monitoring.[9]

Finally, a fourth difference has to do with the disciplinary background of the main authors: criminology (Braithwaite), political science and law (Witteveen), literary studies (Van Klink), and ethics and law (Van der Burg). The more legislation-centric approach of Witteveen and Van Klink may be partly explained by their specific disciplinary outlook, just as my own focus on horizontal and argumentative relationships may be partly traced back to my partial background in ethics.

## 3.3 Interactive Legislation[10]

For a comprehensive understanding of the theory of interactive legislation, it is helpful to understand the context in which it was developed. The context was that of the emerging intertwined disciplines of bioethics and health law in the Netherlands around 1990. At that time, there was very little positive law, let alone statutory law, in the emerging fields of health law and biolaw. If there was, as in the case of euthanasia, it was usually outdated, and medical practice took little notice of it. Healthcare practice developed its own moral and legal standards. (For a more elaborate sketch, see Van der Burg 2014, 127–134.)

In this context, some researchers developed a theoretical approach to describe the emerging processes in health law and biolaw (e.g. Ippel and Vorstenbosch 1994; Van der Burg and Brom 2000). We have named this process interactive legislation, as it builds on broad processes of interaction among a great number of societal actors at all stages of the norm-making and implementing processes.[11] I have argued that, at least in societal fields having a strong ethical dimension, the development of legal norms is shifting from a vertical model in which the legislator authoritatively sets standards for society to a more horizontal, interactive process in which various social actors participate, and among them, the legislator has an important role, though not necessarily a central one. Similarly, the implementation, enforcement, and control of legal norms is shifting from a vertical model in which

[9] Huiskers-Stoop (2015, 445) summarizes her findings as follows:

> It is likely that horizontal tax monitoring, compared to traditional tax monitoring, leads to better tax compliance, greater fiscal certainty, reduction of tax compliance costs and a better relationship with the tax authorities.

Perhaps, the difference may be partly explained by the fact that Stamhuis focuses on the stage in which legal norms are developed, whereas Huiskers-Stoop focuses on the stage in which the norms are implemented.

[10] Some fragments from Van der Burg (2014, 144–148) have been included in this section.

[11] In my view, the best presentations of this approach may be found in Van der Burg (2005, 2014).

government bodies are the main enforcing actor to a more horizontal, interactive process in which various societal actors participate, among which government bodies have an important, though not necessarily central role. Consequently, the separate processes of norm development and norm implementation are merging into one continuous process of norm development and implementation.

In terms of such an interactionist perspective, the role of statutes becomes less central. Moreover, as with legislation in the nineteenth century, the emphasis is as much on codification of norms that have emerged in society as on modification. As a result, the instrumental and the protective functions of law become less important, and two other functions come to the fore: the expressive and the communicative.[12] Statutes often formulate open standards rather than strict rules. On the one hand, these standards may express the common values of a society – or the values of the dominant group – and provide a common normative framework. The statutory formulation then has an expressive function. On the other hand, these standards may provide a common point of reference by which the norm addressees can be guided without being presented with too many detailed and overly restrictive rules. Moreover, the standards invite active interpretation and discussion on those interpretations, both among and between citizens and the various legal authorities. Statutes then also have a communicative function. Because of this continuing process of interpretation and implementation, and the strong focus on more general values and principles, moral norm development and legal norm development are often strongly intertwined.

In my view, a clear shift in perspective is noticeable compared to the traditional interventionist paradigm. The perspective is not that of the legislator, however, but of society at large, and involves a fundamental shift from a vertical to a horizontal relationship. There are vertical elements in this horizontal relationship, as the legislator may codify social norms and the conclusion of societal discussions, and then lend his authority to those norms. Nevertheless, the relationship is fundamentally horizontal.

The instrumental function of legislation is not denied, but is supplemented with communicative and expressive functions. The legislator can still try to use law strategically to further certain aims, although with a more sophisticated view of its functions. However, its view of legislation cannot be called instrumentalist in the usual sense, because the fundamental values and aspirational norms are not formulated by the legislator but have been developed in a broad interactive process in which the legislator provides only a marginal reformulation. In terms of the perspective of interactive legislation, the role of the legislator is much more marginal. Citizens and other stakeholders are respected as co-creators of law.[13]

Rather than being primary actors, the legislator and other state institutions are merely some of the actors engaged in the development of new legal norms. Instead

[12] For a discussion of the various functions of legislation, see Van der Burg (2001); see also Stamhuis (2006, 163) and Chap. 11 in this volume.

[13] There is a strong similarity here to Braithwaite's description of the largely horizontal interaction between the state and various societal actors in his recent work on regulatory capitalism (2008).

of taking the initiative, the legislator follows: he often codifies rather than modifies, and very rarely commands. Finally, an important source of law is legal norm development in society and in specific sub-practices like healthcare; hence, legislation is only one of the sources of law, alongside self-regulation and interactional law.

Let me illustrate this with an example in the field of biolaw. The notion of interactive legislation can be illuminating when one analyses legislation on embryo research, especially in the early stages of legal development.[14] In many Western countries, it has not been the legislature that developed the norms initially, but the research community in dialogue with other stakeholders such as patient organisations and the general public. Through self-regulation and committees, in continuous dialogue with many actors involved, norms were gradually developed and refined. Broad committees were often created by the government to provide advice on legislation and policy recommendations, but these committees usually relied heavily on a dialogue with the research community, ethical and legal experts, interest groups, and the broader public. Even so, in many cases the committees' recommendations were either still too open or too pluralist to be implemented, or they were too controversial. In most countries, it took considerable time before formal statutes were enacted.

Even then, the enforcement of these statutes was left mostly to the research community, through institutional ethics committees or animal experiment committees. The public prosecution rarely played a role in enforcing the norms. For embryo legislation – if enacted in the end – the expressive and communicative functions were very important. Rather than providing detailed norms and instructions, statutes often laid down certain basic principles and general rules. Without defining the concept clearly, such statutes expressed, for example, the intrinsic value of embryos. The concrete balancing of this intrinsic value against other values and interests was usually left to committees in which various stakeholders were represented. These committees were to balance the diverse values and interests at stake, not in general terms but in light of the details of the case at hand. The further development of norms and their implementation thus merged into one continuous process of norm development and implementation. Norm development basically took place in a more case-bound process.

This is just a general sketch. The picture may be slightly different for each individual country, but it seems that in many countries elements of interactive legislation on embryo research may be discerned. Other fields in which elements of interactive legislation may be discerned are biotechnology in general (Brom 2003; Poort 2013; Grotefeld 2003) and non-discrimination law and same-sex marriage (Van Klink 1998; Van der Burg 2005).

A final caveat may be in place here. Proponents of interactive legislation do not defend a general normative thesis that law-making is always better if it is more horizontal and non-instrumentalist, and if it is less state-centric and relies less on the idea that the legislature is the highest legal authority. Proponents merely suggest that there is an emerging interactionist paradigm in which interactional law plays an

[14] The example has been elaborated in Van der Burg (1996).

important role, and that this may sometimes work better. But they do not want to argue that this interactionist paradigm will, or should, replace the interventionist paradigm completely. On the contrary, in many situations we simple have to rely primarily on traditional theories of legislation. The interactionist paradigm may offer additional elements as well as an alternative in some situations, but it cannot replace the traditional paradigm entirely.

What we need is an eye for variation.[15] Sometimes a more interactionist approach works better, while at other times a traditional interventionist approach is to be preferred. Ayres and Braithwaite (1992, 101) argue convincingly 'that there is no such thing as an ahistorical optimal regulatory strategy'. Interactive legislation may be more adequate in areas where ethical controversies are crucial, but I doubt whether it would fully work in the field of tax law. The relationship between the taxing state and the taxpayer is basically of a top-down character, even if horizontal elements can be incorporated into it, especially in the implementation stage, as the experiences in Australia and the Netherlands have demonstrated (Braithwaite 2003; Huiskers-Stoop 2012, 13 and 2015; Gribnau 2015).

## 3.4 Democracy

After this presentation of interactive legislation against the background of a broader interactionist paradigm, it is time to address my second theme. Various concerns have been raised against interactive and communicative legislation. In the traditional paradigm, the focus is on state legislature as the main actor in the field, which is supposed to be democratically elected and therefore legitimate. Putting law-making in the hands of other actors has led to criticisms that this may threaten our democracy. Similar arguments have been made with regard to the rule of law: namely, if state legislation is no longer at the core of our understanding of law, can the rule of law still offer protection?

These concerns have also been addressed by authors advocating communicative legislation (e.g. Witteveen and Van Klink 1999, 137). I have argued that if we develop a different and more substantive understanding of the ideals of democracy and the rule of law, under certain conditions interactionist forms of law can even be seen as more democratic and more supportive of the rule of law (Van der Burg 2000).

An important aspect of democracy is equal participation in decision-making and law-making processes; other relevant aspects concern fair procedures and minority rights. Citizens, and especially those most affected by certain decisions or laws, should have a say in these processes. In the traditional, formal understanding of democracy, this is done through elections. Citizens elect members of parliament and, directly or indirectly, these representatives have an influence on decisions and laws. However, the election of members of parliament is not the only way in which

[15] Van der Burg (2014, 9). I derived this phrase from private conversations with Philip Selznick.

democratic participation can be guaranteed. In fact, the direct participation of all stakeholders in interactive legislation may even be considered more democratic than the merely indirect participation through their representatives in parliament. Of course, interactive legislative processes do not always directly involve all stakeholders personally. However, if procedures are designed well, they may include recognisable representatives of all the relevant interests and views, whereas in parliamentary parties, such a representation of all relevant interests and views is not always guaranteed. Moreover, interactive legislation does better justice to the ideal that citizens especially should have a say in matters that are important to them, because they are only included in the interactive process insofar as they are in fact affected. In parliamentary decisions on, for example, gas extraction in the north of the Netherlands, the votes of those living in that area weigh precisely as much as the votes of citizens living at the other end of the country; in interactive legislation processes, however, it is possible to take into consideration the fact that some citizens are affected more than others.

This certainly does not mean that interactive legislation is always more democratic. We can formulate additional requirements for interactive processes in order to guarantee that they really are democratic.

1. An important requirement is that the processes should truly be inclusive: that is, all stakeholders should be involved. This was clearly not the case in the process leading to the Dutch Corporate Governance Code of 2003 (the Tabaksblat Code), as labour representatives were explicitly excluded from the Tabaksblat Committee (Stamhuis 2006, 139). Inclusiveness is not an easy requirement, because the more stakeholders involved, the more difficult it becomes to reach an agreement. This dilemma is illustrated by the Tabaksblat Code process: if labour interests had been included, the committee would probably not have reached consensus so easily. However, the result of relying on an 'old boys network' was that there was only a 'pretended consensus' (Stamhuis 2006, 153).
2. A second requirement is that the process should be open and allow for the full articulation of all views, rather than being prestructured in a way that favours specific views or conceptual frameworks – and excludes others. An interesting illustration is offered in Margo Trappenburg's analysis of medical ethical debates. Her conclusion was that two of the four debates she had studied – those on organ transplants and on medical experiments with legally incompetent human subjects – were legally structured.[16] The legal framework dominated the debate so much that only arguments that fitted in that framework were effectively allowed during the debates.[17]
3. A third requirement acknowledges that interactive processes are vulnerable to abuse by vested interests. Therefore, there should be adequate checks and balances to protect the interests of minorities and give them a full say. In every

[16] Trappenburg (1993, 340–343). See also Stamhuis (2006, 86).

[17] See also Poort (2013), arguing that too strong a focus on consensus may lead to the exclusion of certain viewpoints.

decision- making process, the protection of minorities can be vulnerable. In parliamentary procedures they can often be ignored because their voices are not important in terms of establishing a majority, or simply because the election system makes it effectively impossible for minorities to be elected at all: for instance, in a winner-takes-all district system. In interactive legislation processes, we may design institutions such that at least every voice can be brought to the table and be heard. For example, by including representatives from all the relevant interest groups, even if they are small, or by structuring consultation processes so that every group can put forward its views.

4. A final requirement is the promotion of a broader orientation towards consensus in the decision-making processes. We need a culture in which a narrow majority is not felt to be satisfactory, and in which all participants are willing to continue the dialogue until a consensus or a compromise has been reached that can be accepted by all or almost all stakeholders – or at least one that is not considered too strongly unacceptable. Such a consensus-oriented culture would fit into the Dutch tradition of consociationalism, of power sharing. However, this tradition should be reinvented and restructured to deal with the usual criticisms that, in its traditional form, it is not inclusive and open enough (see De Been 2012.)

   As both Jellienke Stamhuis and Lonneke Poort have shown, this focus on consensus may lead to the suppression of minority views and dissenters, and to a premature or pretended consensus.[18] I believe their criticisms are justified, but this does not mean that we should skip altogether the orientation to reaching a broad consensus or to constructing a broadly acceptable compromise. Of course, consensus or compromise will often not be possible. Nevertheless, all parties can at least aspire to broaden the majority and to do justice as much as possible to minority viewpoints. It will therefore only work well if this orientation towards consensus and compromise is accompanied by inclusiveness and openness.

Each of these four requirements focuses on a different dimension as regards how interactive processes may go wrong from a democratic point of view: these are lack of inclusiveness; lack of openness for relevant arguments and views; domination by vested interests; and uncompromising attitudes by majorities. The partly informal and unregulated character of interactive processes may make it sometimes more difficult to provide adequate checks and balances and guarantees, but it may also offer opportunities for promoting inclusiveness, openness, and respect for minorities in ways that parliamentary procedures cannot. Moreover, these requirements should not be met only by interactive legislation; they are just as important in traditional democratic decision-making, and are frequently as much at risk there as in the context of interactive legislation.

[18] I fully agree with the critiques by Poort (2013 and Chap. 5 in this volume) that the initial formulations by Frans Brom and me (and by Witteveen and Van Klink as well) were formulated too indiscriminately, suggesting implicitly that consensus should always be the desirable and possible outcome. The positive roles of dissensus and compromise should have been included.

## 3.5 The Rule of Law

A similar critique of the interactionist paradigm has been made with regard to the rule of law. If the rule of law is identified with a *Rechtsstaat*, a state ruled by law as formulated in statutes and constitutions, the interactionist paradigm then provides a very poor image indeed. However, a different understanding of the rule of law is also possible. We need not restrict law to state-made law, nor do we need to apply the rule of law only to state action. Philip Selznick has interpreted the ideal of legality broadly as 'the progressive reduction of arbitrariness' (Selznick 1961, 100). He applies this broad conception of the rule of law to contexts other than the state; legality may thus be relevant to every situation in which there are power imbalances and risks of abuse of power. The advantage of a Selznickian approach is that it provides a broader range of application of the rule of law than does the traditional *Rechtsstaat* model, by including power relations in which the state is not the dominant power but, for example, powerful commercial organisations or unions. Consequently, it also provides guidance in the more horizontal relations of interactional law. Moreover, if we focus on the reduction of arbitrariness as the core meaning of the ideal of legality, we may see that types of law other than state legislation may also help to curb the arbitrary exercise of power. Constitutional customary law, self-regulation, or international interactional law may also be effective means to provide for checks and balances and control of power.

The rule of law is not based on one simple ideal, but refers to various ideals or values. It can be argued that interactive legislation endangers at least two of these ideals: namely finality – the peaceful closure of conflicts – and legal certainty. I will focus on legal certainty, and examine only briefly the issue of finality, since both Nicolle Zeegers and Lonneke Poort discuss it in this volume.

Let me first put the issue of finality into perspective. As representing a value associated with the rule of law, finality and closure are primarily important with regard to court decisions on concrete conflicts. There is nothing in the interactionist paradigm that makes this impossible: courts can still make final decisions and thus end conflicts. The same is true, for example, as regards animal experiment committees: their decision on a concrete experiment is final, or – if it is advice given to a board or to a minister – the latter's decision is final. Only with respect to legislation is there no finality. Whereas in the traditional paradigm the debate is intended to be closed after a statute has been passed, in the interactionist paradigm the debate continues. However, the issue then is whether we truly prefer closure that may be premature – with the result that the law may soon be changed again or that it is ignored in practice – rather than an ongoing debate in which the norms develop gradually. Real closure in legislative debates is only provided if all stakeholders accept the outcome or the norm. I suggest that at least in certain contexts, such as issues with a strong ethical dimension, such a real closure is more likely to happen in interactive legislation than in traditional top-down legislation, where it is always the question of whether all stakeholders will accept the authority of a statute they do not support.

Even so, it may look as if in interactive legislation the debate continues endlessly, and no closure may ever be reached. There are two replies to this critique.[19] The first is that interactive legislation is not the only type of legislation recognised in legal interactionism. Sometimes the legislature simply has to cut through the Gordian knot, because there is no consensus or compromise in sight and legislation simply cannot wait. This is perfectly legitimate. Legal interactionism certainly does not claim that we always have to wait until there is consensus or compromise.[20] Legal interactionism would be extremely naïve to suggest that decisions or statutes should always be made on the basis of consensus. This is often impossible; consequently, a decision or statute will not satisfy all citizens or political parties. That is the condition of politics. Some critics have argued that this is a moment of violence or exercise of power, but to me this seems like a trivial observation rather than a critique, as this exercise of power is inherent in every form of political decision-making in the context of a modern state. So it is not specific to interactive or communicative approaches but a characteristic of politics as such. (See the contributions by Van Klink and Poort in this volume about these critiques.)

The second reply is more positive. Indeed, in some cases there is no consensus, no possibility for compromise, and not even a solution supported by a political majority. No political closure is reached and the debate may linger on. Sometimes, however, that need not be a disadvantage. If the debate persists, this prevents untimely closure. It leaves the dynamics of law open for the future, and it endures pluralism with regard to views of the good law. Perhaps living with that uncertainty is more attractive than living with a sham certainty or an untimely closure.

This brings me to the second value associated with the rule of law: namely, *legal certainty*. It is the ideal that we can know the law and may rely on it, and can predict the behaviour both of officials and of other citizens bound by the law. This formulation of legal certainty relies strongly on there being legal authorities. In an interactionist perspective, the role of these authorities is less central, so we might be tempted to believe that interactive legislation endangers legal certainty. Again, it depends on the perspective as to whether this is true or perhaps even the reverse.

We may discern two dimensions of the ideal of legal certainty. The first is *epistemic or doctrinal certainty*: knowing in detail the content of the law, the legal

[19] Both replies have been inspired by many discussions with Lonneke Poort. She suggests (Poort 2013) that we should look for an alternative approach in public debate, an ethos of controversies, and separate it from legislative and decision-making processes. I am sceptical about this two-track approach because, in my view, the two tracks are too closely connected. I suggest that a more promising perspective to address the criticisms made by Poort is by reconstructing the Dutch tradition of consociationalism. This tradition can and should be revised by orientation towards broadly inclusive consensus and inclusive compromises without denying the importance of dissensus and the frequent need for making decisions and formulating legal norms when consensus or compromise is not or at least not yet possible.

[20] See for a similar point the contribution by Poort in Chap. 5 in this volume, arguing that we must accept the need for temporary decisions.

doctrine. The second is *practical certainty*: being able to predict the behaviour of state officials and other citizens.[21]

Interactive legislation is clearly detrimental to doctrinal certainty. The use of aspirational norms and the reliance on society and various practices for further norm development and specification makes it more difficult to specify in detail the legal doctrine. If we cannot rely on law in the books, it is more difficult to write detailed books about the law. However, except for legal scholars and lawmakers, doctrinal certainty is not a very important value in itself. As a result of their training, lawyers – especially those in the Civil Law tradition – may have a tendency to overestimate the importance of doctrinal certainty, because their education has focused mostly on legal doctrine. I want to suggest that doctrinal certainty is merely a subservient value, in the service of practical certainty. For ordinary citizens, what law in the books tells us is less important than what will happen in practice. Will I be fired or not? How much tax will we have to pay? If I stick to the traffic rules, will I be able to avoid accidents because I can expect other citizens to do so as well?

In my view, when we discuss the rule of law in terms of legal certainty, it is practical certainty rather than doctrinal certainty that is most important. If we want to evaluate whether interactive legislation endangers legal certainty, we should focus on practical certainty and, moreover, we should take a comparative approach. We might ask whether interactive legislation is better or worse than instrumentalist legislation.

If we understand the ideal of legal certainty not in terms of knowledge of certain rules in the statute book but in regard to citizens being effectively able to rely on reasonable expectations regarding the behaviour of state officials, public organisations, and other citizens, we are able to obtain a more positive evaluation of interactive legislation. For if interactive legislation is based on existing patterns in a professional practice, everyone involved in that practice will already be acting according to the new statutory rules before the statute has even been enacted. And if the implementation of the statutory norm is based on an open and inclusive process, every stakeholder will also know what to expect from other actors in terms of more detailed norms. It may be different for non-stakeholders, but then again, the question is why would they be interested? Usually they would not take an interest in those outcomes. If they did, however, the approach would be the same as in the traditional paradigm; they should consult a legal expert – in that case, someone who also knows the field itself – and ask what – according to her expert view – the positive law is.

Moreover, we should not overestimate the degree of practical certainty offered by traditional legislation. The outcomes are often unpredictable if the law has to be applied to specific situations. Ask a lawyer to predict who will win in specific labour or divorce cases, and she will only be able to do so correctly in far less than one-hundred percent of the cases, let alone that a lawyer can predict accurately how

---

[21] Huiskers-Stoop (2015, 40) makes a similar distinction between objective or legal certainty (knowing the formal rules and their correct application) and perceived or fiscal certainty (the feeling that the fiscal obligations of the company are known).

much compensation a court will award to the plaintiff in the case of a labour conflict or how high the fine will be in a criminal case. Of course there are easy cases, but we certainly should not overstate the degree of certainty and predictability the rule of law provides to ordinary citizens. Moreover, the prediction is only based on past official acts of courts and legislatures, and they may revise their view of the law. Courts may overrule precedent, politicians and other officials may decide to no longer enforce a statutory norm, and legislatures may create a new statute. This is especially true in those fields in which interactive legislation has been advocated, such as biomedical ethics, where social and moral norms evolve rapidly, and it is uncertain when or whether the courts and other legal officials will follow.

However, we need not look at those fields to understand why interactionist forms of law may offer a solution regarding the lack of practical certainty. Tax law is a field that is strongly dominated by very detailed and specific rules; there are few aspirational norms. Doctrinal certainty will usually be realised to a high degree. Even so, there is often a crippling practical uncertainty about how tax authorities will actually interpret and apply these rules in complex cases such as the activities of multinational firms. Firms cannot wait for years to know which part of their sales volume and which part of their costs will be taken into account by the Dutch tax office. In order to make investment decisions, they need to know these details in advance. In the 1990s, this need for practical certainty has given rise to a practice of horizontal implementation: namely, rulings. Negotiating with the tax authorities and obtaining rulings about how the authorities will deal with future situations is a means of addressing this type of practical uncertainty.[22] In 2005, the Dutch tax authorities introduced a new, even more horizontal and interactionist tax practice, called horizontal monitoring. In this monitoring practice based on mutual trust, transparency, and understanding, taxpayers can be practically certain that the final tax assessment will be in accordance with the tax return filed. This illustrates that even in a field so strongly dominated by detailed black letter law as tax law, there is not only room but even a need for a more interactionist approach.[23]

This, however, is not a conclusive argument. Again, there are no general arguments on which legislative strategy or which type of law is best. It depends on the specific conditions and problems. Nevertheless, I think that the example of rulings shows that legal certainty in the form of practical certainty may be improved rather than impaired by interactive legislation and interactional law.

It is not a question of whether the traditional or the interactionist paradigm is the best solution in general. Under certain conditions the traditional type of legislation may best serve democracy and the rule of law, while in other situations interactive

[22] Of course, there are also negative aspects to this practice of rulings, especially if they lead to constructions in which multinational firms pay very low taxes. I do not claim that the practice of rulings is *always* good, or that it should not be changed in some respects. I merely use it as an example to show that horizontal interactional law may improve practical certainty rather than be detrimental to it.

[23] For horizontal monitoring in Dutch tax law practice, see Huiskers-Stoop (2012); Huiskers-Stoop and Diekman (2012); Gribnau (2015); and Huiskers-Stoop (2015).

legislation will do this. Which of the two paradigms a scholar prefers may be less an objective issue than a matter of personal character. Some legal scholars may prefer doctrinal certainty, while others opt for openness and dynamics – for both positions there are good albeit inconclusive arguments. In legal scholarship as much as in positive law, we should be aware of the risk of untimely closure. Thus, let the debate continue.

## References

Ayres, Ian, and John Braithwaite. 1992. *Responsive regulation. Transcending the deregulation debate*. Oxford: Oxford University Press.

Braithwaite, Valerie (ed.). 2003. *Taxing democracy. Understanding tax avoidance and evasion*. Aldershot: Ashgate.

Braithwaite, John. 2008. *Regulatory capitalism. How it works, ideas for making it work better*. Cheltenham, UK: Edward Elgar.

Brom, Frans W.A. 2003. The expressive-communicative function of bio-patent legislation. The need for further public debate. In *Patente am Leben? Ethische, rechtliche und politische Aspekte der Biopatentierung*, ed. Christoph Baumgartner and Dietmar Mieth, 117–126. Paderborn: Mentis.

Brunnée, Jutta, and Stephen John Toope. 2010. *Legitimacy and legality in international law. An interactional account*. Cambridge: Cambridge University Press.

Coglianese, Cary. 2003. Does consensus work? A pragmatic approach to public participation in the regulatory process. In *Renascent pragmatism. Studies in law and social science*, ed. Alfonso Morales, 180–195. Aldershot: Ashgate.

De Been, Wouter. 2012. Continuity or regime change in the Netherlands. Consociationalism in a deterritorialized and post-secular world. *Ethnicities* 12(5): 531–555.

Ehrlich, Eugen. 2002 (1936). *Fundamental principles of the sociology of law*. Trans. W.L. Moll. New Brunswick: Transaction Publishers.

Gribnau, Hans. 2015. Taxation, reciprocity and communicative regulation. *Tilburg Law Review* 20: 191–212.

Griffiths, John. 1986. What is legal pluralism? *Journal of Legal Pluralism* 24: 1–55.

Grotefeld, Stefan. 2003. Wie wird Moral ins Recht gesetzt? *Archiv für Rechts- und Sozialphilosophie* 89: 299–317.

Huiskers-Stoop, E.A.M. 2012. Horizontaal toezicht en de eerste uitkomsten van een enquete onder middelgrote ondernemingen. *Maandblad Belastingbeschouwingen* 3: 123–131.

Huiskers-Stoop, E.A.M., and P.A.M. Diekman. 2012. Compliance als hoeksteen voor horizontaal belastingtoezicht. De rol van compliance bij de versterking van het nieuwe belastingtoezicht. In *Jaarboek compliance*, 231–244. Capelle aan den IJssel: Nederlands Compliance Instituut.

Huiskers-Stoop, Esther A.M. 2015. *De effectiviteit van horizontaal belastingtoezicht. Een fiscaal-juridische en empirische analyse*. Dissertation Rotterdam.

Ippel, Pieter, and Jan Vorstenbosch. 1994. De weg naar de wet. In *De Siamese tweeling. Recht en moraal in de biomedische praktijk*, ed. Wibren van der Burg and Pieter Ippel. Assen: Van Gorcum.

Moore, Sally Falk. 1973. Law and social change. The semi-autonomous social field as an appropriate subject of study. *Law and Society Review* 7: 719–746.

Poort, Lonneke. 2013. *Consensus & controversies in animal biotechnology. An interactive legislative approach to animal biotechnology in Denmark, Switzerland, and the Netherlands*. Den Haag: Eleven International Publishing.

Selznick, Philip. 1961. Sociology and natural law. *Natural Law Forum* 6: 84–108.

Senden, Linda. 2005. Soft law, self-regulation and co-regulation in European law. Where do they meet? *European Journal of Comparative Law*, January 9.1. http://www.ejcl.org/91/art91-3.html. Accessed 20 Mar 2015.

Stamhuis, Jellienke N. 2006. *Conflicting interests in corporate regulation. An exploration of the limits of the interactionist approach to legislation in employee participation and corporate governance.* Dissertation Groningen.

Trappenburg, Margo J. 1993. *Soorten van gelijk. Medisch-ethische discussies in Nederland.* Zwolle: W.E.J. Tjeenk Willink.

Van der Burg, Wibren. 1996. Legislation on human embryos. From status theories to value theories. *Archiv für Rechts- und Sozialphilosophie* 82: 73–87.

Van der Burg, Wibren. 2000. Het primaat van de praktijk. In *De overtuigende wetgever*, ed. Bart Van Klink and Willem Witteveen, 105–122. Zwolle: W.E.J. Tjeenk Willink.

Van der Burg, Wibren. 2001. The expressive and the communicative functions of law. *Law and Philosophy* 20(1): 31–59.

Van der Burg, Wibren. 2005. The irony of a symbolic crusade. The debate on opening up civil marriage to same-sex couples. In *Social and symbolic effects of legislation under the rule of law*, ed. Nicolle Zeegers, Willem Witteveen, and Bart van Klink, 245–275. Lewiston: The Edwin Mellen Press.

Van der Burg, Wibren. 2014. *The dynamics of law and morality. A pluralist account of legal interactionism.* Farnham: Ashgate.

Van der Burg, Wibren, and Frans W.A. Brom. 2000. Legislation on ethical issues. Towards an interactive paradigm. *Ethical Theory and Moral Practice* 3: 57–75.

Van Klink, Bart. 1998. *De wet als symbool. Over wettelijke communicatie en de Wet gelijke behandeling van mannen en vrouwen bij de arbeid.* Deventer: W.E.J. Tjeenk Willink. Dissertation Tilburg.

Van Klink, Bart. 2014. Symboolwetgeving. Opkomst, ondergang en wederopstanding van een begrip. *RegelMaat (Journal for Legislative Studies)* 1(29): 5–26.

Witteveen, Willem. 2005. A self-regulation paradox. Notes towards the social logic of regulation. *European Journal of Comparative Law* January 9.1 http://www.ejcl.org/91/art91-2.html. Accessed 20 Mar 2015.

Witteveen, W.J. 2007. Alternatieve wetgeving. De vele gezichten van de wetgever. In *Alternatieve regelgeving* (Preadviezen Nederlandse Juristenvereniging 2007), ed. Witteveen, W.J., Giesen I., and De Wijkerslooth J.L., 1–65. Deventer: Kluwer.

Witteveen, W.J., and B.M.J. van Klink. 1999. Why is soft law really law? *RegelMaat (Journal for Legislative Studies)* 3: 126–140.

Zeegers, Nicolle, Willem Witteveen, and Bart van Klink (eds.). 2005. *Social and symbolic effects of legislation under the rule of law*. Lewiston: The Edwin Mellen Press.

# Chapter 4
# How Law Matters: Sociological Reflections on the Symbolic Dimension of Legislation

**Rob Schwitters**

## 4.1 Introduction

There is no doubt that legal rules do have a symbolic dimension. The laws of a country may, for instance, fit into a narrative about its distinctive cultural traits. One may be proud of the Netherlands as being a liberal and tolerant country, referring to the legislation on prostitution, drugs and euthanasia. Citizens may also defend permissive or prohibitive laws to express themselves as being more or less tolerant. They may advocate more severe punishments and a stricter regulation of immigration to express themselves as hardliners on moral-political issues.

The real effects of law will be less relevant when legal rules are used as position-markers (Sunstein 1996a). In this case, one's preferences for more severe punishments will not be affected by empirical evidence indicating their ineffectiveness. And those demonstrating their approval of conservative positions on ethical issues by advocating a legal prohibition of euthanasia, will not easily shift their position if evidence shows that this prohibition leads to more (hidden) life-terminating practices in which standards of care are violated.

However, the symbolic dimension of law may affect the impact it has, and the symbolic effects may be deliberately used to enhance compliance. For instance, when citizens associate the law with the dignity of the queen, or of judicial authorities, this may induce them to obey the law. And if a repeated violation of the law is seen as characteristic of marginalized citizens, then this will be an extra motive for average citizens to follow the law. A factor which, according to Foucault, was actively incorporated in criminal-policy in the nineteenth century. By imprisoning criminals they were marginalized and excluded from their bonds with the lower classes, in days when social opposition led to violations of the law. It altered the

R. Schwitters (✉)
Faculty of Law, University of Amsterdam, Oudemanhuispoort 4-6,
1012 CN Amsterdam, the Netherlands
e-mail: r.j.s.schwitters@uva.nl



B. van Klink et al. (eds.), *Symbolic Legislation Theory and Developments in Biolaw*, Legisprudence Library 4, DOI 10.1007/978-3-319-33365-6_4

meaning of violating the law and canalized revolutionary tendencies in directions less threatening to the legal order (Foucault 1995).

More idealistic overtones can be discerned in the program of communicative or symbolic legislation as developed by Bart van Klink and Willem Witteveen.[1] In their account the symbolic dimension prevails when the effectiveness of law relies on a persuasion-based compliance.[2] Law can be seen as reservoir of arguments which may be convincing for its addressees. What contributes to voluntary acceptance is that the legislator restricts itself to the drafting of general clauses which are further determined in communicative interaction with citizens, experts and stakeholders (Van Klink 1998; Witteveen and Van Klink 1999; Witteveen 2005).

Van Klink's and Witteveen's account of communicative legislation may be seen as a remedy for the regulationcrisis which became manifest in the 90s (Witteveen 2005). The regulative ambitions of the government could no longer be achieved through the conventional parliamentary democratic procedure and instrumental regulation; a regulation which is based on *top-down* commands which are backed up by sanctions. Communicative legislation should be a more appropriate type of regulation, paralleling the introduction of less hierarchic styles of authority in management, education and in family life (Galanter 2005).

Van Klink and Witteveen contrast their account of legislation, which emphasizes the *positive* symbolic effects, with accounts which attribute the effectiveness of law to a calculative orientation on sanctions (*deterrence*). In this respect they subscribe to a distinction widely adhered to, in which the symbolic effects are supposed to cover a broad category of effects which are not dependent on coercive active enforcement by legal authorities. However, the symbolic effects distinguished by Van Klink and Witteveen, are a specific category of these symbolic effects: they address particularly those effects which are based on a persuasion-based compliance to the behavioral norms embedded in the law.[3] They are especially interested in these symbolic effects, given their low expectations of the contribution of traditional democratic legal procedure to a voluntary compliance with the law.

---

[1] Other labels are used for related approaches such as 'responsive regulation' (Ayres and Braithwaite 1995), 'communicative legislation' or 'communicative approach' (Van Klink 1998, Chap. 2 in this volume) 'interactive legislative approach' (Van der Burg and Brom 2000; Poort 2013 and Chap. 5 in this volume; Van der Burg 2014 and Chap. 3 in this volume).

[2] The terms 'persuasion-based compliance' refer to a voluntary compliance which in Habermas' theoretical framework relies on communicative action: a mutual understanding that can be tested and evaluated in terms of good reasons (Habermas 1997, Ch. 1).

[3] This explains why my definition of 'symbolic' is broader than for instance Van Klink's definition. He considers a law to be symbolic (in the positive sense) 'if a law has acquired an extraordinary meaning within the legal and political community (..). The law is not merely a set of rules, but it is also a symbol for something higher, more valuable, for example human dignity equality or environmental protection' (See this volume). My definition of symbolic does not only refer to the relevance of more fundamental values, but also to other symbolic qualities such as the fact that the law may be seen as an indication of how the majority uses to behave in specific circumstances, or that the law may be seen as the outcome of a democratic procedure which gives the legislated norm a special authority.

How could one not sympathize with an approach to legislation that does not rely on imposing norms top-down and are enforced with sanctions but which relies on persuasion? This idealistic normative approach to legislation shares several insights with the empirical findings of sociologists of law. It builds for instance on the sociological observation that top-down regulation can hardly alter behavior, as long as the imposed behavioral norms are not supported by informal norms embedded in semi-autonomous social fields (Moore 1973). However, in this text I seek to indicate that from a sociological angle there is more to say on the positive symbolic effects.

In sociological empirical accounts of compliance it is an open question what motives and incentives induce people to obey the law. Max Weber, for instance, distinguished custom, tradition, instrumental rational motives, (procedural) legitimacy and value-rational conviction (Weber 1922/1980). Van Klink and Witteveen focus only on a few of the various motives and incentives, merely addressing persuasion and deterrence-based compliance.

I will particularly address the signal-effects, which is a special category of positive symbolic effects which are not acknowledged by the communicative theorists. Compliance with the law has a multifaceted character and cannot simply be explained in terms of persuasion or deterrence. Moreover seeing, as they do, persuasion and coercion as irreconcilable factors, blinds us to the fact that coercion in particular conditions may assist in creating a persuasion-based compliance.

Analyses of compliance may also gain from sociologists' attention to the significance of informal enforcement activities. The willingness of citizens to obey the law is not only dependent on the coercive power of the state but also on the enforcement-activities of citizens (*second order-enforcement)* (Scott 2000). When other citizens enforce a norm, this could cause a shift in the opportunity structure, implying that citizens reconsider their options in response to the negative reactions they anticipate from other citizens. On the other hand, these enforcement activities could also induce them to change their own preferences and motives. The fact that both underpinnings of compliance may have a combined impact implies that we cannot categorically identify compliance as being based either purely on calculation or purely on persuasion.

In this text I will first pay some more attention to Van Klink's and Witteveen's communicative account of law. Next, I will fall back on Habermas to maintain that in current complex societies the coercive power of the state and the formal procedural legitimacy of the law have to be seen as building blocks of communicatively structured compliance. Law-following behavior can hardly be exclusively based on persuasion. In the last part I will address some more concrete illustrations of effects of legislation which cannot be exclusively attributed to deterrence or to persuasion but which build on the combined effects of a calculative orientation and internalization. The first category of these effects concerns the signal-effects, effects which rely on precise and clear norms and are not dependent on an actively motivated adherence to the legislated norms. A lukewarm acceptance is sufficient. Second, I will address the significance of second-order enforcement for the willingness of the addressees to obey the law. Finally, I will indicate that especially in circumstances in which the law contributes to overcoming problems of collective action, its effect will rely on the simultaneous impact of deterrent effects and persuasion-based effects. It is another illustration of the fact that the coercive power of the state and the formal procedural legitimacy of law may function as assistants to foster a persuasion-based acceptance of the law.

## 4.2 Communicative Legislation: The Superior Status of Legislator's Arguments

What is new in the communicative approach of legislation, as formulated by Witteveen and Van Klink, is not so much the discovery of symbolic effects as the articulation of how particular symbolic effects of law may be used as a remedy for problems of legislation in contemporary society. These communicative theorists attribute these problems to the deficiencies of a classical instrumentalist model of legislation.[4] When it concerns objects of regulation prevailing in current societies, which are characterized by complexity, dependency on expert knowledge and conflicting ideological commitments, the instrumentalist device to rely on clear and distinct legal directives, backed up by sanctions, is inadequate (Witteveen and Van Klink 1999). According to the communicative theorists, effective regulation of those domains requires the active engagement of citizens in the process of legislation. The legislator has to restrict itself to the creation of a legal framework which embodies general clauses. The interpretation and specification of these clauses has to be left to the cooperative activities of the legislator, addressees of law and to legal and other experts. In this cooperative endeavor, dialogue and persuasion play a central role (Witteveen and Van Klink 1999).

The prominent place of dialogue and persuasion in the communicative account of legislation reflects an increasing democratization and spreading of education and the more urgent task of integrating diverging life styles and opinions. Moreover in complex and technologically advanced societies the legislator often simply lacks the expertise to create detailed legislation. It has to fall back on the knowledge of experts.

The question can be raised whether the capacities of the legislator in this bottom-up approach of legislation are not reduced to the articulation of norms which are already prevailing in semi-autonomous social fields? Do legislated norms have a distinct quality which differentiates them from informal norms? Is the legislator able to alter informal norms?

The representatives of the communicative approach do not seem to have many doubts about the transformative-potentials of the legislator. According to them "symbolic law offers a vocabulary that affects the way in which legal and political actors perceive reality". It offers "a source of arguments". Moreover "these arguments or *topoi* can be used as trumps" (Chap. 2 in this volume). However, it is not

[4] For a more elaborate exploration of instrumentalist and communicative legislation, see Chap. 3 in this volume.

easy to conceive what they regard as the basis of this persuasive weight of legislator's arguments. What constitutes their 'trump status'? Why are they authorized to do more than just advise (Cotterrell 2005)?

A possible explanation might be found in the *constitutional* function which Van Klink and Witteveen ascribe to communicative law. A communicative law constitutes an interpretive community, that is a group of legal and political actors and citizens who are participating in the interpretation and application of the law. The legislator has to restrict itself to the introduction of general clauses.[5] These clauses express certain principal values and aims to which the addressees of law are already committed or to which they will become gradually committed, while they are engaged in the interpretative process. Two qualities which can be derived from this constitutional function can be distinguished: first, the participation of experts and stakeholders will improve the quality of the law. Second, participation of addressees in the process of legislation will have a positive effect on their obedience. It commits them to the legislator's cause.

It seems quite plausible that the legislator's arguments will be more convincing when these may rely on the input of engaged citizens, experts and stakeholders. However, the extra weight of these arguments cannot be exclusively explained in terms of this responsive quality. Only the fact that the legislator is able to constitute interpretative communities and win the engagement of citizens and stakeholders implies that the legislator is imputed a special authority.[6] This has to be attributed to the formal rules constituting its authority and the coercive power it can fall back on to enforce the legislated norms. It is impossible to deny the relevance of factors that the communicative theorists have disqualified as belonging to a top-down perspective.

## 4.3 Coercion and Positivity as Building-Blocks of Dialogue (Habermas)

As stated, Van Klink and Witteveen tend to contrast their legislative ideal with top-down instrumental legislation.[7] This may explain why they emphasize that the effectiveness of communicative legislation relies on debate and persuasion and not on procedural legitimacy and enforcement backed up by sanctions.

Especially in a sociological empirical account the multi-faceted bases of compliance have to be acknowledged. In this respect a one-dimensional focus on persuasion-based acceptance ignores the multitude of motivations and incentives

[5] The legislation on embryo research, Van der Burg refers to (Chap. 3 in this volume), is a good illustration of experts and stakeholders being engaged with the process of norm development.

[6] In later publications Van Klink acknowledges that the coercive dimension of legislation cannot be ignored, see e.g. Van Klink (2005).

[7] They have also described the differences in terms of 'soft law' versus 'hard law', see Witteveen and Van Klink (1999) and Witteveen (2005).

underpinning the obedience to law that prevail in social reality, and underestimates the significance of the formal procedural qualities of law and the coercive power of the state. Both factors are appreciated in Habermas' account of law, which nevertheless, as Van Klinks and Witteveens account, articulates the communicative dimension. He considers the formal procedural qualities of law and the coercive power of the state to be necessary building blocks of a persuasion-based obedience. The communicative dimension of law is optimally realized when there exists an appropriate balance between communicative legitimacy, positivity and coercive power (Habermas 1997).

Habermas sees law as a product of modern complex societies, and as a necessary device to realize a norm and value-based coordination of behavior and social integration.

According to him, the integration of societies is dependent on communicative intersubjective orientations. He distinguishes communicative action, as a necessary basis of social integration, from strategic action. In strategic action, actors do not strive for mutual understanding but try to realize individual aims. Communicative action is more demanding because it is successful merely insofar as cooperation is based on a consensus between the actors regarding the reasonableness of their aims (Habermas 1997).

Although social integration can to some extent be based on strategic forms of action (e.g. success on the market, efficiency), societies are stable over the long run only if the social order is perceived as legitimate and in accordance with what is true, right and good. It requires the grounding in consensual norms, which assumes actors to be orientated towards reaching understanding (Habermas 1997).

In modern societies this grounding in communicative action is a special accomplishment. In pre-modern societies the social order was based on shared norms and values being taken for granted. But current societies cannot fall back on the integrating force of perceived norms because these are too differentiated and pluralistic. Understandings which used to be shared and taken for granted are doubted and contested. With modernization, social interaction comes to depend more on communicatively actively accomplished consensus as opposed to consensus prescribed in advance by tradition. Only those norms that can meet with the approval of those potentially affected are considered to be valid. Integration has to rely on the active exchange of arguments through communicative action that relies on the compelling force of the better argument. Habermas sees it as the task of social institutions and the law to facilitate these rational discourses (Habermas 1997).

Modern societies need more advanced modes of normative orientation to cope with complex interdependencies. This complexity is mainly the consequence of the differentiation of domains that are based on systemic rationales such as the economic and political-administrative domain. While in pre-modern societies conduct is predominantly normative and symbolically structured, modern complex societies have domains that are governed by non-linguistic media such as money (economic domain) and power (state/administration). These media disentangle economic and administrative activities from religion, family-relations and traditional bonds and values (Habermas 1997).

Money and power enable the coordination of complex systemic interdependencies. What counts in the 'systems' is effectiveness and success. Coordination is not based on shared understandings or agreement but it takes place behind the actor's back (e.g. the rationality of the market) (Habermas 1997). It is the prevalence of these complex system that facilitates the productivity and wealth in our societies.

As mentioned earlier, according to Habermas, a robust coordination and integration cannot rely on the systemic rationale of the market and political administration but requires communicative underpinning. The institutionalization of communicative processes is urgent since the systemic rationalities embedded in the market and bureaucracies have a tendency to crowd out these consensual forms of integration ('colonization of the life world') (Habermas 1987, 1997). All norms become vulnerable to being assimilated into the strategic rationality of social subsystems of finance and administrative power.

To counter the erosion of communicative action, this process of a growing systemic complexity requires a post-conventional mode of normative integration. A mode of social integration which relies on discursive processes in which only the best argument counts. In Habermas's perspective this amounts to the proper place for instrumental and strategic orientations and the right balance between these orientations and communicative orientations.

In pre-modern societies actors were able to derive their mutual expectations of behavior from a framework of perceived norms. Once a discursive orientation on norms becomes predominant this steady framework is no longer available. Law is able to integrate complex societies because it has some mechanisms that make the integration less dependent of actual consensus. It grants citizens a private domain where they are allowed to follow their own preferences and motivations (*private autonomy*) (Habermas 1997). And in contrast to morality, law leaves the motives for compliance open while demanding law-following behavior. It does not require citizens to comply with the law for the right reasons. It may rely on voluntary rational adherence but it may also rely on identification with others, the fear of punishment or the negative reactions of fellow citizens (Habermas 1997). This does not mean that it should not be an endeavor of the legislator to win the assent of the addressees of law. But it is a functional sociological observation that thanks to the formal procedural qualities of law (rules imposed by recognized authority) and its coercive character, law is able to deal with the contingency of norms in complex societies. Law is the medium which in modern societies helps to ease the burden of social integration that falls on moral discourse and communication. Norms and values incorporated in a legal framework can be doubted and discussed because the law entails a mechanism to define which norms should be followed (positivity) until further notice and which norms can be enforced (Habermas 1997).

Habermas observes that in complex societies, such as ours, features of the legislation-process also imply that you cannot expect all legislation to rely on the voluntary assent of all citizens. This would be too ambitious given the range of relevant arguments and expert-knowledge which plays a role in processes of legislation and the pluralism prevailing in current societies. The bulk of political decision-making includes, for instance, pragmatic issues in which empirical knowledge is relevant. Moreover, the principle of certainty reduces the options for the selections of norms. Further, time pressure sets limits on open discussion in which only the better argument counts. Within the actual process of legislation, compromises, bargaining and the majority rule play an important role. Therefore Habermas reformulates the idea that normative decisions have to be able to win the assent of all citizens, when it applies to matters of legislation. Legislated rules only have to be indirectly legitimized by the universal assent of those affected, which means that the assent merely has to apply to the democratic character of the procedure (Habermas 1997).

A law is legitimate when, besides protecting private autonomy, it can win the assent of members of the legal community because it is the product of a formal decision-making body which is based on deliberation and discourse. Apparently this concept of democratic legitimation brings Habermas very close to accepting the prevailing institutions as sufficient guarantees of his discursive rationale and seeing the formal procedural qualities of law as a sufficient guarantee of the communicative underpinning of law. However, if this should be an appropriate description of his position, it would seriously diverge with the ambitions of Van Klink's and Witteveen's project of communicative legislation.

Although it has to be admitted that there is a lot of institutional realism in Habermas' perspective, a formal democratic procedure does not suffice to accomplish his discursive ideals. The discursive underpinning is only realized when citizens are able to regard themselves as the authors of law (*public autonomy*) (Habermas 1997). Democratic procedure is only one requirement. The other is that there has to exist a fruitful interplay between the deliberation and decision-making in governmental institutions and informal discussions among ordinary citizens. He sees it as the role of citizens and mass media to create well-considered public opinions. It is the role of legislative bodies to be receptive to the information, arguments and suggestions which are developed in the discursively structured public domain. These considerations bring him close to the ambitions of the communicative theorists.

To summarize, what Habermas shares with the defenders of communicative legislation is the appreciation of the discursive underpinning of legislation. But where in the design of communicative legislation the significance of procedural legitimacy and coercion is downplayed (as if legislation is just a matter of a horizontal dialogue between legislators and addressees) Habermas regards these factors as building blocks for a persuasion-based compliance. In a complex and pluralistic society social integration and coordination of behavior are not only dependent on the persuasion-based adherence with legislated norms but also on these factual dimensions of law; a dimension which is based on formal legal qualities, coercive power and which allows a calculative orientation of actors. This factual dimension releases the integration of the necessity of an actually prevailing consensus and facilitates debate and argumentation.

## 4.4 Alternative Bases of Compliance

The compliance with the law is mediated through various social phenomena which cannot be identified in terms of the persuasion versus coercion contrast. Seen through a sociological lens, in particular the impact of citizens' mutual expectations of law-following behavior has to be taken into account. Illustrative are the signal-effects. These effects rely especially on the information the law gives the addressees of other citizen's preferences and how they will react when the rule is violated. A refined sociological account has to acknowledge that coercion is not only the outcome of state-imposed punishment but also of the negative reactions of fellow citizens (*second-order enforcement*) (McAdams 1997; Scott 2000; Griffiths 2003). The contribution of law to solving problems of collective action offers a good illustration of this.

### *4.4.1 Signal-Effects*

There is a symbolic dimension of legal rules which is not explored by the advocates of communicative legislation. It concerns the *signal-effects*, effects which require clear and precise legal rules. These effects are extensively discussed in the dissertation by the Norwegian sociologist of law, Vilhelm Aubert (1954). This exploration has never received the attention his study of regulation of the position of Norwegian housemaids has received, which is regarded as *the* illustration of *negative* communicative legislation. Negative, because the law just creates the illusion of social reform, in fact lacking the enforcement apparatus to have any social impact (Aubert 1961). This negative concept has long been the standard interpretation of symbolic legislation.

Instead, the assumption underpinning Van Klink's and Witteveen's concept of communicative legislation is that symbolic effects may contribute to law's effectiveness. The same applies to the signal-effects which can be addressed in the footsteps of Aubert. However, contrary to the effects of communicative legislation, signal-effects do not rely on actively motivated conviction but on a lukewarm passive compliance with the law. This compliance is based on the law's significance for addressees' mutual expectations of behavior. People do not merely see the law as a set of prescriptive norms but also as a description of the rules the majority adheres to (Aubert 1954). These rules give people an indication of the expectations they may anticipate in various situations, in which informal norms are diffuse. People often adapt their behavior to these laws without much passion or conviction but just because it makes life easier.

It is not difficult to understand that signal-effects are relevant in situations in which the content of the norms is not morally precarious or not contested. Whether you have to drive on the right side or the left right of the road is for instance morally irrelevant. A legal prescription informs car drivers what they may expect from each other and this signal-effect in combination with the drive of actors not to risk life and limb, may be seen as the major source of compliance.

This does not imply that signal-effects merely prevail when it concerns morally neutral regulations. For instance doctors may see legal rules prohibiting euthanasia as a reflection of standard practices within the medical profession. This may explain why in many countries with laws prohibiting euthanasia, the legal authorities are reluctant to actively enforce these rules. Bringing doctors to trial would reveal that euthanasia is a common practice. That might undermine the repressive policy, since doctors practicing euthanasia on severely suffering terminal patients, may count on the sympathy of many citizens. They may be seen as heroes instead of as criminals. For legal authorities which are reluctant to allow euthanasia it may be wise policy to ignore the violations of the prohibitive rules.[8] To summarize, a slack enforcement may contribute to law's effectiveness when this contributes to the illusion that prohibited behavior is not practiced (Aubert 1954).

McAdams explains the signal-effect as the consequence of the significance of informal (non-legal) sanctions for the effectiveness of law (McAdams 1997). People's compliance with the law is often more dependent on rule enforcing reactions of fellow citizens and colleagues than on formal enforcement activities. People are likely to act in accordance with their perceptions of what other people expect and appreciate. This impact of others' opinions is especially relevant when it concerns the willingness of people to enforce norms. Since the costs of these informal reactions are not very high the inclination to react is related to what there is to win or lose in esteem. People will be reluctant to react when they anticipate that others will not adhere to the norm that informal reactions are required in the prevailing situation (McAdams 1997; Scott 2000).

A legislated norm may empower those who adhere to the content of the law to use informal sanctions against those violating the law. This applies especially for those with weak preferences for the norm. They are less likely than those with intense preferences to address those violating a norm when they are not certain about the support they may receive from others (scared to be seen as busy-bodies or dummies). They will be more willing to sanction violators of the norm, the more they assume that others are also willing to sanction (Scott 2000). An increase in the number of potential norm-enforcers reduces the expected costs of conflict of any single enforcer and enforcement-activities may then even contribute to one's reputation. A legislated norm teaches the community about the majority's opinion on the prevailing norms and the rights or duties to enforce these norms (Scott 2000). It also reduces the inclination of those violating the norm to react aggressively to an attempt to shame them.

The success of the anti-smoke legislation in the USA can be seen as an illustration of the signal-effects. The prohibitive law was considered to be very effective, while lacking an active enforcement policy. The explanation being that the law changed the perception of what the majority's sensitivities were towards smoking.

[8] Following this line of thought I explained the in Norway prevailing enforcement-policy, see Schwitters (2005).

Before the introduction of the law there were already many citizens who opposed smoking. But as long as they were uncertain about the opinions of others they were reluctant to complain directly to smokers in public places. The ordinance that was introduced, which imposed clear rules about where smoking is permitted, gave them an indication of the norm upheld by the majority and stimulated them to enforce the rule against violators (McAdams 1997; Kagan and Skolnick 1993).

The signal-effects may be seen as a distinct category of symbolic effects. These do not have to rely on the motivated acceptance of the content of the law and these effects do not directly depend on the sanctioning power of the state. However, the fact that legislated rules are seen as indications of majority's beliefs and practices presupposes that legislated rules are given extra authority. This distinctive authority cannot be explained without acknowledging the (indirect) effect of coercive power and the formal procedural basis of law.

### *4.4.2 Law as an Assistant to Change Behavioral Norms*

In analyses that emphasize the relevance of informal reactions on violations of norms, the effectiveness of law may be seen as the outcome of a shift in the opportunity structure. Those considering disobeying the law will respect the law because they fear the negative reactions of others. Coercive power is not merely embodied in state sanctions but also in informal sanctions. This implies that Holmes' 'Bad Man' figures not only in the judicial domain but in the non-legal domain as well.

Behavior that is determined by the opportunity structures might in the long run be governed by the feeling of duty, which reflects an internalization of a norm (McAdams 1997; Scott 2000). There will be a more fluent transition when the opportunity structure relies on informal norms than on norms embedded in the law. Within informal social relations a rich variety of sanctions prevails, including subtle reactions such as furrowing your brows, while the repertoire of legal sanctions is quite limited. In addition, informal sanctions can be flexibly adjusted to the context and the features of the norm-violating individual.

Within the domain of non-legal relations there are more supplementary mechanisms which stimulate addressees to adhere to the norms of desirable behavior. Processes of identification for instance, may urge them to internalize norms and values. People are likely to copy the behavior of those they are dependent on or respect. Legislation may be more or less effective to the extent it is adequately intervening in the structures of status and identification. If for instance, repeated violation of the laws is a propensity of marginalized groups in society and more respected people tend to obey the law, the adoption of norms in law may have a positive effect on following these norms.[9]

[9] This rationale makes it understandable that competing groups in society are often eager to have norms they cherish, articulated in legislation, see e.g. Gusfield (1967).

A refined sociological account of compliance has to acknowledge the combined effect of shifts in the opportunity structure and internalization. For instance an ordinance prohibiting littering on the streets may inform individual citizens about the sensitivities of the community. The immediate effect may be a change of their behavior because they anticipate informal sanctions. In the long run the external norm which prohibits littering may be internalized and in that case a new normative framework prevails which influences individual citizens. The relation between effects which have to be attributed to changes in the opportunity structure and internalization, is rather complex. Both underpinnings of compliance are interrelated and overlapping.

Legal economists are inclined to exclusively focus on opportunity structures and ignore the fact that the law may alter and modify preferences. In their perspective effective enforcement is based on cost/benefit calculations. According to this rationale civil liability for instance enhances safety because it confronts actors with the financial costs of violations of standards of care. But when this device, appealing to the calculative orientations of the addressees of law, is adopted in legal practice it may crowd out the informal enforcement mechanisms which may be invoked by imposing a duty to obey legally imposed standards of care. If the reaction to socially undesirable behavior is limited to confronting the violators with the financial costs of their behavior, the sanction becomes a license that permits behavior as long as one is willing to pay for it.[10] A price tag on behavior will create a completely different appreciation of regulated behavior than an imposed duty, which may promote informal enforcement and (as a consequence) internalization.

Internalization is especially likely when law solves problems which have their origin in a conflict between narrow individual preferences and deliberate political choices (well-considered interests). Cass Sunstein observes that people are often kept from following their deeper convictions and commitments because they obey reputational norms that stimulate them to ignore these convictions and commitments. A majority may oppose dog owners not cleaning up after their dogs but refrain from enforcing this norm for fear of the reactions of others (to be seen as a busy body). People's private convictions may diverge greatly from public appearances. Hockey players may for instance prefer not to wear mouth guards as long as it is seen as cowardice but would use this protection if the reactions were not so negative. Following the rationale of Sunstein, the introduction of an ordinance prescribing mouth guards may then be a proper device to overcome the adverse effects of the unreflected reputational norms (Sunstein 1996b). The introduction of a legal rule may in the end foster a persuasion-based adherence to the prescribed norm (Sunstein 1996b; McAdams 1997; Scott 2000).

This rationale that the law may be contributing to overrule unreflected private preferences with deeper concerns can also be recognized in problems of collective action (Sunstein 1996a). While individual preferences may cause *free-rider behavior* among actors, legal intervention may create a situation that corresponds better

---

[10] Along this lines I have criticized economical accounts of non-pecuniary damages, see Schwitters (2012).

with their real (well-considered) preferences. Free-riding, which in first instance may be reduced by the coercive power of the state to impose a norm, may soon be reduced because actors will voluntary adhere to the imposed norm. Moreover, legislation may alter mutual expectations and reactions and bring about a decisive shift in informal enforcement activity (Sunstein 1996a; Scott 2000).

Even when the difference between narrow interests and well-considered interests is not as clear as within the framework of the problem of collective action, legislation may be persuasion-based when it reflects well-considered interests. Sunstein refers to legal bans on the sale of sexual and reproductive capacities. Law may fortify norms regarding the permissible use of money (restrict commodification) (Sunstein 1996a). One may also consider practices of sex-selection (enabled by IVF and sperm-sorting). What might be a desirable policy on the basis of the direct interests of parents, might not be wise policy when wider social implications are taken into account. To submit the gender of children to human intervention might lead to a contested gender policy: to provoke the government to correct or influence the decisions of individual parents when their aggregate decisions result in an unbalanced population. A legal ban on sex-selection might be a devise to avoid this politization of sex-selection and be more in line with well-considered interests.

It is the essence of political deliberation and democratic processes to substitute narrow short-term preferences for deliberated preferences. People's political judgments are not a product of their narrow self-interest. People may in their role of political actors favor altruistic or other aims which are not reflected in their actual daily behavior. Narrow self-interests or reputational norms may keep them from acting in accordance with these aims. In their political judgments they may opt for legal rules and institutions which alter their immediate preferences. They may seek the assistance of law to create a social order that they appreciate more than the prevailing order. Political judgments may reflect second-order preferences (wishes about wishes) (Hirschman 1984; Sunstein 1996b; Habermas 1997). A legislated rule, which brings behavior more in tune with reflected interests, may produce an important change in behavior and while first being dependent on coercive force, soon relies on communicative underpinning.

## 4.5 Final Remarks

In this text I have delved deeper into the question: how law matters. More particularly, I have addressed the symbolic effects of law. In the program of communicative legislation the effectiveness of law is explained in terms of the participation of citizens, stakeholders and experts. Their participation should improve the quality of legislated norms and contribute to their acceptance of these norms. This program has developed in response to the deficiencies of traditional, democratic *top down* legislation. It assumes more horizontal relations between the legislator and the citizen. Compliance is seen as based not on commands backed up by sanctions, but on the positive symbolic effects of persuasion.

Taking a sociological stance, I questioned some presumptions of the communicative program. I followed Habermas, who shares the normative ambitions of the communicative theorists, but acknowledges the functional significance of the coercive power of the state and the formal procedural legitimacy of the law, for a persuasion based compliance in complex societies. I also suggested that compliance may be based on symbolic effects other than persuasion. The *signal*-effects are an illustration of this. Finally, I showed that especially in circumstances in which the law contributes to overcoming problems of collective action, its effect will rely on the simultaneous impact of deterrent effects and persuasion-based effects. This is another illustration of the fact that the coercive power of the state and the formal procedural legitimacy of law can help to foster a persuasion-based acceptance of the law.

## References

Aubert, Vilhelm. 1954. *Om straffens sosiale funksjon*. Dissertation, Oslo.

Aubert, Vilhelm. 1961. The Housemaid. An occupational role in crisis. In *Sociology, the progress of a decade*, ed. S. Lipset and S. Melser. Englewood Cliffs: Prentice-Hall.

Ayres, Ian, and John Braithwaite. 1995. *Responsive regulation. Transcending the deregulation debate*. Oxford: Oxford University Press.

Burg, Wibren van der. 2014. The dynamics of law and morality. A pluralist account of legal interactionism. Farnham: Ashgate.

Burg, Wibren van der and Brom, Frans W.A. 2000. Legislation on ethical issues. Towards an interactive paradigm. *Ethical Theory and Moral Practice* 2000:57–75.

Cotterell, Roger. 2005. Legal effects and moral meanings. A comment on recent debates on approaches to legislation. In *Social and symbolic effects of legislation under the rule of law*, ed. Nicolle Zeegers, Willem Witteveen, and Bart van Klink, 339–359. New York: The Edwin Mellen Press.

Foucault, Michel. 1995. *Discipline and punish. The birth of the prison*. New York: Vintage Books.

Galanter, Marc. 2005. Preface. The perplexities of legal effectiveness. In *Social and symbolic effects of legislation under the rule of law*, ed. Nicolle Zeegers, Willem Witteveen, and Bart van. Klink, XVI–XX. New York: The Edwin Mellen Press.

Griffiths, John. 2003. The social working of legal rules. *Journal of Legal Pluralism* 48: 1–69.

Gusfield, Joseph R. 1967. The Symbolic process in public designation of deviance. *Social Problems* 15(fall): 175–188.

Habermas, J. 1987. The theory of communicative action. In *Lifeworld and system, a critique of functionalist reason*, vol. 2. Boston: Beacon Press.

Habermas, J. 1997. *Between facts and norms*. Cambridge, UK: Polity Press.

Hirschman, Albert O. 1984. Against parsimony. Three easy ways of complicating some categories of economic discourse. *Psychological and Sociological Foundations* 74(2): 89–91.

Kagan, R.A., and J.H. Skolnick. 1993. Banning smoking. Compliance and enforcement. In *Smoking policy. Law, politics and culture*, ed. R. Rubin and S.D. Sugarman, 49–69. Oxford: Oxford University Press.

Klink, B.J.M. van. 1998. *De wet als symbool. Over wettelijke communicatie en de Wet gelijke behandeling van mannen en vrouwen bij de arbeid*. Deventer: W.E.J. Tjeenk Willink

Klink, Bart van. 2005. An effective-historic view on the symbolic working of law. In *Social and symbolic effects of legislation under the rule of law*, ed. Nicolle Zeegers, Willem Witteveen, and Bart van Klink, 113–145. New York: The Edwin Mellen Press.

McAdams, Richard H. 1997. The origin, development, and regulation of norms. *Michigan Law Review* 96(2): 338–433.

Moore, Sally F. 1973. Law and social change. The semi-autonomous social field as an appropriate subject of study. *Law and Society Review* 7: 719–746.

Poort, Lonneke. 2013. *Consensus & controversies in animal biotechnology. An interactive legislative approach to animal biotechnology in Denmark, Switzerland, and the Netherlands*. The Hague: Eleven International Publishing.

Schwitters, Rob. 2005. Signaalwetgeving. Positief symbolische effecten in het werk van Vilhelm Aubert. *Recht der Werkelijkheid* 26(1):57–64.

Schwitters, Rob. 2012. Non-pecuniary damages, symbol or financial incentive? Bert Niemeijer and Rob Schwitters (eds). *Recht der Werkelijkheid* 33(2) (Tort, behavior and social context), p 48–64.

Scott, Robert E. 2000. The limits of behavioral theories of law and social norms. *Virginia Law Review* 86(8): 1603–1648.

Sunstein, Cass R. 1996a. On the expressive function of law. *University of Pennsylvania Law Review* 144(5): 2021–2053.

Sunstein, Cass R. 1996b. Social norms and social roles. *Columbia Law Review* 96(4): 903–968.

Weber, Max. 1922/1980. *Wirtschaft und Gesellschaft. Grundriss der Verstehenden Soziologie, Fünfte, Revidierte Auflage besorgt von Johannes Winckelmann*. Tübingen: J.C.B. Mohr.

Witteveen, Willem. 2005. Turning to communication in the study of legislation. In *Social and symbolic effects of legislation under the rule of law*, ed. Nicolle Zeegers, Willem Witteveen, and Bart van. Klink, 17–45. New York: The Edwin Mellen Press.

Witteveen, W.J., and B.J.M. van Klink. 1999. Why is soft law really law? A communicative approach to legislation. *Regelmaat* 14(3): 126–141.

# Chapter 5
# The Tension Between the Functions of Law: Ending Conflict Versus Dynamics

**Lonneke Poort**

## 5.1 Introduction

When regulating complex issues such as medical and biotechnological developments, communicative approaches to law are presented as an alternative to the traditional instrumental approach to law. While this latter approach seeks to establish clear-cut rules to end conflict and guide behaviour, it can be extremely difficult, if not impossible, to set concrete rules for dealing with complex issues. Communicative approaches to law leave room for alternative, more general clauses that can then be concretised in practice. The use of general clauses comprising open norms creates *dynamics*, which are a characteristic element of communicative approaches to law. *Dynamics* are best explained as a process of ongoing norm development. The use of open norms is suitable for addressing complexity and situations in which there is a lack of knowledge (Van Klink 2016, Chap. 2, this volume: 8); leaving the law and the legislative process open in this way creates scope for supplementing this knowledge.

A second characteristic element of communicative approaches to law is *interaction*. This can be seen in the focus on horizontal decision-making or, in other words, in the involvement of relevant actors in developing norms on a horizontal level. This involvement ensures a responsiveness to developments in the field and can improve the adequacy of the norms developed, especially in the case of complex ethical dilemmas such as animal biotechnology and embryo selection (Van der Burg and Brom 2000). An additional benefit is that horizontal processes may contribute to greater acceptance of norms, given that all actors were able to have a say in developing them.

L. Poort (✉)
Department of Sociology, Theory and Methodology, Erasmus University Rotterdam, Erasmus School of Law, Burgemeester Oudlaan 50, 3062 PA Rotterdam, the Netherlands
e-mail: lonnekepoort@gmail.com



B. van Klink et al. (eds.), *Symbolic Legislation Theory and Developments in Biolaw*, Legisprudence Library 4, DOI 10.1007/978-3-319-33365-6_5

In recent years, the basic elements of communicative approaches to law and the extent to which they can be applied have been studied both by proponents and opponents, especially in the Netherlands. Theoretical models, such as symbolic legislation (Witteveen and Van Klink 1999), the interactive legislative approach (Van der Burg and Brom 2000; Stamhuis 2005; Poort 2013) and responsive regulation (Ayres and Braithwaite 1995; Black 1998) build on the idea that regulation will function more adequately if it adapts to norm development in the field (Van der Burg 2016, Chap. 3, this volume).[1] At the same time, these communicative approaches have attracted considerable criticism and their functionality questioned (see, for example, Griffiths 2005). Besides questions concerning these approaches' lack of authority, several critical scholars have highlighted the tension between law's basic function of ending conflict and the incentive in communicative approaches to law to leave norms open for further debate (Griffiths 2005; Van Klink 2014), given that the existence of open norms may conflict with the rule of law.

Communicative approaches to law have been criticised not only from an external, but also from an internal perspective. Proponents of an interactive legislative approach have challenged, for example, its practical functionality in specific cases (Stamhuis 2005; Poort 2013). Both Stamhuis (2005) and Poort (2013) have criticized the interactive approach for having a too strong focus on reaching consensus. Consensus plays a dominant role in this approach, both in structuring horizontal processes and in justifying the use of open, symbolic norms. Consensus is seen as a way of overcoming the lack of enforcement (see Poort 2013: Chap. 3). The search for consensus is, thus, a strong means of countering external criticisms of a communicative approach to law. I have demonstrated, however, that consensus-thinking counteracts the basic elements of the communicative approach to law in that it risks foreclosing the debate, while this approach specifically sets out to leave debate open for further development. Although the alternative I have presented – an *ethos of controversies* (Poort 2012, 2013) – cannot be used to justify norms, it may contribute to promoting an open-ended debate by focusing on controversies rather than on seeking to reach consensus as soon as possible.

Although the ethos of controversies may offer a solution for avoiding premature closure of debate, it has not yet been used to address the general criticisms of communicative approaches to law. Whereas consensus could serve as justification for countering these criticisms, the ethos of controversies would seem to increase the tension referred to above between law's basic function of ending conflict and the incentive that communicative approaches to law provide to leave debate open for further development of norms.

This chapter seeks to address the tension between ending conflict and stimulating dynamics, while at the same time avoiding a repetition of arguments from ear-

[1] Explanation of differences in the use of communicative approaches to law in this chapter in contrast to their use by other authors in this volume.

lier debates between opponents and proponents of communicative approaches to law. It starts by exploring the general discussion about communicative approaches to law and then elaborates on the difficulties regarding consensus-thinking in communicative approaches to law, while also discussing the possibility of an ethos of controversies. The third and final part addresses the main purpose of the chapter, which is to analyze the ethos of controversies in the light of the tension, and argues that the tension can be resolved by an understanding of the ethos of controversies in a two-track approach to norm development (Poort 2013). Instead, therefore, of explaining the tension, this chapter presents a different way of looking at it, and one which in turn helps to reduce it.

## 5.2 Communicative Approaches to Law

As I explained earlier, there are several theoretical models of communicative approaches to law that start from the assumption that regulation will function more adequately if it adapts to norm development in the field.[2] To that extent, these approaches identify certain additional functions of law alongside the traditional instrumental and constitutional functions. These include an emphasis on the communicative and interactive functions of the law, whereby the law is seen as a framework in which communicative structures are institutionalised and in which vocabulary for communication is introduced. The communicative approaches to law also underline its expressive or symbolic function, with the open norms incorporated into the legal framework expressing a certain aspiration or value.

Section 5.2.1 explores one of the communicative approaches to law, being the interactive legislative approach, and discusses the basic grounds of this approach, inspired by the ideas of Selznick on responsive law and Fuller on interactionism, as well as my criticism of the functioning of the interactive legislative approach in practice. As all the communicative approaches to law share a strong focus on the communicative dimensions of law-making and several of these dimensions' characteristics, these basic grounds need to be discussed in order to provide an understanding of communicative approaches to law in general. In Sect. 5.2.2, I briefly examine the criticism that has been expressed with regard to these basic grounds, with a specific focus on the tension between ending conflict and stimulating dynamics.

[2] For a further analysis of the differences and similarities, see Van Klink and Van der Burg (both in this volume).

## *5.2.1 Interactive Legislative Approach*

### 5.2.1.1 Basic Grounds[3]

This chapter takes the ideal-typical model of the interactive legislative approach, as reconstructed in my earlier research, as its starting point. This model builds on the ideas of Selznick and Fuller, who have both had a major influence on the development of this approach and on other theories of symbolic and communicative law. The interactive legislative approach reflects two main elements that are also visible in the ideas of Selznick and Fuller: interaction and dynamics.

The term 'interaction' relates to the need for a process of interaction between the legislature and society or relevant sectors within a society (Selznick), as well as both horizontal and vertical structures of communications (Fuller). According to the latter, the vertical structures are embedded in the horizontal ones, with both structures considered relevant for guiding interactive processes.

The term 'dynamics' refers to an ongoing process of norm development that ensures that legislation does not continually have to play catch-up with society (Selznick). The use of open norms and the interplay between interaction and ongoing norm development – which contributes to concretisation of these norms – define dynamics. These elements are similar to the general elements characteristic of all communicative theories of law, including the emphasis on the communicative dimensions and functions of law, the use of aspirational norms and the importance of interaction (Poort 2013).

Selznick (1992) argues that legislation developed through dialogue responds better to society's needs and aspirations. He argues that, in law-making, the problem itself plays a more prominent role than formal rules that guarantee validity. If the problem itself is put at the forefront, rules must be open for revision to ensure that they continue to reflect general values and principles in society (Selznick 1992). In summary, characteristic elements of the responsive approach to law are the use of open, flexible norms and the need for continuing dialogue (Poort 2013).

Fulller's works *The Morality of Law* and *The Principles of Social Order* have strongly influenced the communicative approaches to law. The relationship between the effectiveness of the legal system and human interaction is an important feature in his works. Fuller argues that legal decision-making finds its basis and justification in human interaction. Human interaction originates in relationships built on the principles of reciprocity and shared commitment and, according to Fuller, the legal system requires both in order to be effective.

The principle of reciprocity can be related to vertical structures of decision-making, such as those flowing between an association and its members. The members have a duty to follow the basic rules that keep the association together. At the same time, this duty sets the basic rule for a social order (Fuller 1969); reciprocity in that sense stipulates a minimum for responsible interaction.

---

[3]The explanation of these basic grounds builds on Chap. 3 of my dissertation (Poort 2013). For a more in-depth analysis of Fuller and Selznick, see Sect. 3.2 of this dissertation.

The second principle holding human interaction together is the principle of shared commitment that results in a sense of co-operative effort. 'Shared commitment' refers to a form of horizontal interaction in which actors search for common ends or purposes (Fuller 2001). Shared goals ensure that actors are more willing to co-operate. This willingness to co-operate is required in order to enable responsible interaction. At the same time, shared goals and a willingness to co-operate cannot in themselves function as a good basis for legal rules. Consequently, vertical structures are equally important. Fuller argues that, to that extent, vertical and horizontal structures are both needed for responsible interaction.

#### 5.2.1.2 Consensus Versus Controversies

I tested the ideal-typical model of the interactive legislative approach by analysing the regulations on animal biotechnology in Switzerland, Denmark and the Netherlands. One of the major outcomes of these case studies is discussed here.

What is remarkable is that despite the open, interpretable character of the law in the above countries and the various tools for interaction, debate on the subject was prematurely closed; in other words, dynamics were not established. An explanation for this premature closure can be found in the strong focus on reaching consensus, which plays a dominant role both in structuring horizontal processes and in justifying the use of open symbolic norms in the theoretical models on communicative law, and specifically in the ideal-typical model of the interactive approach.[4] This strong focus on consensus is counterproductive as, instead of setting the scene for rational debate, consensus-thinking risks foreclosing the debate. Let me explain this (see Poort 2012, 2013).

Consensus-thinking implies structuring legal decision-making around a search for consensus. The use of consensus in structuring legal decision-making builds on two theoretical conceptions: consensus as the ideal outcome, and consensus as a regulative ideal. Consensus as the ideal outcome can be traced back to Habermas' notion of the *ideal speech situation* (Habermas 1996): under ideal circumstances, all discussants can participate in debate with equal opportunity and equal power, and without any constraint. Habermas claims that this situation can be established when striving for consensus, while also arguing that if all parties can consent to an outcome of a debate, this outcome represents the ideal outcome.

Consensus as a regulative ideal is, however, a more dynamic concept than consensus as the ideal outcome. The regulative ideal involves an orienting aim for rational debate, in which actors use rational arguments for the purposes of persuasion. Proponents even acknowledge that consensus may never be reached. Instead of emphasizing the outcome, this idea focuses on the benefits of aiming for consensus as a method for structuring deliberation. Gutmann and Thompson (1996), for

[4] Van der Burg and Brom have acknowledged this criticism of their ideal-typical model of the interactive legislative approach. The ideal-typical model described by Van der Burg in this volume does no longer cohere, therefore, with the ideal-typical model criticised in my earlier work.

example, regard aiming for consensus as a method for structuring debate that reduces disagreement and ensures an openness to other people's points of view.

In the above case studies, a strong focus on consensus seems to have resulted in a premature closure of debate.[5] A clear example can be found in the case of Switzerland (Poort 2012, 2013: Chap. 6), where the concept of the dignity of living beings was one of the incentives for regulation and was consequently discussed in detail during the regulatory debate. Although the relevant actors agreed that the dignity of living beings was an important concept, they were diametrically opposed on the question of its meaning. The Swiss Ethics Committee on Non-Human Gene Technology (ECNH) played a leading role in the conceptualisation of the dignity of living beings in the regulatory context. It emphasized that there was not yet any consensus on the meaning of dignity; the committee members themselves could not even agree on the preliminary conceptualisations. It also stated that the dignity of living beings required further elaboration and crystallisation in regulative practice.

In this example, consensus was not reached, but instead functioned as a regulative ideal. In the meantime, the preliminary conceptualisations of the dignity of living beings started to function as a basis for the regulatory framework in practice. In contrast, however, to earlier statements on the need for further elaboration of the concept of this dignity, both the ECNH and the responsible authorities adopted and applied the preliminary conceptualisations without questioning their meaning. The concept therefore started to function as if it were based on consensus, while in reality there was no consensus at all on its fundamental meaning. Indeed, the concept was still highly disputed. The preliminary conceptualisations presented dignity as a gradual concept, thus bypassing a fundamental distinction between human beings and non-human beings, even though this distinction is crucial for certain people, particularly those with Christian beliefs. These differing viewpoints were not reintroduced into the debate and were even silenced, despite their remaining part of the problem. The presentation of dignity in this case as a gradual concept eventually led to heated debates between members of the ECNH and theologians (Poort 2013: Chap. 6), and it is doubtful whether attitudes continued to remain positive once the debates had reached this stage (Honig 1993). The preliminary conceptualisations of the dignity of living beings started to function as an end or the ideal outcome. There was then no need for further fundamental debate on the dignity of living beings since the ideal outcome had already been achieved. This case study illustrates that instead of establishing dialectic, the relationship between consensus as a starting point, consensus as an ideal outcome and consensus as a regulative ideal seems to have worked against any further crystallisation.

This case study shows that taking consensus as both a regulative ideal and an ideal outcome has certain consequences. To a certain extent, consensus-thinking may structure debate and encourage positive attitudes in the form of a willingness to co-operate. At the same time, however, it has its limits. To begin with, having a common goal of reaching consensus does not ensure that actors agree on the contents of this consensus or on how to reach this common goal. Secondly, as soon as

[5] For a more in-depth analysis of the risks of consensus-thinking, see Poort (2012, 2013).

a temporary decision has been reached, debate is closed or at least not encouraged to continue as the temporary decision is easily perceived as being built on consensus (in other words, the ideal outcome). To that extent, there is no need for further discussion, and debate consequently stagnates. Thirdly, consensus-thinking risks excluding controversial views as it does not leave scope for pluralism and its characteristic of undecidability (Mouffe 1999: 757). Actors who do not accept the definition of the problem or the directions to be taken risk being excluded. Consequently, both problem-definition and debate are restricted. Eventually, however, the failure to address the real problem will backfire on the decision-makers. The Swiss case provides a good example in this respect: debates that become heated do not contribute to the rationality of debate in which actors have positive attitudes and are willing to co-operate (Rescher 1993). It is doubtful, therefore, whether attitudes will remain open and positive. We can also question whether an outcome really reflects the ideal outcome if certain viewpoints are excluded. Instead, such an outcome risks addressing the wrong problem (Hisschemöller and Hoppe 1995).

An ethos of controversies offers an alternative (Poort 2012, 2013: Chap. 10). In a nutshell, this ethos structures decision-making processes by exploring controversies, with a primary focus on providing a structure for decision-making on complex issues, when aiming for consensus is premature. These complex issues can be seen as unstructured problems characterised by uncertainties, both about norms and facts (Hisschemöller and Hoppe 1995). These uncertainties mean it is not possible to arrive at concrete decisions, based on consensus. To that extent, such an ethos, which confronts differing viewpoints, concerns and preferences, helps to stimulate norm development and decision-making on such issues.

An ethos of controversies is rooted in Waldron's notion of respect for disagreement Waldron (1999) and in agonistic theories such as the ideas of Mouffe (1999). Waldron acknowledges that a decision has to be made, but at the same time that a conflict can linger on. Awareness that disagreement still exists demonstrates greater respect for the views that have been excluded from the decision-making. Agonistic theorists emphasize the conflicting nature of opinions in pluralistic democracies. Mouffe argues that the lack of resolvability that comes along with this conflicting nature has to be recognised. She claims that recognition of the conflict will do greater justice to the conflicting nature of pluralism (Mouffe 1999: 757). Furthermore, in my view, this will also help to establish a more realistic regulatory framework.

In contrast to Mouffe, however, the ethos of controversies does not contravene the deliberative promise as it continues to emphasise the need for and possibility of dialogue and offers a means of structuring debate within the decision-making processes. To this extent, the ethos of controversies has a different design from that of consensus-thinking. The former focuses primarily on problem-definition and norm development, whereas the latter also seeks to legitimise decisions. The ethos of controversies thus offers only a partial alternative.

The ethos of controversies can be explained in the context of three stages: articulation, confrontation and awareness (Poort 2013: Chap. 10), with the interplay between these three stages ideally promoting an ongoing process of legal norm development.[6] The first stage involves defining the problem by articulating the various viewpoints, concerns and preferences. Instead of searching for commonalities, this stage of articulation involves taking stock of the variety of viewpoints, concerns and preferences. It is argued that by focusing on the variety, the problem will be defined more comprehensively: instead of the conflict being simplified, insights into it will be provided.

In the second stage, the variety of viewpoints confront each other. Actors acknowledge differences in reasoning, with the debate being structured through the confrontation between the different viewpoints, concerns and preferences. In this way, actors are forced to explain, to think through and perhaps even to reconsider their ideas. As a result, these viewpoints, concerns and preferences are no longer loose statements, but reflected opinions. In that sense, confrontation may contribute to the further development of norms.

In the end, however, decisions have to be made. And it will, of course, be much easier to make decisions if everyone agrees. We cannot legitimize decisions built on disagreements, even if the latter are clearly articulated and crystallised through confrontation. An ethos of controversies cannot, therefore, legitimize decisions. Nonetheless, the ethos of controversies ensures, in the third stage, that these controversies at least have a role in decision-making. It does so by promoting an awareness of the conflict that still exists after decisions have been made. To that extent, this ethos can do more justice to the differing viewpoints. This final stage relies on acknowledging the temporary status of the compromise (political or otherwise) laid down in the legal decision. Zeegers explains the ethos of controversies as a specific type of deliberative democracy, in which dynamics are the key element (2016, Chap. 15, this volume). To that extent, she identifies certain requirements that shape the last stage and argues that, providing certain requirements are met, it is legitimate for decision-making processes to be closed off to diverging viewpoints.[7] The first requirement is for the diverging viewpoints to have been addressed at an earlier stage of the norm development process. The second requirement prescribes the "acknowledgement that the outcome of decision-making is a 'temporary political achievement' that rests on a compromise" (Chap. 15, this volume). Only then it is legitimate to close off the decision-making process.

[6] Here, a distinction is made between legal decision-making and legal norm development. Legal decision-making and lawmaking can address a current situation or conflict. Legal norm development does not necessarily end, however, when legal decisions have been made or regulations implemented. Regulations and legal decisions influence legal norm development and are strong directives towards a certain path of development. However, norm development continues, given that its context continues to develop (responsive character). See also Van der Burg (2014).

[7] Zeegers does not make a fundamental distinction between decision-making and norm development. Both terms are therefore used to explain her argument.

### 5.2.2 Criticism of Communicative Approaches to Law

The previous sections describe the interactive legislative approach as an example of communicative theories of law. Some of the characteristic elements described give cause to criticise communicative theories of law in general. These criticisms relate specifically to the lack of enforcement (Westerman 2005), the tension with law's basic function of ending conflict (Van Klink 2005), and the lack and misunderstanding of authority (Lindahl 2000; Van Klink 2005). This chapter focuses on one characteristic element that is open for contestation: dynamics, which are heavily criticised for the tension they cause with law's basic function of ending conflict.

One question that I am often asked when presenting or defending my research and arguing for the need for dynamics concerns how decisions should be made. Although ongoing debate can seem attractive, decisions ultimately have to be made; in other words, law's basic function of bringing conflicts to an end has to be allowed to operate. Van Klink, for example, points out some serious conceptual flaws in the theory on communicative law, with one of these flaws concerning the role of violence. This latter aspect is closely related to the tension between dynamics and the ending of conflict. At some point, decisions have to be made; since we cannot make decisions that build on controversies or ongoing debate, an element of violence is inevitable. According to Van Klink, communicative approaches to law ignore this violence. The concept of *temporary* decisions can be raised in defence of the interactive legislative approach. An argument for dynamics or ongoing debate is not so much a matter of postponing decisions, but instead of being aware of the decisions' temporariness. In this context, Lindahl (2000) highlights the problem of violence in communicative approaches to law, also with regard to temporary decisions. Dialogue and persuasion form the foundations of the legal framework in communicative approaches to law, and are even contrasted to violence and commands. Lindahl argues, however, that the context of the dialogue is already built on a process of inclusion and exclusion, with this process automatically requiring a moment of violence against those who are excluded. In terms of the communicative law theory, this constitutes an interpretive community of those who recognize the regulatory framework (Van Klink 1998: 93). This community is thus constituted by law. Lindahl's argument is that, to this extent, the communicative approach differs less fundamentally from the instrumental approach. The constitutional moment of the interpretive community is a command issued through law. The dialogue and interaction within the interpretive community are thus based on a moment of force and violence that occurs when constituting the community.

Although the interactive legislative approach does not recognize an interpretive community or constitutional moment of inclusion and exclusion, it acknowledges that decisions need to be made and that norms set by the legal framework can be open. To a certain extent, force and violence against viewpoints seem to occur here, too. One way or another, decision-making involves a moment of inclusion and exclusion. Here we can recognize law's basic function of ending conflict: at a certain point, decisions have to be made. It is the intention of law-making to make those decisions and, thus, to include or exclude. How does that relate to the need for dynamics? Here, the tension between ending conflict on the one hand and the need for dynamics on the other comes to the fore.

An obvious response `in terms of countering this tension would be to raise the possibility of consensus-thinking as an alternative to violence. Based on the outcomes of my case studies, however, I am highly critical of consensus-thinking as seen in the interactive legislative approach. Another obvious response would be to point to the additional functions of law that the opponents of communicative approaches seem to ignore. This reply, too, will not bring the debate any further, with opponents most likely to reply that even if they acknowledge these additional functions of law, decisions have to be made and violence will inevitably occur. This chapter will then end up in a repetition of arguments and a ritual dance between opponents and proponents around the value of these functions and the institution of law and authority. Van Klink tries to bring the discussion further in this volume by explaining symbolic legislation as an essentially political concept. Proponents and opponents of the communicative and interactive approaches to law adopt different political stances, in which the political refers to their different understandings of regulatory power, their different ideas about the division of regulatory power and their different views on the role of actors in the regulatory process (Chap. 2, this volume).

I will now seek to shift this discussion in a different direction by taking these criticisms seriously and analyzing them in the light of my argument in favour of an ethos of controversies. To this extent, I will address the tension that, at first sight, would seem to be worsened by this ethos, given that focusing on controversies does not readily lead to concrete decisions. In doing so, however, I will show that rather than worsening the tension, analyzing the tension in the light of an ethos of controversies may even contribute to addressing it. The interactive legislative approach seeks to cause less violence to the various viewpoints in decision-making; my aim, therefore, is to show that adopting an ethos of controversies within such an approach can create scope for openness and flexibility that, along with the temporariness of decisions, can do justice, instead of violence, to the controversial dimensions of an issue at stake.

## 5.3 Tension

The tension between ending conflict and the need for dynamics would seem to be worsened by an ethos of controversies. Focusing on the controversies characterizing an issue instead of searching for consensus as a basis for decision-making does not encourage efforts to put an end to conflict. On the contrary, thinking in terms of controversies emphasizes the conflicting nature of the issues to be addressed by the regulatory framework. An ethos of controversies is presented as a method to promote dynamics and would seem, therefore, to contradict efforts to bring conflict to an end. Although this ethos is a response to the failures of consensus-thinking from an internal perspective, the external criticism needs to be addressed in order to enable debate about the communicative theories of law and to shift this debate in a different direction. Exploring an ethos of controversies against the background of the tension will also help to develop this ethos and to concretize its basic elements. The following subsection attempts, therefore, to address this tension by understanding the ethos of controversies within a two-track approach.

### *5.3.1 Addressing the Tension*

The interactive legislative approach is especially suited for dealing with issues in specific circumstances, such as issues of an ethically sensitive nature. Dynamics are, then, a way of dealing with the conflicting nature and the dimension of undecidability characterizing such issues. At the same time, an ethos of controversies acknowledges that decisions also have to be made on these issues, even if the parties do not yet agree. Dynamics in terms of ongoing debate should also not be taken to mean the opportunity to change law whenever we want. According to Zeegers, certain requirements need to be meet when re-opening norm development, with legal safeguards such as legal certainty and equality consequently still being of value (2016, Chap. 15, this volume). A first question that comes to mind, therefore, is whether the tension is in fact as great as is presented.

In my view, this tension cannot be addressed solely within the legal track. Instead, a two-track approach, involving additional channels, is required. This two-track approach was previously developed to address a different flaw in the interactive approach, which is the dominant legal discourse (Poort 2013: Chap. 10). Even though, the interactive approach seeks – against the background of an ethos of controversies – to open debate on controversial positions, that debate continues to be framed by the legal discourse. By adopting a two-track approach, however, we can broaden the debate to include other discourses of relevance to the issue at stake. Given that the previously mentioned issue of animal biotechnology is a complex legal problem and this complexity is predominantly due to moral dimensions, the tracks in this approach focus, on this occasion, on law-making on the one hand and the development of moral norms on the other.[8]

The interplay *between* these tracks is also relevant if we are to address the tension. The legal track involves decision-making, even if the parties' views are still diametrically opposed. In this track, legal safeguards can be protected and decisions

[8] This two-track approach is not necessarily exclusive and limited to these two tracks. The point made here is that, for the sake of norm development, the process should not be limited to a legal discourse, but should also be encouraged in other discourses. The moral discourse is most prominent in the case studies addressed in this research. However, other discourses may be relevant for norm development in other specific fields. In this paper, however, I have chosen to focus, for the sake of the argument, on a two-track approach. Elaborating on other potentially relevant tracks would never be comprehensive and would distract from the focus of the points I am seeking to make here.

can be made (thus bringing conflict to an end), while the moral track allows opposing viewpoints to be re-introduced and discussed in further depth (thus creating dynamics). While, as Van der Burg claims, "the most obvious institution for control is the law", there will always be a dimension of undecidability that, in itself, requires further exploration and development (Van der Burg 2009: 64). An interplay between the legal track and the ongoing debate in the moral track can address this dimension by ensuring that the ongoing debate, in which controversial viewpoints are fully acknowledged, can feed into further development of legal norms.

Here, the requirements for an ethos of controversies, as introduced by Zeegers, become relevant (2016, Chap. 15, this volume). As explained in Sect. 5.2.1, Zeegers defines two requirements that need to be met if the closing-off of legal decision-making processes to new viewpoints is to be regarded as legitimate: (1) These viewpoints must already have been addressed in the earlier process of norm development; and (2) The decision that has been made should be regarded as a temporary political achievement, based on a compromise.

I would propose reformulating certain aspects of these requirements. To start with, given that dynamics are a key element of the interactive legislative approach, I would suggest defining requirements for *re-opening* debate and cohering with the need for dynamics rather than setting requirements for *closing off* debate.

A first requirement for re-opening debate would, therefore, be that a viewpoint has not been addressed earlier, fully or equally during the process of norm development. Rather than defining the latter as a requirement, I would suggest defining it as a basic condition if an ethos of controversies is to function adequately in all three stages. When dealing with controversial issues with a strong moral impact, the regulation might not represent the end of the moral conflict, and neither will it guarantee that the existing legal framework will not prove problematic in the face of new developments in the future. In other words, even if legal rules are set or legal decisions made, legal norms may still continue to develop. Dynamics and rational debate are possible only if the parties involved acknowledge that the outcomes of decision-making rely on a compromise in the form of a temporary political achievement. In my view, therefore, the basic condition to be met if an ethos of controversies is to function adequately is to acknowledge both the temporary political achievement and the conflict that still exists or may re-arise.

I would also propose stipulating a second requirement as there may be an additional reason for re-opening the development of legal norms: new developments in the field may change the context in which the issue at stake should be understood. Future developments in biotechnology, for example, may make it possible to select an embryo based on eye colour or on talents the future child may develop. Such developments will change the moral impact of the technology and may result in new or changing viewpoints. As well as diverging viewpoints being excluded from norm development, certain viewpoints may never previously have been introduced at all. Therefore, imposing a second requirement will allow a debate or process of norm development to be re-opened if new developments give rise to new moral or social dilemmas.

Following on from this second requirement, a third requirement can also be distinguished. In addition to new developments in the field that may result in new or changing viewpoints, society itself or viewpoints themselves may change.[9] Changes in society, for example, have resulted over time in equal rights for women and in same-sex marriages. These changes broke through previous barriers and required a re-opening of the legal framework.

It should be noted that these three requirements are cumulative, with each in itself creating an incentive to re-open the debate. Although a reason for re-opening the debate may also arise without the basic condition (i.e. the first requirement) being met, there is a risk in situations in which the basic condition or either of the above requirements is not met that dynamics will create an incentive to change law whenever a member of society wants it. Let me explain how the introduction of this basic condition and the other two requirements can overcome this risk in the context of an ethos of controversies. First of all, the basic condition allows both tracks to operate alongside each other by creating scope both for decision-making and for further debate. Furthermore, it allows both tracks to interact. An interplay between the tracks can be established only, however, if it is acknowledged that the legal decision does not necessarily constitute resolution of a moral (or other) conflict.

In addition to the basic condition, certain requirements have to be set so as to ensure that law functions as an institution of control and cannot simply be changed on a whim (which would undermine the legal safeguards). An interplay between the two tracks should be possible only if one or both of these requirements are fulfilled. Even though the moral debate may continue after decisions have been made, this does not necessarily mean that the process of developing legal norms should be re-opened. The diverging viewpoints should already have been balanced in the decision-making process. Only if a viewpoint was excluded from this process, or new developments give rise to new dilemmas requiring reconsideration, should there be scope to re-open norm development. Not every moral debate, however, gives cause to re-open norm development. If that were the case, the legal safeguards and the basics of law would be meaningless. By setting these requirements we can create room for conflict and diverging viewpoints, while also adapting to the institutions of law.

In my view, understanding an ethos of controversies through a two-track approach is a way to address the tension between seeking to end conflict on the one hand and stimulating dynamics on the other hand in that it shows that the two functions do not necessarily have to conflict and can operate alongside each other. This approach thus reduces the tension and ensures there is room both for the legal track and for the moral track. An interplay between these two tracks ensures an awareness that a legal decision does not necessarily signify the end of a conflict. In other words, the two-track approach makes continual norm development possible.

Nonetheless, the design of the legal track still requires a specific understanding of law. The ethos of controversies, as well as the two-track approach, is designed from the perspective of the interactive law approach. We can still question whether

---

[9] I owe this point to Wibren Van der Burg, who commented on an earlier version of this chapter.

opponents of communicative approaches to law will accept this perspective, the basic conditions applying to the ethos of controversies and this renewed interpretation of the tension. I have attempted to explain and strengthen the functioning of an ethos of controversies by identifying its basic conditions and the requirement for dynamics and hope that this will feed into a continual debate on communicative approaches to law in general.

## 5.4 Discussion

This chapter has sought to address the tension between law's basic function of ending conflict and the need for dynamics in a communicative approach to law. This tension becomes visible in such an approach, particularly in one of its theoretical models, the interactive legislative approach. This approach is claimed to be especially suitable for addressing complex issues because it facilitates horizontal structures of law-making and uses open, symbolic or aspirational norms. Both of these elements combine to ensure the interpretative character of the law and, therefore, allow dynamics to prevail. At the same time, law's basic function in traditional approaches is to end conflict, and this would seem to contradict the interactive legislative approach's aim of creating dynamics. Although this tension would seem to be worsened by the adopting of an ethos of controversies, which was introduced into the interactive legislative approach at an earlier stage in order to stimulate dynamics, my explanation of how an ethos of controversies can operate in a two-track approach represents an attempt to resolve this tension. This two-track approach creates room for ongoing development of the moral norms by making a distinction between moral and legal norm development (Poort 2013: Chap. 10).

The two-track approach offers a tool for putting the renewed interpretation of the tension between ending conflict and stimulating dynamics into practice. This renewed interpretation, which demonstrates that the tension is not as great as often presumed, can be visualized by making a distinction between a legal and a moral track. This distinction, in turn, creates scope in the legal discourse for ending conflict and scope in the moral discourse for acknowledging that conflict still exists. In an ethos of controversies, the interplay between the two tracks ensures an ongoing debate that takes account of the variety of viewpoints still existing even after a decision has been taken. The fact that the debate remains open for further norm development in the moral track may encourage an awareness in the legal track that the conflict still exists. In other words, both dynamics and the ending of conflict can establish a position in legal decision-making when an ethos of controversies operates in a two-track approach. This interplay, however, is possible only when the basic condition and at least one of the two specified requirements is fulfilled. The basic condition involves acknowledging that the decision being made is a temporary political achievement, based on a compromise. These three requirements (i.e. the two requirements and the basic condition) ensure that the debate can be re-opened only if diverging viewpoints were previously excluded from norm development, if new developments in the field give rise to new dilemmas requiring further debate, or if society and existing views in society have changed.

In conclusion, this chapter contributes to the debate on communicative approaches to law by elaborating on the ethos of controversies in the light of a two-track approach. I hope, therefore, to have created scope in legal decision-making processes both for dynamics and for ending conflict. My argument, however, does not fully counter the criticism of communicative approaches to law as it leaves several essential questions about the functioning of an ethos of controversies and about symbolic law in general unanswered, including questions on the authority of law and guidelines on how to operate an ethos of controversies and how to arrive at decisions when focusing on bringing conflict to an end.

Van Klink may have a point in stating that the differences in the interpretation of communicative approaches to law will continue to exist. He clarifies these differences by considering a symbolic or interactive legislative approach, based on a pre-scientific political choice (Chap. 2, this volume). I have doubts, however, as to whether this explanation is sufficient and what the consequences of this pre-scientific choice for legal scholarly debate are or should be. I believe, therefore, that the story needs to continue and would suggest that a next step would be to develop tools for putting an ethos of controversies into practice. For a sneak preview of such tools, I refer to the research of Castle et al. (2013), who tested a method of contested exchange in a process of environmental decision-making comparable to an ethos of controversies and who brought all the relevant actors together in a workshop on climate change. The focus of the workshop was on exploring differences in opinions and views, with the actors being required to talk these differences through. By the end of the workshop, they had gained more respect for each other's views and the discussion became less polarized. The positive outcome of this experiment could be a good starting point for further elaboration of the ethos of controversies.

## References

Ayres, Ian, and John Braithwaite. 1995. *Responsive regulation. Transcending the deregulation debate*. Oxford: Oxford University Press.

Black, Julia. 1998. Regulation as facilitation. *The Modern Law Review* 61: 621–660.

Castle, David, and Keith Culver. 2013. Getting to 'No'. The method of contested exchange. *Science and Public Policy* 40: 34–42.

Fuller, Lon L. 1969. *The morality of law* (revised edition). New Haven: Yale University.

Fuller, Lon L. 2001. *The principles of social order*, selected essays of Lon L. Fuller. Edited, with an introduction by Kenneth I. Winston. Oxford: Hart Publishing.

Griffiths, John. 2005. Do laws have symbolic effects? In *Social and symbolic effects of legislation under the rule of law*, ed. Nicolle Zeegers, Willem Witteveen, and Bart van Klink, 147–161. Lewiston: The Edwin Mellen Press.

Gutmann, Amy, and Dennis Thompson. 1996. *Democracy and disagreement*. Cambridge, MA: Harvard University Press.

Habermas, Jürgen. 1996. Three normative models of democracy. In *Democracy and difference. Contesting the boundaries of the political*, ed. Seyla Benhabib, 21–30. Princeton: Princeton University Press.

Hisschemöller, Matthijs and Rob, Hoppe. 1995. Coping with intractable controversies. The case for problem structuring in policy design and analysis. *The International Journal of Knowledge Transfer and Utilization* 8(4): 40–60.

Honig, Bonnie. 1993. *Political theory and the displacement of politics*. Ithaca: Cornell University Press.

Lindahl, Hans. 2000. Communicatieve wetgeving en volkssoevereiniteit. In *De overtuigende wetgever*, ed. Bart van Klink and Willem Witteveen, 173–188. Deventer: W.E.J. Tjeenk Willink.

Mouffe, Chantal. 1999. Deliberative democracy or agonistic pluralism? *Social Research* 66: 745–759.

Poort, Lonneke. 2012. An ethos of controversies. A critical analysis of the interactive legislative approach. *Legisprudence* 1: 35–55.

Poort, Lonneke. 2013. *Consensus and controversies in animal biotechnology. An interactive legislative approach to animal biotechnology in Denmark, Switzerland, and the Netherlands*. The Hague: Eleven International Publishing.

Rescher, Nicholas. 1993. *Pluralism. Against the demand for consensus*. Oxford: Clarendon.

Selznick, Philip. 1992. *The moral commonwealth. Social theory and the promise of community*. Berkeley: University of California.

Stamhuis, Jellienke. 2005. Communicative law. A quest for consensus. In *Social and symbolic effects of legislation under the rule of law*, ed. Nicolle Zeegers, Willem Witteveen, and Bart van Klink, 277–297. Lewiston: The Edwin Mellen Press.

Van der Burg, Wibren. 2009. Law and bioethics. In *A companion to bioethics*, ed. P. Singer and H. Kuhse, 49–67. Oxford: Blackwell Companion Guide.

Van der Burg, Wibren. 2014. *The dynamics of law and morality. A pluralist account of legal interactionism*. Farnham: Ashgate.

Van der Burg, Wibren. 2016. The emerging interactionist paradigm and the ideals of democracy and rule of law. In *Symbolic legislation theory and new developments in biolaw*, ed. Bart Van Klink, Britta Van Beers, and Lonneke Poort, 37–53. Cham: Springer.

Van der Burg, Wibren and Frans W.A. Brom. 2000. Legislation on ethical issues. Towards an interactive paradigm. *Ethical Theory and Moral Practice* 57–75.

Van Klink, Bart. 1998. *De wet als symbool. Over wettelijke communicatie en de Wet gelijke behandeling van mannen en vrouwen bij de arbeid* (diss. Tilburg). Deventer: W.E.J. Tjeenk Willink

Van Klink, Bart. 2005. An effective-historical view on the symbolic working of law. In *Social and symbolic effects of legislation under the rule of law*, eds. Nicolle Zeegers, Willem Witteveen and Bart van Klink, 113–145. Lewiston: The Edwin Mellen Press.

Van Klink, Bart. 2014. Symboolwetgeving. De opkomst, ondergang en wederopstanding van een begrip. *RegelMaat: Journal for Legislative Studies* 29:5–26.

Van Klink, Bart. 2016. Symbolic legislation. An essentially political concept. In *Symbolic legislation theory and new developments in biolaw*, ed. Bart van Klink, Britta van Beers, and Lonneke Poort, 19–35. Cham: Springer.

Waldron, Jeremy. 1999. *Law and disagreement*. Oxford: Clarendon.

Westerman, Pauline. 2005. Some objections to an aspirational system of law. In *Social and symbolic effects of legislation under the rule of law*, ed. Nicolle Zeegers, Willem Witteveen, and Bart van Klink, 299–316. Lewiston: The Edwin Mellen Press.

Witteveen, Willem, and Bart van Klink. 1999. Why is soft law really law? *RegelMaat (Journal for Legislative Studies)* 3: 126–140.

# Chapter 6
# Symbolic Legislation and Authority

**Oliver W. Lembcke**

## 6.1 One (Too) Simple Story

Studies on the concept of symbolic legislation are usually part of a broader discourse on legal effectiveness (e.g. Friedman 1975; Bryde 1993; Isensee 1999: 33). In this context symbolic legislation is widely perceived as an example of ineffective law-making, more or less affirming the legal status quo. In addition, it is often held that symbolic legislation represents a lack of political will to actually change the legal status quo. In this sense symbolic legislation does not only appear to be the opposite of substantive law-making, but it is also seen as an attempt to pretend actions by the legislator, where in fact there was no reform of the status quo intended. Instead of using law making as ways and means of political problem solving, the ineffectiveness of an intended piece of legislation is *ex ante* and voluntarily chosen by the key actors of the law-making process (Voß 1989; Blankenburg 1977). In addition to the element of ineffectiveness (of changing the legal status quo) it is the element of unwillingness (of the main political actors in the law-making process) that accounts for the centerpiece of a view on symbolic legislation that can be dubbed here as the 'common view'.

On this basis of this common view, referred to in this volume as the symbolic legislation in the negative sense,[1] the relationship between authority and symbolic legislation may be a simple story: In a nutshell, symbolic legislation is then a sign of political crisis. One way to present this story is David Easton's "Systems Analysis of Political Life" (1965). In Easton's framework the lack of effectiveness equals to a lack of policy output, which can/or will be seen as insufficient by the 'environ-

[1] See the chapters of Van de Burg and Van Klink.

O.W. Lembcke (✉)
Institute of Political Science, Friedrich Schiller University Jena,
Carl-Zeiss-Str. 3, D-07743 Jena, Germany
e-mail: oliver.lembcke@uni-jena.de

 
B. van Klink et al. (eds.), *Symbolic Legislation Theory and Developments in Biolaw*, Legisprudence Library 4, DOI 10.1007/978-3-319-33365-6_6

ment' (the public, voters, citizenry, etc.). Insufficient policy output raises the probability of negative responses by the environment via 'feedback loop' towards the political system, expressed in a growing level of 'demand' compared to the level of 'support'. If the political system (because of the inability or unwillingness of its political actors) ignores this feedback for a longtime or continues to produce 'symbolic legislation' the disappointment may grow. This can lead to the consequence that the former level of 'specific support' (that is support related to the particular policy, their actors etc.) will decline. Robust democratic systems may be able to compensate the reduced level of 'specific support' by 'diffuse support', viz. the people's belief in the legitimacy of the political system. However, the constant flow of negative feedback can also exhaust the level of 'diffuse support' – which ultimately affects the legitimacy and stability of the political system *in toto*. Taken together, demands which are not met by effective legislation over a significant period of time may eventually result in disenchantment with politics – which undermines in any case political authority. An even stronger version of this argument would be: If political authority is responsible for such ineffective legislation, then this may well be seen as a symptom of a crisis.

This paper wants to challenge the common view of symbolic legislation as sign of political crisis based on two grounds. First it aims to present an alternative view by introducing the symbolic dimension as an integrative element of politics. Within this perspective symbolic legislation appears to be not simply a sign of political insufficiency, but is portrayed as a strategy of integration by means of legislation that seeks to establish or reinforce a common ground of the political community. To be sure, this paper does not dispute the empirically well-documented relationship between the output capacity of political system (performance) and the support of (and trust in) the political institutions by 'the environment', to use Easton's terminology (see, for example, Newton and Norris 2000; Wang et al. 2006). It claims, however, that the relationship between symbolic legislation and political authority is more complex than the simple story suggests, because symbolic legislation entails a dimension of integration; and this dimension is different from – and in some ways the opposite of – symptoms of political disenchantment or political crisis (sub 2.).

Second, this paper challenges the authority concept of the common view. The latter presupposes a concept of authority that depends on the performance of the political actors. Here, political authority more or less equals to output legitimacy. In contrast, this paper introduces a different notion of authority claiming that authority should be less strictly tied to legitimacy then the common view suggests. This claim may already find some response in Easton's concept of 'diffuse support' itself. By introducing the notion of 'diffuse support', Easton indicates that political legitimacy is less understood as a clear-cut product of cause and effect depending on the policy output. In a similar fashion the case can be made – and, actually, has been made in the literature on authority – that political authority is not directly linked with legitimacy, e.g. in the way that legitimacy is the 'ground' ('reason' or 'source') of authority (e.g. Garthoff 2010). If this holds true than the lack of effectiveness as one of the alleged main characteristics of symbolic legislation may still result in disenchantment with politics. It may also still have a negative effect on political legitimacy. However, it does not need to undermine political authority. Against this background, this contribution calls for a more elaborate analysis of the concept of authority and its implications for the relationship between authority and symbolic legislation. In this sense it is the purpose of this paper to throw some light on different conceptual elements of authority. It also tries to show that these elements are not easily integrated into a coherent theory of authority. In order to substantiate this thesis the paper discusses briefly Joseph Raz's concept of legal authority, which is perceived as one of the most important theoretical attempts in this matter within recent years (sub 3.).

Instead of a coherent theory of authority this paper aims for a topographic mapping of the different conceptions of authority. Preliminary steps toward such an approach are presented in the section following the discussion of the different elements of authority (sub 4.). The main purpose of this section is to use the analytical framework to cover different aspects of the relationship between political authority and symbolic legislation. To be sure, these are rather low rungs on the 'ladder of abstraction'. However, in the absence of an integrative theory of authority it seems to be conducive to distinguish, at least, four configurations in which the different models of authority are related to the symbolic dimension of law and politics. There are may be more. This article does not claim that these four configurations already provide a complete scheme of every possible configuration. It does, however, claim that the four configurations are identified through a systematic application of the analytical framework. In addition, it also claims that two of the four configurations – here dubbed as 'contested authority' and 'the authoritative and its actors' – are of particular relevance for the policy field of biolaw which is typically a field of value-loaded conflicts'. The other two configurations are located on a more abstract level. At first glance, they are less relevant in terms of policy, but they can be read as the theoretical basis for the other two more specific policy examples. Building on these four configurations the final section tries to shape the different perspectives on the relationship between authority and symbolic legislation (sub.5). It confirms the view that the latter is not necessarily undermining the former, unless it transforms authority into power.

## 6.2 Two Dimensions of Symbolic Legislation

In recent years symbolic legislation has become increasingly a subject of research again. This is partly due to the renewed interest in the symbolic dimension of politics starting already in the 1960s (Edelman 1964) and expanding in the wake of neo-institutionalism since the 1980s (March and Olsen 1984). As a result of this development the literature on symbolic legislation is somewhat split, at least in terms of the evaluative dimension. One aspect of the literature focuses on the deficiencies of symbolic legislation (see already Arnold 1935). The critique is directed to the ineffective law-making on the one side and the legislator's intention to accept *ex ante* this kind of ineffectiveness on the other. Taken together, this combination results in a kind of "alibi legislation" (Kindermann 1989), potentially motivated by dodgy political strategies of deceiving or even manipulating the public (Dwyer 1990; Campbell 1993).[2] In contrast, another approach with a more affirmative attitude towards symbolic legislation stresses the argument that legal regulations in general are typically based on value-oriented ideas of the legislative intention which can or even should find an adequate symbolic expression (Gusfield 1967; Voß 1989). In this line of reasoning the symbolic dimension plays a role in addressing and communicating[3] certain values of a political community – more as a 'moral appeal' than a 'political alibi' (Noll 1981: 357 ff.).

How to make sense of these different dimensions of symbolic legislation? This paper suggests a conceptual matrix that resembles in some ways Newig's differentiation between the dimensions of "legal-material" and "political-strategic" legislation (Newig 2010: 304 f.). According to Newig the first dimension covers the intended changes of legal status quo (as far as they are stated in the legislation), whereas the second dimension refers to the political purposes 'behind' it. Yet, already at first glance, this differentiation seems to be somewhat flawed. One objection is that legal regulations, at least in democratic settings, will always be subject to political compromise and in this sense subject to the strategic dimension. Therefore it seems to be rather futile to distinguish between law and politics in such a way. Moreover, Newig remains somewhat vague about the second dimension, which is the political-strategic legislation. His example (the political strategy of dodging an issue) suggests that the second dimension is related mainly or exclusively to political power. If that holds true, his approach simply reproduces the critical position of the discourse on symbolic legislation: As inherently instrumental – here for acquiring or maintaining political power – symbolic legislation would look similar to the type of alibi legislation, symbolizing an empty box of broken promises.

Building on Newig's distinction this paper pursues a different conceptual strategy by introducing the notion of integration.[4] Integration, in general, means 'being part of a unity'. Of the many notions that follow from this rather abstract definition in the realm of politics neo-institutionalist approaches have reminded us in particular of the importance of institutional embeddedness. According to this perspective political order and its means of decision-making (including law making) depend to a large extent on underlying routines, templates, ideas etc. that provide the necessary framework of interpretation and self-understanding. In this sense integration can be seen as a process and, at the same time, a product of combining order and orientation.

[2] The constitutional implications are discussed by Meyer (2009) and Lübbe-Wolff (2000) with reference to environmental law.

[3] For the importance of the communicative function see Zeegers et al. (2005).

[4] Newig (2010: 305) himself mentions briefly the concept of integration in the context of symbolic legislation, yet without any further elaboration. From his standpoint the notion of integration is, of course, less fitting to the label 'political-*strategic* dimension'.

This inherent 'process-product ambiguity' of integration[5] is of particular importance with respect to political institutions and their decision-making process. Political institutions typically embody a certain status of integration and, at the same time, contribute to the process of integration by their (collectively) binding decisions. These two ways of 'being part of a (political) unity' are connected with two modes of representation in which political institutions operate: delegation and symbolic representation. Both modes of representation share the same core purpose that according to Pitkin is the making present of something that is absent (Pitkin 1967: 8). Yet the modes differ in their representational logic (Göhler 1994): Delegation is voluntaristic in its nature. The key elements here are interests, responsiveness, and accountability. Symbolic representation is different. It contains the expression of self-images and their underlying values of a particular community. This form of representation is less directly linked with interests; it's more about identity and the function of integration that comes with it. Both modes can be analytically differentiated but, as a result of the product-process ambiguity, they are empirically interwoven with each other.

What follows from this theoretical sketch? First, the symbolic dimension is directed to integration and therefore an integral part of political representation by political institutions and their decision-making processes. Symbolic legislation is therefore characterized by its purpose of promoting and/or maintaining (socio-political) integration by political means. This kind of legislation may be ineffective in terms of changing the legal status, but in order to qualify as symbolic legislation it has to be related – following the conceptual suggestion of this paper – on the symbolic dimension of the political order. Second, the scope and content of the symbolic dimension can be made subject of empirical analysis, guided by the following questions that build on the old-fashioned distinction between instrumental and symbolic functions of legislation (e.g. Carson 1974; Dwyer 1990): Does the intended legislation change the legal and/or social status quo? Does it have enough institutional 'bite' to be effectively enforced (dimension of effectiveness)? And who does benefit from the legislation (dimension of integration)? The former is measured by the effective instruments (viz. in terms of changing the status quo as in enforcement mechanisms, agencies etc.) within the legislation, the latter by the intended scope of integration. – The following graph may illustrate the underlying conceptual ideas related to the concept of symbolic legislation.

[5] As product (or result) integration refers to status of units being integrated, whereas integration as a process refers to the mechanism or media in which integration (or disintegration) comes about.

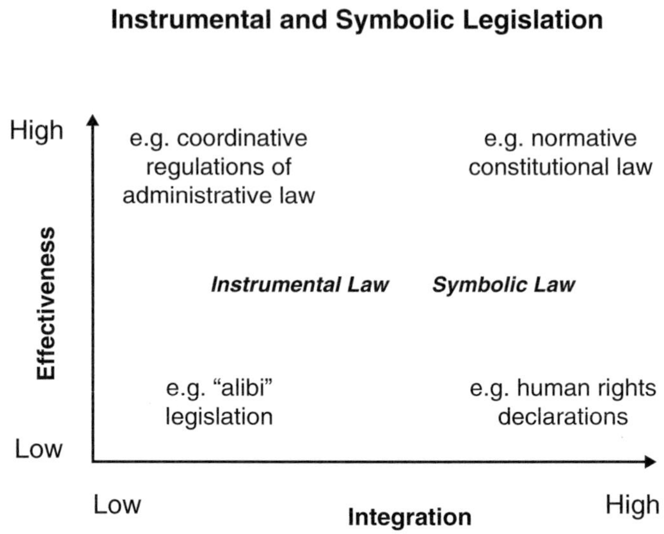


The illustration may also help to distinguish briefly four 'prototypes' of legislation based on the two dimensions of 'effectiveness' and 'integration':

- An example of the *first prototype* is the already mentioned "alibi legislation". It lacks not only legal effectiveness, but also any meaningful integrational perspective. It is usually the result of high political pressure which is met by the political actors in charge aiming at a strategy of pretence. That is to say they pretend to be responsive to public concerns and to the main interests of the majority (measured by public opinion polls), without actually being responsive at all. Among others Newig (2010: 310–314) has given an intriguing example by his insightful comparative analysis of two case studies taken from German environmental policy. In its pure form so-called alibi legislation is instrumental in character and self-referential in terms of political power.
- Like alibi legislation, non-binding human rights declarations do not really change the legal status quo. However, they provide for an integrational perspective because of their universally inclusive claim.[6] In this sense they can serve as an example for the *second prototype*. A historically important case in point is the French *Déclaration des Droits de l'Homme et du Citoyen* (1789/91).
- The history of human rights may also illustrate the underlying dynamic of legal principles once they have become part of an enacted legislation. Their normative status may be weak at first, but they can develop into a powerful political and legal tool over time, e.g. via interpretation by courts.[7] The emergence of human rights regimes in constitutional democracies (and beyond) is an example of the

[6] For the ambivalent strategy of "symbolic constitutionalization" see Neves (1999).

[7] Stone Sweet (2007) discusses several examples which he frames as juridical coup d'états. This 'success story' is not restricted to the constitutional arena of human rights or basic rights. In some countries equality laws provide another series of examples.

*third protoype* which combines effective legislation with a high standard of integration.

- The integrational perspective is typically very limited if regulations of administrative law serve the core purpose of coordination. For this reason, administrative regulations often provide the legal material for the *fourth prototype*: One of the classic examples is the rule of keeping to the right side of the road. No doubt, it is a necessary regulation, highly effective, but also at the same time highly contingent – simply because the other side of the road would also be possible.

In sum, symbolic legislation is characterized by a political claim of socio-political integration.[8] This claim does not need to be transformed into effective legal instruments, but it is typically linked with value-orientations and/or self-images of the political community as the addressee of this legislation.

## 6.3 Three Dimensions of Political Authority

This paper argues for a rather 'strong' concept of symbolic legislation. It is strong in the sense that politics, at least political institutions, cannot do without a symbolic dimension. For this reason it would be ill-fated to understand the concept of symbolic legislation simply as an impotent version of political authority. But this perspective begs, of course, the question of what is meant by political authority. And this question is even more justified, since the discourse on authority is multifarious and fragmented. To answer the question, this section makes use of a typology that analytically distinguishes between three ideal types (Friedman 1973; Flathman 1980). The section does not claim that this typology is exhaustive; nor does it claim to come up with a coherent theory of authority. But it does claim that these types cover a lot of empirical phenomena which are relevant in the political realm (Sect. 3.1). Moreover, it seems that these types have not been brought into a coherent theory so far; and maybe that is too much to ask for anyway. Among other challenges, such a theory would have to be able to cope with the possible transformation of each type into a different type of authority. The example of Raz's approach – which is briefly discussed at the end of this Sect. (3.2) – should demonstrate that this is a demanding task.

[8] The claim of social integration underlines the actor's perspective within the process of law-making. Whether these claims succeed or whether they are meaningful from the perspective of the observer inspired by system theory is different question. See in this respect Priban's contribution to this volume.

### *6.3.1 Three Types of Authority*

The first ideal type refers to the Platonic model of an expert who is recognized as 'being *an* authority'. Based on his/her superior knowledge the expert gives advice to the layman. He/she delivers reasons for belief, but does not impose duties on others; at most the expert's advice specifies what a person's duty is.[9] The office holder is the second ideal type of authority, for it refers to Hobbes's model of 'being *in* authority'. Those who possess this kind of authority are "entitled to require the performance of their claims without either the approval of them by those subject to their authority, or any acknowledgement of the rational or persuasive nature of their declarations" (O'Sullivan 2003, 44). The *Leviathan* has been influential in portraying the core of political authority as being 'in authority' (Lev., ch. 16.). Hobbes argues for a strong connection between authorization and authority with the idea that the main feature of authority is that it refers to a right to issue directives which obligate.

What distinguishes the two types from each other is the difference between content and source. Unlike the expert-type authority the political authority in a formal sense is supposed to give reasons for action and not reasons for belief.[10] A well-known expression of this understanding is expressed in Hobbes's dictum *'auctoritas sed non veritas facit legem'*. In this sense, political authority should be as content-independent as possible (Flathman 1987: 29), because it is the source of the authority (and the source alone) that guarantees that the office holder has the competence to issue rules and directives which obligate.[11] Michael Oakeshott has picked up on Hobbes's concept of political authority as a basis for his concept of a 'civil association' – in contrast to an 'enterprise association' (Oakeshott 1975b: 108–112; 114–118): A civil association is a moral association based on the authority of non-instrumental rules that impose on its associates the obligation to respect the conditions which are prescribed in the law (Van Klink and Lembcke 2013). Similar to Hobbes the authority here is strictly tied to the notion of office, while the enterprise association, according to Oakeshott, replaces this kind of formal constitutional commitment with a substantive commitment to a particular vision of the good society.

Oakeshott's attempt to build his theory of civil association on the formal concept of authority has been criticized, mainly for the reason that role conceptions related to 'being in authority' cannot "form hermetically sealed compartments or windowless monads that are or could be altogether isolated from another" (Flathman 1980:

---

[9] For a powerful analysis of the 'Platonic' model see Arendt (1961: 107–115).

[10] For a critique see Hurd (2001).

[11] "As it features in liberal democratic societies, the most striking characteristic of such authority is that one who holds it can demand compliance even when the laws or policies in question are not regarded as rational, persuasive or substantively acceptable by those who acknowledge their validity. This is because the authority, being formal, does not demand substantive consensus" (O'Sullivan 2003: 46).

108). Instead they must presuppose a background of shared values and beliefs[12] – and if that is true, then this kind of authority is acknowledged because of rules and procedures which themselves are perceived as *'the authoritative'*. In this line of reasoning 'the authoritative' is the third model of authority. It contains the very rules and procedures that settle questions concerning the decision-making competence as much as questions concerning the extent, scope and content of the office holders – similar to Hart's *rule of recognition* (Hart 1994) –, and by that provides ultimately the ground for exercising the power of being 'in authority' (Flathman 1980: 20 ff., 159 ff.).

### 6.3.2 Coherent Theory of Authority

The three different models raise the question of a unified theory of authority. However, whether such a theory actually exists is a contested issue within the discourse on authority. One of the most elaborated approaches – which is, however, restricted to the concept of *legal* authority – was developed by Joseph Raz.[13] His concept is based on the axiomatic difference between *de facto* and legitimate authority. In general, each government claims to have morally legitimate authority with every legal directive it sets, but, as Raz points out, not all of these directives actually possess this kind of legitimacy. Therefore, he "denies the existence of a general obligation to obey the law even in a reasonably just society" (Raz 1986: 70). The individuals can be obliged to obey the law only if the law is "justified by considerations which bind them" (ibid.: 72). Raz suggests two standards of justification which have to be met by the authority's claim in order to be legitimate. The first standard refers to the "dependence thesis" which says that "all authoritative directives should be based on reasons which already independently apply to the subjects of the directives" (ibid.: 42). The second standard is covered by the "normal justification thesis" which determines the scope of the individual's obligation to obey the law:

> It claims that the normal way to establish that a person has authority over another person involves showing that the alleged subject is likely better to comply with reasons which apply to him (other than the alleged authoritative directives) if he accepts the directives of the alleged authority as authoritatively binding and tries to follow them, rather than by trying to follow the reasons which apply to him directly (ibid.: 53).

Where the first standard has the purpose to provide for context-sensible decision-making that takes into account the characteristics of the problem, as much as the circumstances covered by the directive,[14] the second standard aims for an independent assessment which enables the individual to act according to his or her own

---

[12] See Fuller (1987) as a response to Flathman's criticism.

[13] For the purpose of this article the reference to Raz is restricted to his study "The Morality of Freedom" (1986).

[14] This meaning is underlined by Raz's usage of the 'arbiter' as a metaphor.

judgment. Taken together, these two standards in effect come close to the idea of the first model of 'being an authority': The law's authority is embedded in reasonable grounds mediating the (legal) relationship between the individual and the state, while at the same time the reasonability of these grounds serve as the source of legitimacy through 'trust' of the government's expertise expressed in its legal directives. This relationship is, however, fragile in principle, because from the point of view of the normal justification thesis the claim of authority, as much as the potential challenge of this claim, depends on a case by case assessment of individual reasoning. But this ultimately undermines the concept of authority, as Raz ' himself notes:

> [the] whole point and purpose of authorities [...] is to preempt individual judgment on the merits of a case, and this will not be achieved if, in order to establish whether the authoritative determination is binding, individuals have to rely on their own judgments of the merits (ibid.: 47 f.).

In order to stabilize law's claim of authority, he therefore introduces a third standard expressed by the "pre-emption thesis". Its upshot is that legal directives have an exclusionary effect on an individual's choice of action: "[T]he fact that an authority requires performance of an action is a reason for its performance which is not to be added to all other relevant reasons when assessing what to do, but should exclude and take the place of some of them" (ibid.: 21). In other words, it is not the content any longer that matters here, but the source – in a very similar way compared to the model of 'being in authority'. And if legal norms have the authority to exclude competing individual reasons, the following questions – in analogy to the formal model of authority – need to be answered: On what grounds is this exclusive status of legality based on? And what source can provide for this exclusionary effect that "trumps" the potential or even perpetual challenge by individual reasoning? These questions are *in nuce* questions about the relationship between the pre-emption thesis and the two other theses, in particular the normal justification thesis. It seems that Raz's concept not only lacks a convincing answer to these questions, but also suffers from an inherent tension between two different models of authority (Martin 2010): one that is close to the expert model of authority, the other one that is more or less identical with the formal concept of authority.

## 6.4 Four Analytical Configurations

In the absence of an integrative theory of authority it seems to be conducive to distinguish between four configurations in which the different models of authority are related to the symbolic dimension in law and politics. The following analysis tries to show that two basic insights can be gained from this section: First, there seems to be a close connection between the concept of formal authority and the symbolic dimension of politics. Second, this link to the symbolic dimension becomes somewhat more complex with the introduction of 'the authoritative' into the realm of politics.

(1) *Formal Authority and its symbolic dimension*: It has been mentioned already that in the history of political ideas, Hobbes has given *the* classical example of a formal concept of authority. His key idea is to cut off every possible external source that could possibly justify or legitimize the *Leviathan's* authority. The authority's right to impose duties on the subjects of the legal order should rather not be subject of any (ideological) interpretation fueled by values or benefits outside the political order. In this sense the legal order of the State is not an instrument to the final *telos* (peace), but to the peaceful mode of co-existence among the individuals. This mode is a moral practice, and the morality of this practice is represented by each of the different individual wills that are directed to the existence and maintenance of the political order as such. Does this concept of formal authority exclude the symbolic dimension of law and politics? Surely not, since Hobbes himself has given one of the best-known examples in the history of political ideas for that he thinks the opposite being true: he invented the powerful metaphor of the 'mortal God' and he also commissioned the frontispiece of his *Leviathan* which is widely seen as a congenial illustration of his thought. In this respect it also seems to be worth remembering that his strategy of visualization does not only encompass the notion of political sovereignty (the sword together with the bishop's crozier), but also the integration of the multitude of individuals. This multitude serves, metaphorically speaking, as the chest protector of the 'mortal God'. It has been mentioned already that the individuals only become a unity, called 'the people', via representation. At the same time, the frontispiece also expresses the importance of social integration as the origin of modern statehood. Both principles – sovereignty and integration – are interrelated with each other; neither of these two principles can substitute the other: political sovereignty depends on social integration as much as social integration depends on political sovereignty. This interrelation is the first message of Hobbes's visualization strategy; it is connected with a second message, which says that both principles are 'real' only through representation (Skinner 1999). Applied to political authority, each directive (ordered by the authority) implies a claim of representing 'the people' (integration) and their authorization of power that allows for speaking in their name (sovereignty) – which is the same as acting as a representative of the State. In this sense it belongs to the legacy of Hobbes's political theory that formal authority needs to be accompanied by symbolic references to the otherwise only fictional existence of an integrated 'body politick' or commonwealth. Moreover, his conception of the commonwealth is an example for the connection between the two functions of symbolic legislation in the positive sense, the expressive and the constituting function: by giving expression to the political unity the political order itself is constituted.

(2) *Contested Authority*: Oakeshott has called Hobbes's *Leviathan* an example of a 'civil association' (Oakeshott 1975a). A civil association is a mode of association based on moral practice, which is a practice free of any purposes other than the existence of the political order itself (Oakeshott 1975b: 149–158). Its non-instrumental character allows citizens to freely choose for themselves among their own preferences and to pursue their own interests. Neither preferences nor interests are determined by the civil association, only the way of conduct is regulated by the

rule of law. For Oakeshott it is the "adverbial" character of the law ('*lex*') that makes the difference between a civil association and an enterprise association: *Lex* is directed to the rules of conduct; it is about 'how' to live a life according to the law ("lawfully"), but not about 'what' to do in life. Building on this differentiation, Oakeshott believes that authority and freedom are not necessarily opposite poles. But if the mode of association is of different nature, they may start to contradict each other: In contrast to the civil association, the enterprise association is instrumental in nature. It is about reaching a goal or fulfilling a purpose (e.g., general welfare, public good etc.). Once installed, this goal or purpose determines the necessary allocation of resources with the effect, though, that government's authority becomes more or less involved in a power game, because power is needed to coordinate and decide upon 'who gets what, when, where, and how' (as the well-known phrase by H.D. Lasswell says) for the purpose of the common enterprise.

Oakeshott would agree with Hobbes that formal authority within a civil association is inherently refering to the symbolic dimension that encompasses the relationship between sovereignty and integration. However, the more authority turns into power (or 'de facto authority'), in Oakeshott's eyes an indication for an enterprise association, the shallower this reference becomes. One of the consequences seems to be that authority does not rest any longer on its own unchallenged legitimacy; instead, as 'de facto authority', it becomes depending on 'de facto legitimacy' (or legitimation) which transforms authority in an instrumental value, measured by the degree of 'support' or 'demands' as the result of 'the public's' evaluation of the performance or output of the actual political authorities ('being in authority'). Another consequence is that the symbolic dimension loses its direction. It is not directed to authority any more but to the factual (and actual) consensus or dissent that is reflected by the degree of support or demands as the new challenging input for the political system.

However, this is not to say that symbolic legislation (even in the negative sense) is necessarily a 'bad' kind of legislation. On the one side it tends to commit itself to ideological visions of the good society under the conditions of an enterprise association. Whether or not this affects the status of the political authority in a given legal order is, of course, an empirical question. Authority can fade away, if the one being 'in authority' is not respected as 'being an authority'. On the other hand it can play an important role in highly contested policy-fields, as it is so often the case in the field of biolaw. Here the type of conflict is usually value-loaded (e.g. sanctity of life vs. right of self-determination). This *"dissensus"* type of conflict (Aubert 1963) is a real challenge for political actors, because the common strategies to reach compromises are doomed to fail: No politics of *'do ut des'*, no distribution, or even redistribution. Under these circumstances symbolic actions (decisions, legislations) can provide a *modus vivendi* between the conflicting parties, but to reach such an agreement is a difficult task. It involves among other things the capacity of establishing a common language (or vocabulary) which allows for a mutual understanding of the problem(s) involved, while enabling, at the same time, different interpretations of the alleged same understanding. In the field of biolaw, the legal concept of human dignity might serve as a case in point,[15] for it has, at least partly, proven to be suited for such a purpose of conflict management by means of legal jurisdiction. The secret of the success is mainly its strong inclusive formula, which is normatively based on the reference to every human being (and mankind), and conceptually on the integration of different (sometimes even contradicting) value-orientations (Lembcke 2013b).

However, both elements, the normative significance as well as the integrating scope of different conceptual networks, belong to the fabric of which 'essentially contested concepts' are typically made of: "To say that such a network [of concepts – OWL] is essentially contestable is to contend that the universal criteria of reason, as we can now understand them, do not suffice to settle these contests definitively" (Connolly 1993: 225). Because of their inherent ambiguity the reference to these concepts can produce double-edged externalities on the political discourse: On the one hand, concepts like human dignity may have the power to build a common ground of mutual better understanding. On the other hand, they can also invite contestation or promote an ongoing political struggle. In short, as much as politics builds on concepts which are essentially contestable the outcome may be self-contradicting in effect.

(3) *The Performative Dimension of Authority*: Arendt suggests a different reading of Hobbes than Oakeshott. In her view, the *Leviathan* presents not only the draft for an authoritarian political order, but his concept of authority is authoritarian, because it is based on a foundationalist conception of 'the authoritative'. It represents the attempt to substitute politics as political acting through hierarchy in which political acting is reduced to applying rules. For this reason foundationalism is already authoritarian, since it destroys the performative part of politics and by that politics as such. According to Arendt, Hobbes's political theory represents this line of thinking that has ultimately led to the vanishing of authority – a loss that has become visible during the revolutionary attempts to establish a *constitutio libertatis* in America and France at the end of the eighteenth century.

In her study *On Revolution* she is quite outspoken in her comparison of the American and the French Revolution (Arendt 1963): In short, the French got it all wrong, because they tried to establish a *sovereign* political order based on the idea of 'the nation' as the highest ideal and empowered by force of the *'pouvoir constituant'*. In Arendt's view this absolute power was simply an unfortunate inversion of the absolutism during the Ancien Régime with more or less the same effect in the end. It destroyed the freedom it wanted to achieve, not the least, because it transformed authority into a monstrous program of legitimized power justified to use its force against every enemy of 'the nation' in the name of universal freedom. By contrast, the Americans performed rather well, because they built a *political* order which reflects the open (i.e. contingent) situation of the revolution itself which calls for agreement, promise, or forgiving – voluntary actions with the potential to create *political* power, which is power in concert, not sovereign power. In Arendt's view, a precondition for such a 'powerful' joint venture is to understand the inherent perfor-

[15] For an elaboration of this concept, see for instance Van Beers' chapter.

mative dimension of politics which is endangered (if not destroyed) by establishing 'the authoritative' as a foundation of order which then serves as 'the absolute' or 'the constitutive'. But 'the absolute' (value) silences us: Speech becomes dispensable, silence preferable; all that matters is the isolated acquiescence to a truth; if everything is clear, nothing needs to be said and done; power turns out to be useless, or turns into violent force to break the resistance of the non-complier.

Arendt's fierce argument in favor of the American Revolution was countered by Derrida's somewhat different reading: In the end, Arendt's reading of the revolutions may be only another "fable" (Derrida 1986: 10). However, "fabulous" indeed is her idea of a performative dimension of authority which keeps the revolutionary moment alive and open for the contingency and renewal that comes with it. As for the relationship between authority and symbolic legislation two points are at least of interest here: The first point refers to the expressive function of symbolic legislation (in the positive sense) with respect to the constitution. The case of American Revolution shows, according to Arendt, the expressive power of the constitution and its various procedures of the constitution-making processes by symbolizing the contingent political space that followed the rupture. Second, because of its contingency the newly created political space symbolized by the constitution-making process is open for what Arendt calls 'augmentation'. The Latin origin of this word points in her view to the core of political authority: the capacity to enrich the performance that leads to the foundation of a political order which is (like every political institution, and ultimately like every human being) incomplete and, hence, because of its permanent transformation, 'augmentable'.[16] Taking together, it seems that for Arendt the expressive function of the symbolic legislation (in the positive sense) matches well the self-constituting dimension of political authority which cannot be derived from higher values but only expressed by performative actions and interpretations.

(4) *'The Authoritative' and its Actors*: The German discourse about the constitution and its principles is an intriguing example for the interplay between the *'Grundgesetz'* (GG) as 'the authoritative' and the Federal Constitutional Court (FCC) as the institution 'being in authority'. It has become widely accepted that the German Constitution is more than a mature legal document from which specific practical directives can be extracted by means of proper legal techniques. It establishes ground rules for political order, but the constitution does more than just that: it also gives expression to fundamental principles of social co-existence (human dignity, democracy, individual rights, equality, etc.), principles whose authority is meant to vouch for the legitimacy of political action. These guiding principles are

[16] If this capacity (to augment) actually resides within the US Supreme Court, as Arendt has argued (Arendt 1963: 201), is a somewhat sober and maybe not so fabulous end of her narrative – but an end that makes her fable pretty real. For an illustrative example see Kahn's discussion of the Supreme Court "as the guardian of that symbolic order" that is built on the "belief in American citizenship as participation in a popular sovereign that expresses itself in and through the rule of law". For this reason the Supreme Court "wields an unquestionable power. So much so that it can tell us that the President of the United States is the person who lost the popular vote", as it was the case in the Bush v. Gore decision (Kahn 2003: 2686).

'value-loaded' and symbolize different guiding ideas (*ideés directrices*) underlying the legal document as an expressive-evaluative dimension and supplying the legal validity with moral legitimacy. This context may already explain in part the 'success' of the German constitution, measured by the standing and prestige of this document among the German population. However, this development is in large parts 'fabricated' by the FCC and its effective strategy to establish itself as the 'guardian of the constitution' – and this strategy is all the more effective the less obvious' the precarious nature of the 'fabricated' validity of FCC's decisions appears (Lembcke 2007). In sum, the symbolic dimension may become 'constitutive' because of the interplay between 'the authoritative' GG and the FCC as the actor 'being in authority'. The German example may be a prominent case in point, yet it is by far not the only case (e.g. Kahn 2003; Lembcke 2013a; Stone Sweet 2007). In more general terms, in modern states political institutions as actors being 'in authority' may have various reasons to ground their claim on a normative foundation called 'the authoritative'.[17] However, it seems that Hobbes's insight still remains true: 'the authoritative' itself exists only through representation by the actor 'being in authority'. Empirically speaking, this might be a strong incentive for the representative to switch roles, and to ground the claim of 'being in authority' in (the claim of) the superior knowledge of 'being an authority'.

## 6.5 Crisis of Authority?

So, in the end, what can be said about the relationship between symbolic legislation and political authority? Is the former a sign of a crisis of the latter? In the light of the previous section: not necessarily. Additionally to the seemingly close connection between the concept of formal authority and the symbolic dimension of politics, the argument has suggested two potential developments: First, symbolic legislation may become instrumental, e.g. for conflict management of *dissensus* conflicts'; second, actors 'being in authority' might be able to switch roles in order to ground the representative claim in superior knowledge of 'being an authority'. These potential developments do not need to give reason to mark them as phenomena of a political crisis, at least not in the sense of an ineffective authority.

Most likely, however, Hannah Arendt would disagree, albeit for a different reason. In her opinion the concept of authority is not only in a stage of permanent crisis, she went so far as to argue that "[p]ractically as well as theoretically, we are no longer in a position to know what authority really is" (Arendt 1961: 92). It has "vanished" from the modern world and was replaced by external forms of power, either by "coercion by force" or "persuasion through arguments":

> The authoritarian relation between the one who commands and the one who obeys rests neither on common reason nor on the power of the one who commands; what they have in

[17] Some of these reasons may have something to do with the mode of an enterprise association.

> common is the hierarchy itself, whose rightness and legitimacy both recognize and where both have their predetermined stable place (ibid.: 93).

According to her, if symbolic legislation is used to create or maintain a higher normative order, it is just another ill-fated attempt to implement the force of reasoning – and by that to cover the very fact that authority's time in the realm of politics has come and gone. To Oakeshott (and others) this assumption is prone to exaggeration (Oakeshott 1962: 88–90), which is in many ways the result of her 'essentialist' linkage between authority and the founding act of a public order.[18] Yet he shares Arendt's position that authority is different from power (even legitimized forms of power) and he confirms that it is best understood as a particular hierarchy among equals, which is not imposed (by power) but accepted voluntarily (without reasoning or justification). From this perspective symbolic legislation becomes a sign of crisis if it enables or promotes the transformation of authority into power – e.g. by narrowing the scope of integration.

The analysis of Raz's approach might also provide for another answer: As mentioned, the example shows that an integrative theory of the three different types of authority proves to be a difficult undertaking. One cue that can be taken from Raz's failure to integrate the different types is that the 'crisis of authority' is, at least in part, a crisis of the theory of authority. The more political theory – or in Raz's case: legal theory – tries to spell out the criteria distinguishing between de facto and legitimate authority, the more (or so it seems) it transforms the concept of authority into the concept of legitimacy. Yet the latter is morality, and the former is power – *ergo*: in both ways theory contributes to the 'vanishing' of authority. In this respect the tension itself that exists between the formal and the substantial type of authority may represent not so much the underlying tension between power and morality but the theoretical insight into the complexity of the concept. In a similar fashion this might also be true for the conceptual tension between the different dimensions of symbolic legislation. In the light of these complexities it does not come as surprise that the relationship between the two concepts is not easy to structure: Neither does symbolic legislation in a negative sense cause the crisis of authority nor does symbolic legislation in a positive sense promote authority. However, as this paper has tried to argue, it is worthwhile to study their relationship. For this purpose we need to sharpen our analytical understanding of the different (sometimes contradicting) elements of the concepts and our empirical tools to comprehend the various configurations in which they interrelate with each other.

## References

Arendt, Hannah. 1961. What is authority? In *Between past and future*. New York: Viking.
Arendt, Hannah. 1963. *On revolution*. New York: Viking.

[18] See for a detailed critique of Arendt's position Flathman (1980): 71–78. Honig (1993: 96–104) tries to shed new light on Arendt's argument and by that defending it against Flathman's critique.

Arnold, Thurman. 1935. *The symbols of government*. New Haven: Yale UP.
Aubert, Vilhelm. 1963. Competition and dissensus. Two types of conflict and conflict resolution. *Journal of Conflict Resolution* 7: 26–42.
Blankenburg, Erhard. 1977. Über die Unwirksamkeit von Gesetzen. *Archiv für Rechts- und Sozialphilosophie* 63: 31–56.
Bryde, Brun-Otto. 1993. *Die Effektivität von Recht als Rechtsproblem*. Berlin: Gruyter.
Campbell, Robert M. 1993. Symbolic regulation. The case of third-party regulation of Canada post. *Canadian Public Policy – Analyse de Politiques* 19(3): 325–339.
Connolly, William E. 1993. *The terms of political discourse*. Oxford: Blackwell.
Carson, W.G. 1974. Symbolic and instrumental dimensions of early factory legislation. A case study in the social origins of criminal law. In *Crime, criminology and public policy. Essays in honour of Sir Leon Radzinowicz*, ed. Roger Hood. London: Heinemann.
Derrida, Jacques. 1986. Declarations of independence. *New Political Science* 15: 7–15.
Dwyer, John Patrick. 1990. The pathology of symbolic legislation. *Ecology Law Quarterly* 17: 233–316.
Easton, David. 1965. *A framework for political analysis*. Englewood Cliffs: Prentice-Hall.
Edelman, Murray. 1964. *The symbolic uses of politics*. Urbana: University of Illinois Press.
Flathman, Richard E. 1980. *The practice of political authority. Authority and the authoritative*. Chicago: Chicago UP.
Flathman, Richard E. 1987. Authority. In *Blackwell encyclopaedia of political thought*. Oxford: Blackwell.
Friedman, Lawrence M. 1975. *The legal system. A social science perspective*. New York: Sage.
Friedman, Richard B. 1973. On the concept of authority in political philosophy. In *Concepts in social and political philosophy*, ed. Richard E. Flathman. New York: Macmillan.
Fuller, Timothy. 1987. Authority and the individual in civil association. Oakeshott, Flathman, and Yves Simon. In *Authority revisited*, Nomos 29, ed. Roland J. Pennock and John W. Chapman. New York: NY UP.
Garthoff, Jon. 2010. Legitimacy is not authority. *Law and Philosophy* 29: 669–694.
Göhler, Gerhard. 1994. Politische Institutionen und ihr Kontext. Begriffliche und konzeptionelle Überlegungen zur Theorie politischer Institutionen. In *Die Eigenart der Institutionen*, ed. Ibid. Baden-Baden: Nomos.
Gusfield, Joseph R. 1967. Moral passage. *Social Problems* 15: 175–188.
Hart, Herbert Lionel Adolphus. 1994. *The concept of law*, 2nd ed. Oxford/New York: OUP.
Honig, Bonnie. 1993. *Political theory and the displacement of politics*. Ithaca/London: Cornell UP.
Hurd, Heidi. 2001. *Moral combat. The dilemma of legal perspectivalism*. Cambridge: CUP.
Isensee, Josef. 1999. *Vom Stil der Verfassung*. Opldaen/Wiesbaden: Westdeutscher Verlag.
Kahn, Paul W. 2003. Comparative constitutionalism in a new key. *Michigan Law Review* 101(August): 2677–2705.
Kindermann, Harald. 1989. Alibigesetzgebung als symbolische Gesetzgebung. In *Symbole der Politik, Politik der Symbole*, ed. Rüdiger Voigt. Opladen: Leske & Budrich.
Lembcke, Oliver W. 2007. *Hüter der Verfassung. Eine institutionentheoretische Studie zur Autorität des Bundesverfassungsgerichts*. Tübingen: Mohr Siebeck.
Lembcke, Oliver W. 2013a. Autorität der Verfassungsgerichtsbarkeit – eine Skizze in vergleichender Absicht. In *Politik des Verfassungsrechts. Interdisziplinäre und vergleichende Perspektiven auf die Rolle und Funktion von Verfassungsgerichten*, ed. Christian Boulanger and Michael Wrase. Baden-Baden: Nomos.
Lembcke, Oliver W. 2013b. Human dignity – a constituting and constituted legal principle. Some perspectives of a German discourse. *Archiv für Rechts- und Sozialphilosophie, Beiheft* 129 (Human Dignity as the Foundation of Law) 207–231.
Lübbe-Wolff, Gertrude. 2000. Verfassungsrechtliche Grenzen symbolischer Umweltpolitik. In *Symbolische Umweltpolitik*, ed. Bernd Hansjürgens and Gertrude Lübbe-Wolff. Frankfurt am Main: Suhrkamp.

March, James G., and Johan P. Olsen. 1984. The new institutionalism. Organizational factors in political life. *The American Political Science Review* 78(3): 734–749.
Martin, Margaret. 2010. Raz's morality of freedom. Two models of authority. *Jurisprudence* 1(1): 63–84.
Meyer, Stephan. 2009. Die Verfassungswidrigkeit symbolischer und ungeeigneter Gesetze. Die Normenwahrheit – ein neuer Verfassungsrechtsbegriff und dessen Folgen für ein altes Problem. *Der Staat* 48: 278–303.
Neves, Marcelo. 1999. *Von der symbolischen Gesetzgebung zur symbolischen Konstitutionalisierung. Ein Überblick.* (IfS-Discussion Papers, Department of Government Studies). München: IfS.
Newig, Jens. 2010. Symbolische Gesetzgebung zwischen Machtausübung und gesellschaftlicher Selbsttäuschung. In *Wie wirkt Recht?* ed. Michelle Cottier, Josef Estermann, and Michael Wrase. Nomos: Baden-Baden.
Newton, Kenneth, and Pippa Norris. 2000. Confidence in public institutions. Faith, culture, or performance? In *Disaffected democracies. What's troubling the trilateral countries*, ed. Susan Pharr and Robert Putnam. PUP: Princeton.
Noll, Peter. 1981. Symbolische Gesetzgebung. *Zeitschrift für Schweizerisches Recht* 100: 347–364.
Oakeshott, Michael. 1962. Review of 'between past and future' (Hannah Arendt). *Political Science Quarterly* 77:88–90.
Oakeshott, Michael. 1975a. *Hobbes on civil association*. Indianapolis: Liberty Fund.
Oakeshott, Michael. 1975b. *On human conduct*. Oxford: OUP.
O'Sullivan, Noël. 2003. Power, authority, legitimacy. In *Understanding democratic politics. An introduction*, ed. Roland Axtmann. London: Sage.
Pitkin, Hanna F. 1967. *The concept of representation*. Berkeley: University of California Press.
Raz, Joseph. 1986. *The morality of freedom*. Chicago: Clarendon.
Skinner, Quentin. 1999. Hobbes and the purely artificial person of the state. *The Journal of Political Philosophy* 7: 1–29.
Stone Sweet, Alec. 2007. The Juridical Coup d'État and the problem of authority. *German Law Journal* 8(10): 915–928.
Van Klink, Bart, and Oliver W. Lembcke. 2013. What rules the rule of Law? A comparison between Michael Oakeshott and Hans Kelsen. In *Politik und Praxis – Michael Oakeshott im Dialog*, ed. Michael Henkel and Oliver W. Lembcke. Tübingen: Mohr Siebeck.
Voß, Monika. 1989. *Symbolische Gesetzgebung. Fragen zur Rationalität von Strafgesetzgebungsakten*. Ebelsbach: Gremer.
Wang, Zhengxu, Russel J. Dalton, and Shin Doh Chull. 2006. Political trust, political performance, and support for democracy. In *Citizens, democracy and markets around the Pacific Rim. Congruence theory and political culture*, ed. Russel J. Dalton and Doh Chull Shin. Oxford: OUP.
Zeegers, Nicolle, Willem Witteveen, and Bart van Klink (eds.). 2005. *Social and symbolic effects of legislation under the rule of law (Studies in political science)*. Lewiston: Edwin Mellen.

# Chapter 7
# On Legal Symbolism in Symbolic Legislation: A Systems Theoretical Perspective

**Jiří Přibáň**

## 7.1 Introduction

The concept of symbolic legislation is controversial and typical of normative, critical and aspirational meanings. Some scholars refer to symbolic politics and legislation as a disguise of real forces operating within the rule of law. According to these critics, the function of law is to politically stabilize and legitimize existing power constellations and to psychologically comfort the powerless. Criticising the acts of symbolic legislation, therefore, is part of a more general ideological critique of law and politics.

Other scholars use the concept of symbolic legislation to contrast the effectively enforced legislative acts to the acts primarily recognizing and reasserting specific societal values beyond the legal enforcement. This distinction draws on the difference between the instrumental and value rationality of law, yet often involves criticisms of symbolic legislation as a weak form of law resigning to the regulative function of law and merely recognizing values and moral aspirations of a political society under the rule of law (Dwyer 1990). In this context, symbolic legislation signifies dysfunctional acts of legislation unable to effectively regulate and govern political society.

However, some recent legal and socio-legal theories have adopted the concept of symbolic legislation as a concept contrasting the instrumental use of legislation to the legislation preferring 'soft' enforcement and stimulating social interaction and moral reflexivity of society as a community of values (Zeegers et al. 2005). In this theoretical context, the concept of symbolic legislation acquires positive meaning and the legislator is expected to weaken the instrumentalist top-down policy backed by the system of sanctions by adopting a policy of social engagement and interaction

J. Přibáň (✉)
Cardiff Law School, Cardiff University, Museum Avenue, CF10 3AX Cardiff, UK
e-mail: priban@cf.ac.uk

 
B. van Klink et al. (eds.), *Symbolic Legislation Theory and Developments in Biolaw*, Legisprudence Library 4, DOI 10.1007/978-3-319-33365-6_7

steered by morally reflexive and aspirational legislative acts. Symbolic legislation is considered part of a legislative policy reasserting common values and ideals and preserving political society's moral unity (Van der Burg and Takema 2003).

Despite the variety of theoretical uses and critical meaning, the concept of symbolic legislation shows that the system of positive law is impossible to reduce to the instrumental rationality of legal regulation and accommodates the symbolic rationality of values and principles promoting society's self-reflection as moral community. In this chapter, I, therefore, ask if symbolic legislation can facilitate social cohesion and normatively protect commonly shared values. My answer to this question is negative because, unlike, for instance, Wibren van der Burg who describes 'the integrationist paradigm' promoting the ideals of democracy and the rule of law in this volume, I argue that no social system can guarantee social cohesion in functionally differentiated modern society. However, I also hope to answer the other question, namely why the system of positive law is irreducible to its instrumental rationality and adopts the specific mode of symbolic communication stretching far beyond policy goals of symbolic legislation.

I, therefore, distinguish between the policy-oriented concept of *symbolic legislation* and the concept of *legal symbolism* as a more general concept describing the symbolic function of the system of positive law. I am not interested in a critique of law as a symbol of government. I am equally uninterested in either criticizing, or praising specific acts of legislation for their symbolic recognition and promotion of common values. Instead, I am interested in understanding whether these symbolic acts of law actually can transform modern pluralistic societies into morally unified communities and the symbolic rationality of law is perceived as 'an essentially political concept' (see Chap. 2 analysis of symbolic legislation in this volume) and the ultimate force of social integration.

This chapter, therefore, is not constrained by topics of legal policy and normative aspirations of symbolic legislation. Instead, it primarily focuses on the tension between legality and a specific form of symbolic communication facilitated by the system of positive law. Apart from being a functionally differentiated system, law involves expressive and evaluative symbolism which aims at collective self-reflection combined with attempts at self-perfection. The instrumental rationality of legality is accompanied by the evaluative symbolic rationality of collective identity and shared culture. Although moral values are not the 'foundations' of the legal system, they cannot be entirely eliminated from it.

Similarly to Oliver Lembcke's Chap. 6 in this volume, I draw on the duality of the instrumental and symbolic rationality of law. However, the problem of authority is not a central frame of reference of my study. Instead, I argue that the concept of legal symbolism signifies the operations whereby the legal system internalises the moral 'absolutes.' Morality is described as a loose system which does not have the institutional and operative capacity of law and, therefore, uses the legal system to stabilise itself in modern society. Against the Durkheimian and Parsonsian views of social integration and value unity, which are still highly influential in contemporary sociology of law, I argue that the symbolic rationality of law cannot have the general integrative function and only internalizes the specific function of culture as a temporal reservoir of collective moral aspirations and identifications. Legal symbolism is part of the process of functional differentiation and no piece of symbolic legislation, therefore, can provide access to the moral values and principles guaranteeing the normative unity of modern pluralistic society. It would be too ambitious to expect that the contemporary system of positive law can answer the philosophical question and grant our access to a shared universe of symbols which would normatively reflect on the meaningful existence of society and its members.

Unlike scholars promoting the concept of symbolic legislation, I am sceptical of the law's capacity to promote social cohesion by recognising moral values and encouraging social discussion and endorsement of these values in the process of interaction and civic activism between the legislator and different segments of society. Social cohesion founded on the symbolic rationality of commonly shared moral values is impossible to achieve in functionally differentiated society. Any symbolic legislation, rather, needs to be analysed as part of the functionally differentiated symbolic communication permeating different social systems including the system of positive law.

I introduce my general argument by briefly addressing Ernst Cassirer's philosophy of a symbolic universe of human ideals materialised in and differentiated into different social sectors, such as religion, art, science and law. Instead of treating law as a normative order enforcing commonly shared values and ideals and therefore functioning as a guardian of social cohesion and moral unity, this philosophical view considers law one of many forms of symbolic communication of human ideals which are internalized and therefore cannot operate as the law's foundations.

I subsequently discuss the sociological functionalist view of law as an external index of social solidarity. This method, most famously used by Emile Durkheim, makes the problem of symbolisation part of immanent social communication and removes it from the realm of philosophy contrasting social immanence of norms to the transcendental ideals of humanity. However, Durkheim's sociological broad brush effectively turns the complex system of positive law into a mere symbol of values safeguarding social solidarity. Its main contribution to the study of symbolic communication and law thus paradoxically consists of a relativistic perspective of morality and values. Unlike romanticising theories looking for lost foundations of authentic social life and *Volksgeist* and considering the first function of law to be a symbol and guardian of this spirit of a people, Durkheim's sociological theory firmly analyses the general function of law to be a symbol of social solidarity and totality of social life.

One of the most important modern legal and social philosophical categories, the spirit of the people, is subsequently interpreted in light of this functionalist perspective. This critical interpretation opens the possibility of understanding legal symbolism within the context of modern social and sociological theories of law analysing the law's dual claim to legal and moral authority. However, law cannot secure general moral authority in functionally differentiated society and the paradoxes emerging between the instrumental rationality of legal regulation and the symbolic rationality of law as an expression of moral community cannot be de-paradoxified by claiming the supremacy of social solidarity and commonly shared values and principles.

In the final sections of the chapter, I, therefore, engage in analysing legal symbolism as part of functional differentiation of the systems of law and morality which leads to the conclusion that symbolic communication, instead of being treated as a foundational and/or external determination of legality's meaning and content, runs as specific communication internalised by the legal system and using the code of legality to refer to what is morally 'good' and 'bad.' Through this moral self-reflection, it establishes our identity in terms of functionally differentiated cultural symbolism distinguishing between 'us' and 'them' in terms of time and space. I, therefore, conclude by highlighting the basic fact about functionally differentiated society, namely that any symbolic communication of ideals and values in law, rather than building ultimate moral commonwealth and social cohesion, only contributes to further differentiation between the instrumental and symbolic rationality of law.

## 7.2 In Search of a *vinculum functionale*: On the Philosophy of Symbolic Forms

There is a profound difference between philosophical and sociological perspectives of social symbols and symbolic communication. The philosophical perspectives emphasize what might be called 'a new dimension of reality' (Cassirer 1944: 24) or 'a fifth dimension' (Elias 1991: 47) of human existence which reshapes the four social dimensions of space-time into a shared universe of symbols communicating the meaningful existence to both individuals and societies. Its primary purpose is to examine social symbols as an expression of human nature and/or the media communicating and searching for the meaning of human existence.

The fifth dimension of social symbols constitutes the cultural life of human beings which, since early mythologies, has incorporated both anthropological explanations of the origins of human existence and cosmological explanations of the origins of the world. If there still remains 'a clue to the nature of man' (Cassirer 1944: 23) in the modern world of science and functionally differentiated society, it is to be found in a symbolic universe of humankind responding to the physical world by the active and complicated process of thinking.

In this respect, Ernst Cassirer, for instance, states that 'instead of defining man as an *animal rationale*, we should define him as an *animal symbolicum*. By so doing we can designate his specific difference, and we can understand the new way open to man – the way to civilization' (Cassirer 1944: 26). According to this anthropological philosophy, symbols do not have actual existence in the physical world, yet they have a 'meaning' and thus make a clear distinction between actual reality and possibility. The difference between things and symbols constitutes human culture as a realm of the difference between facts and ideals. The general function of symbolic thought is thus the establishment of ideals, which, by definition, are impossible to materialize. They are in the state of potentiality which is both a necessary and indispensible part of our social reality.

This function of symbolic communication involves the capacity to treat ethical ideals as if they can materialize in the form of moral facts (Přibáň 2007). The symbolic system thus has an important function of constantly reshaping the human universe according to the human thought. Nevertheless, Cassirer's anthropological philosophy of symbolic forms does not recognize a human metaphysical 'essence' and considers language, myth, religion, art, science and other constituents of thinking and acting as sectors of the circle of 'humanity.' These sectors are held together by a common bond. But this bond is not a *vinculum substantiale*, as it was conceived and described in scholastic thought; it is rather a *vinculum functionale*. It is the basic function of speech, of myth, of art, of religion that we must seek far behind their innumerable shapes and utterances, and that in the last analysis we must attempt to trace back to a common origin (Cassirer 1944: 68).

The state, law and other institutions are thus expressions of both communal existence and the common human origin. According to the anthropological philosophy, the functional value of the state and law is its symbolization of the common human origin and the capacity of human thinking to turn the natural world into the image of the social world. The unity of human existence is not given by a metaphysical substance existing in itself and known by itself. Instead, it needs to be conceived as a functional unity and multiplicity of the fundamental human task and power of man 'to build up a world of his own an "ideal" world' (Cassirer 1944: 228).

This view sharply contradicts the Hobbesian perspective according to which the natural world is irrational and brutal and social institutions, such as law and the state, control and civilise natural instincts and tame aggressiveness and destructive atavisms of the human nature. Against the Hobbesian version of political society and civilisation as the nature's opposites and controlling mechanisms of the humanity's dark and destructive side (Gehlen 1964: 105), the philosophy of symbolic forms identifies the commonality of human thought in all symbolic forms from myths and religion to the science and art.

All these forms eventually are symbolizations of human existence and its achievements. The symbolic thought is the common root of laws of human evolution and constitutes 'a final community of function' (Cassirer 1946: 84). Language, religion, art, politics and law are the symbols of recognition and remembrance of humankind manifested in its culture and elevating the human 'spirit' above the physical 'life.' This appears to be a philosophical solution of the paradox of the human consciousness of modern culture presenting human beings with new life chances and gifts, yet seemingly denying them their enjoyment.

The human capacity to transcend the physical world and create a new world of metaphors and symbols leads to the liberation of human existence from natural limits including human natural instincts. The human thought equally fights against the chains of natural world and pursues its emancipation from social institutions. Social symbols thus do not merely strengthen the stability of institutions and often communicate calls for their reconstruction, if not revolutionary destruction and/or replacement.

Instead of being a source of inhibition, society channels impulses of inspiration and aspiration of humankind. Normative aspirations of philosophy significantly contribute to this symbolic communication of human emancipation in modern society and social reconstruction of the human condition. Nevertheless, there is an important lesson to be learned from Cassirer's philosophy of symbolic forms, namely that law is not the ultimate symbol of cultural recognition protecting humankind from falling into the natural world's anarchy. It, rather, is one of many examples of symbolic communication and its function as a 'sector of the circle of humanity' does not coincide with the moral integrative function of modern society. The philosophical *vinculum functionale* of a common bond actually rules out the possibility of societal institutionalisation and legal codification of such a bond. Cassirer's philosophy of symbolic forms thus serves as a lesson in both common humanity and societal differentiation of its sectors with their specific communication and points of reference. Symbolism in the legal sector cannot be treated as its normative foundation. It, rather, is independent and emancipated from constraints of legal normativity and therefore can equally contribute to its stabilization and destabilization, confirmation and change.

## 7.3 The Law's Symbolic Function: On the Sociological Functionalism

Unlike normative claims of philosophers and their explorations of the human nature and the function of symbols and values carved in stone of social institutions, the sociological perspective considers social symbols and symbolic communication a function of general social processes and developments. Instead of treating normative contradictions and different aspirations as an element of human thought and its capacity of symbolic expression, it claims to explain them as particular responses to general social 'facts.'

Instead of using anthropological and ethnological methods to demonstrate the general capacity of human thought to express itself through different symbolic forms, the sociological functionalism treats social symbols as expressions of social reality and collective representations. Within this tradition of sociological positivism, Emile Durkheim's functionalist view of law as an 'external index' of social solidarity which is its 'visible symbol' (Durkheim 2013b: 57) continues to play an important role in contemporary general sociology and sociology of law (Veitch 2010: 189).

Describing social facts as the facts of moral life and treating them as 'phenomena like any others' (Durkheim 2013a: 3), Durkheim pursued the goal of the positive science of morality observing, describing, classifying and explaining these facts as collective representations. He treated social facts as mental and inter-subjective beliefs and sentiments contrasted to the physical world. His concept of social solidarity thus was a specifically sociological response to the simple, yet persisting question of what are the bonds uniting men together (for critical reflections, see Hart 1968: 1–13). While social solidarity cannot be directly observed and positively measured, law with its specific measures and different forms of regulation can be treated as its external index.

Legal institutions, such as the sanctions of the criminal law, are symbolic in the sense that they express and function as a reinforcement of dominant collective representations (Garland 1983). The legal evolution from repressive to restitutive laws, for instance, informs the sociological distinction between two different types of solidarity – mechanical and organic.

According to Durkheim, law 'symbolizes the special solidarity engendered by the division of labour' (Durkheim 2013a: 54). This symbolism of law is part of the profound methodological innovation made by Durkheim, namely the circular explanation of social phenomena by social phenomena. This circularity of the social explained only by the social (Luhmann 2008: 19) is what makes Durkheim one of the sociology's founding fathers (Lukes 1972) with profound influence on the sociology of law (Clarke 1976: 246).

Furthermore, this circularity defines symbols as social function. In this respect, Durkheim clearly states that 'all law … is a social function' of symbolizing the special social solidarity (Durkheim 2013a: 55) and the function of law thus involves the maintenance of the cohesion of a society by sustaining its common consciousness. Law is not a technique of social control but an expression of moral collectivity shaped by common values and social relationships. Society is a site of the sovereign power while politics is only one of its systems and politicians operate as the social solidarity's functionaries. Like politics, law derives and expresses this solidarity and moral ideas, therefore, 'are the soul of the law' (Durkheim 1987: 150).

Unlike Weber and other sociologists of his time, Durkheim virtually ignored the actors and institutions making, interpreting and enforcing the law. The role of legal professions, specialists and traditions are unimportant vis-à-vis Durkheim's metatheoretical assumption that law is but a symbol of social solidarity (Cotterrell 1991). Following the evolution of law from repressive to restitutive regulation, Durkheim considers social evolution a complex development from less to more advanced forms of solidarity, and from passive collectivism of submission to active individualism of consensual cooperation (Turkel 1979: 721).

Nevertheless, Durkheim's symbolic function of law is profoundly different from the romanticizing sociological tradition contrasting community to society and using organic metaphors to contrast idealized communal bonds of the past to the present alienating structures of industrial society. Unlike Tönnies and other scholars searching for the possibility of revitalizing the community's authentic sense of belonging in the modern social life, Durkheim used the concept of organic solidarity to signify the functioning and interdependence of increasingly differentiated modern society. His distinction between the mechanical and organic solidarity is thus profoundly different from the community/society distinction drawing on the spontaneity/organization and authenticity/artificiality distinctions typical of the romanticized sociological critique of modernity.

While moral behavior is shaped by society and thus disinterested and motivated, rather than by individual self-interest, by devotion, Durkheim acknowledges that the moral foundations of modern law, if not legally enforced, may be weakened and lead to a kind of moral unraveling. Unlike the sociological romanticism mourning the loss of community in modern society and the philosophical functionalism treating the symbolic forms as different functions of the human capacity to re-invent the world, Durkheim's sociological functionalism is morally relativistic and attributes different sets of moral ideas and values to different societies. Despite this positivistic relativism, Durkheim, nevertheless, admits that societal evolution and a system of collective beliefs and practices of modern industrial society leads to the specific morality of individualism typical of pluralistic and diverse societies (Turner 1993). This system of collective practices and beliefs in individualism, however, is not a matter of individual rational choice but a societal 'religion' focusing on the human person as a sacred object and thus sanctioning its protection by special collective authority.

## 7.4 The Historical Legal Metaphor of 'the Spirit of the Laws,' Its Symbolism and Critique

Durkheim's functionalist distinction between the mechanical and organic solidarity is a lot more complex than a simple contrast between *old* community and *new* society which is typical of evaluative distinctions between nature and reason, particularism and universalism or affectivity and rationality (Tönnies 1957: 34). Despite some intellectual inspirations and conceptual overlapping, such as the status/contract distinction, the functionalist legal symbolism outlined by Durkheim profoundly differs from early attempts at historical positivism of legal and social theories interpreting law as a mode of expression of a collective *spirit* unifying and arranging manifold social structures into one totality.

As Hegel remarked, it was Charles Montesquieu who first recognised the true philosophical position by demanding that legislation should be examined as a variable moment in the social totality (Hegel 1967: 16). According to Montesquieu, the spirit of the laws is the result of mutual social influences that may have only relative validity and are related to the life of specific nations and external conditions such as climate, geography, and space. In this respect, Raymond Aron commented that the general spirit of the laws is a product of the 'totality of physical, social, and moral causes … which enables us to understand what constitutes the originality and unity of a given collectivity' (Aron 1965: 46).

Montesquieu considers the spirit of the laws 'the unifying principle of the social entity.' This synthesising meaning makes the concept of the spirit of the laws close to the concept of *culture* because it signifies specific folkways and mores of a people persisting through its history (Aron 1965: 21). Historical jurisprudence and early sociological theories were, indeed, inspired by Montesquieu's historical methodology and sense of the importance of historical particulars (Meinecke 1972). However, romantic thinkers such as Herder gave a normative twist to Montesquieu's concept of the spirit of the laws by contrasting alienating modernity to the authenticity of the past communal life of different peoples. The spirit of the laws became identified with the spirit of the people expressing itself in customs, traditions, language, folk tales and other experiences of the peoples' allegedly authentic existence increasingly threatened by societal structures of modern industrial society.

Drawing on Herder's romantic philosophy, it was F.K. von Savigny who used the 'deeply mystical idea' (Cotterrell 1992: 21) of *Volksgeist* in his critique of A.F.J. Thibaut's proposals to introduce a codified civil law for all German states. Against these legislative proposals, Savigny argued that legal codifications primarily must reflect and recognise the original spirit and common consciousness of the nation *(Volksbewusstsein)* (Savigny 1975: 27–28). In order to be recognised by its subjects, the codified law needs to *express* the nation's ethos detectable in its history, mythology, religion, customs, or folk tales. The legislator's primary function is to follow the spirit of the nation and respect legal customs and traditions (*Gewohnheitsrecht*).

The central role of law in expressing a historically and morally unique cultural system – the spirit of the laws – informed the idealistically romantic historical jurisprudence of Savigny or Gierke as well as many different forms of Hegelian historicism in the social sciences in the nineteenth and twentieth centuries. The legal system was depicted as a system of a limited number of principles rooted in the totality of culture. Romantic legal scholars and philosophers like Savigny believed that 'law is the totality of life' seen from a specific viewpoint (Timasheff 1974: 343).

The unifying concept of the spirit of the laws was considered to be both an analytical concept of general positivistic laws and a specific cultural totality governing the life of political society and making it clearly different from and incommensurable with other societies. The spirit of the laws was the historical spirit presenting itself in every field of human culture and society including law and politics. The methodology of historical positivism synthesised specific findings of philology, ethnography, jurisprudence, political theory and other sciences to identify the fundamental features and laws of a particular society and its culture. Historical jurisprudence thus established new comparative methods and genealogical perspectives accommodating ethnographic knowledge of folkways, linguistic findings, and political and social developments of the state and its institutions.

The spirit of the laws became a leading metaphor for what the collective identity and values of society in its totality should be like. The legal system's function was to faithfully mirror the historical, cultural and political uniqueness of a people. It was considered an aspect of cultural tradition and experience of the overwhelming and persisting domain of culture. At the same time, law was expected to codify a cultural code for society and symbolize its unity.

## 7.5 Legal Symbolism Beyond Its Spirit: On the Paradox of the Law's Dual Appeal to Political and Moral Authority

Drawing on Durkheim's sociological functionalism, Roger Cotterrell identifies the duality of law's function by stating that 'law is the regulation and expression of community' (Cotterrell 2006: 28). While law regulates and settles social conflicts by its procedures, the same procedures and decisions reaffirm values and the moral integration of community. Furthermore, legal regulation is differentiated itself and this differentiation cannot be contained by an essentialist morality operating as the legal system's foundations.

It is impossible to essentially ground the functionally differentiated subsystems of modern society in the superior system of common moral values. Nevertheless, the legal regulation of different sectors of modern society often leads to the constitution of different communities in the most diverse forms from commerce and care to employment and environment. In Cotterrell's sociological theory of law, community is not associated with nostalgia for things past destroyed by alienating modern social structures. Instead, community is both an outcome and precondition of law's functionality.

According to this view, law is not merely a self-referential and self-evolving subsystem exclusively communicating through the medium of legality. It equally responds to the external need for social solidarity, moral justification and symbolic collective expressions. While delineating culture from the essentialist concept of communal identity and relieving law from the job of legislating for this identity, law responds to the first imperative of communal relationships, namely mutual interpersonal trust (Cotterrell 1995: 330). Law's communities, therefore, need to be perceived as communities of trust specifically established and guaranteed by means of legality.

This duality of law's function in modern society is replicated by the paradox of law's dual appeal to moral and political authority. While political authority claims to be ultimately grounded in moral authority, it also is increasingly divorced from any kind of moral foundations and tends to exclusively operate as a political technique channeled through the medium of legality. Law is thus both an instrument of power and a symbol of its moral justification (Cotterrell 1995: 337).

The law's paradoxical appeal to both moral and political authority had already been reflected in Thurman Arnold's classic treatment of law as a symbol of government. According to Arnold, 'the function of law is not so much to guide society, as to comfort it' (Arnold 1935: 34). Though the concept of the rule of law may be the moral background of revolt, it ordinarily operates to induce acceptance of things as they are. It does this by creaing a realm somewhere within the mystical haze beyond the courts, where all our dreams of justice in an unjust world come true. Thus in the realm of the law the least favored members of society are comforted by the fact that the poor are equal to the rich and the strong have no advantage over the weak (Arnold 1935: 34–35).

The symbolic function of law thus coincides with legitimation of political power by the appeal to the fundamental principles of justice and the rule of law. However, instead of the revolutionary or rebellious appeal of these principles, their symbolic invocation by law leads to the acceptance of the social and political state of things and their institutional framework.

The symbolic function of law as the moral justification of political order is possible because of the capacity of social symbolic communication to accommodate the most contradictory concepts and trends, such as obedience and revolt, and to recognize the most diverse social claims and actions as equally justified despite all social differences and inequalities. It is paradoxically the symbolic rationality of law and not its capacity to operate as a political power instrument what constitutes its strength in society and makes it a source of social stability.

The legal system includes the capacity of legality to both operate as the internal code of communication and escape from reality and create its fictional symbolic universe of ideals, values and beliefs in legitimacy of power and moral respect of government. Law as ‘a great reservoir of emotionally important social symbols’ (Arnold 1935: 34) legitimizes inequalities of political power, economic conditions and social privileges as manageable and justifiable by the symbolic communication turning the particular differences into the general unity of society. The absolute moral claim of legal symbols turns them into a power instrument of political ideology while constituting the moral status of society as community of values beyond power politics.

However, the law’s capacity to constitute its community of values and bonds of trust does not mean the capacity to codify and impose a specific form of life and culture through legality. It, rather, is the capacity to transform political legislation into a specific culture of communal bonds without completely eradicating communal particularisms by societal universalism (Cotterrell 1995: 325). Nevertheless, this symbolic function of law communicating and expressing the community of moral values and bonds hardly can be fully understood without taking into account the paradoxical detachment of law from both morality and culture.

Legal symbolism described by the Durkheimian functionalist tradition as a kind of mutual reinforcement between the instrumental rationality of legal regulation and the symbolic rationality of moral community, therefore, needs to be critically revised and its paradoxes de-paradoxified by analysing both overlaps and differences between the systems of positive law and morality in modern society. This methodological step requires re-thinking the differentiation between law, morality and culture in particular and the very process of functional differentiation of modern society in general.

## 7.6 The Semantics of Legal Symbolism: On Morality, Culture and Legality

Symbolic communication imagines modern differentiated society as a unity and thus enhances moral reflections of social cohesion. It maintains the identity of a collectivity, its social boundaries and internal development (Augé 1982; Firth 1975; Turner 1967). Legal communication is not immune from this fundamental desire for social unity and collective identity. Apart from the instrumental rationality of formal legality, the legal system thus makes the symbolic rationality of communal bonds, collective identity, and unity part of its communication (Tushnet and Yackle 1997: 84).

The concept of legal symbolism signifies the operations whereby the legal system internalises the moral 'absolutes' and manipulates them within its internal temporal horizon. It thus informs political society about its moral fabric and identity (Ricoeur 1985: 6, 106). Morality is to be understood as a system using the good/bad binary coding to communicate social processes, actions and even systemic operations and evolution in terms of their praiseworthiness. It considers society as a system of evaluative communal bonds constituting friendship and normative commitments or expectations, such as formulated by the classic fiction of the social contract leading to the establishment of civil society or the discussed romantic version of the spirit of the laws establishing harmony and congruence between the legislated norms and folkways of a people.

Nevertheless, moral concepts, such as natural law and rights, civil society or the spirit of the people, do not exercise any sort of normative superiority and sovereignty over other social systems in functionally differentiated society. Morality is not the ultimate system guaranteeing social integration because of its expressive power to symbolise the totality of society. It, rather, needs to be considered a *loose* system because it does not have the autopoietic operative capacity of law, politics, economy and other social systems. Autopoiesis of law, politics, economy, education and other social systems actually means a higher a-morality of these systems (Luhmann 2008: 163) operating independently of what is described as good or bad by the system of morality. Unlike these a-moral autopoietic systems, morality does not have self-constituting and self-referential institutions closing its good/bad binary coding. Its communication subsequently depends on internal operations of law or politics. Legal symbolism is an example of such internalisation of the good/bad coding of morality by the system of positive law.

Modern society is typical of semantic overproduction leading to the possibility of evaluating all sorts of social phenomena in terms of their good or bad 'nature' and 'impact', yet without general steering capacity of the system of morality. There is both permanent oversupply and shortage of the moral coding and society experiences both too much morality and its absence. An example of this coeval shortage and oversupply of moral norms are various codes of professional ethics emerging in the autopoietic social systems of economy, law, politics, science etc. Lawyers, economists, politicians, scientists and others are expected to act according to their codes of ethical conduct, yet these expectations are specifically related to the systemic rationality of law, economy, politics, science etc. They do not have the capacity to operate as general moral expectations applicable to the whole of society. These codes and other evaluative interventions of social operations are further evidence that morality does not have its autopoietic capacity and therefore uses autopoietic social systems including the system of politics and positive law to stabilise itself in modern society.

Furthermore, the symbolic rationality of law internalizing the moral binary code good/bad has its temporal dimension. It thus benefits from the cultural communication coded in the past/present distinction and adopts the concept of memory as part of collective identity. However, this identity is not a matter of substance but function of culture as latency related to the operations of specific autopoietic systems.

Culture is a specific subsystem communicating temporal differences and thus constituting specific self-referential concepts of legal culture, political culture, economic or any other systemic culture drawing on the recursive self-identification of these systems through the difference between their past and future. These legal internalizations of moral and cultural communication contribute to the most diverse juridical processes of 'dealing with the past' and critical evaluations of collective history, present and future.

Internalizing the latency of culture, the symbolic rationality of law selects and evaluates past experiences and future expectations and turns them into moral praiseworthy absolutes and values which, by definition, are considered non-negotiable and unquestionable. Societal themes of shared traditions, collective identity and memory are critically assessed from the moral perspective and the temporality of culture gets transformed into the 'absolute' moral evaluation. These moral symbols of collective identity, however, are not fundamental in the sense that they would form a common basis or spirit of society. As Luhmann comments: '[S]ociology does not … find the way to what Hegel had called "spirit" [*Geist*]. It does not belong to the humantities [*Geisteswissenschaften*]' (Luhmann 2013: 331).

The moral code of good/bad, like any other social systemic communication, is subject to the temporal change and thus communicates permanent 'transvaluation of values.' Due to the cultural system's use of collective memory and identity, the moral 'absolutes' lose their absolute meaning and become historically and socially relative evaluations. Modern society is thus constantly reminded of the genealogy of its morality.

The legal system consequently facilitates a symbolic communication network semantically independent of instrumental legality and dominated by the value-based strategic rationality which stabilises the system of morality. However, this network cannot reconcile the systemic divorce between the instrumental rationality and cultural symbolism of law. Instead of breaking the iron cage of modernity and achieving social and epistemological unity, it merely confirms the communicative and operative pluralism of functionally differentiated modern society in which 'some bits of information always may escape' (James 1978: 81), even within the domain of law.

## 7.7 Concluding Remarks: On the Legal Symbolism of the Functionally Differentiated Society

Legal symbolism is best understood as the legal system's specific reflection of social expectations of communal togetherness, goodness and justice. It is a mode of legal communication originating in the symbolic communication of moral values of political community, its cultural unity and collective identity. Despite its external origin, symbolic communication is an internal part of the legal system that is also involved in the process of legal self-reproduction and self-reference. It constitutes a specific kind of legal semantics contributing to the external symbolic rationality of morality and culture.

The richness and persuasive force of legal symbolic semantics is particularly strong in the context of constitutionalism and constitution-making, yet may be identified in other areas, such as criminal law and biolaw. Revolutions and other events of social discontinuity invoke the symbolic rationality of law to facilitate new modes of morally communicating the common good and to specify which values and principles need to be legally codified to secure the unity of an emerging polity. The legal system is morally expected to represent the totality of society and the foundation of its values. The law's role seems to be that of identifying and expressing the internal boundaries of political society and its essential cultural fabric. The legal system is thought of as representing cultural traditions, ideological expectations and essential forms of moral discourse.

The symbolic rationality of law obviously is both historical and prospective, transmitted and shared by members of a polity. Like any living organism, society is never located in a single instant and is constituted by its past, present and future which cannot be split up and considered as individual elements of societal life. History is thus always already taken into consideration by the present society which equally cannot describe itself without referring to its future. Collective memory and its legal symbolization are never just simple returns to the past events and their repetition. It rather is a rebirth of the past selected, recollected and synthesised by the present symbolic communication. Instead of repetition of past experiences, collective memory provides for their reconstruction and reconstitution.

In this respect, it is noteworthy that the term 'prudence' (*prudentia*) is etymologically connected with 'providence' (*providentia*) and signifies the present ability to foresee and prepare for future needs (Cassirer 1944: 54). This ability is obviously linked to the ability of learning from past experiences and adjust future expectations according to them. At the same time, the future also functions as a symbolic ideal and the process of formulating future goals, therefore, is an ethical task driven by normative expectations. No wonder law constantly reconstitutes the past and future of society to stabilize its present symbolic communication.

Law has the symbolic power to condemn the past and its injustices and thus creates the new symbolic universe of a changing political society. The system of positive law selects parts of society's past and future and thus contributes to the synthesis of the present identity of *us* opposed to the past or present identity of *them*. It codifies collective identity by symbolising the transcendental ethical ideals of society and the specific meaning of the past, present and future. However, the legal system is unable to ultimately codify collective identity by effectively dealing with the morally unjust past and the just future. Despite the processes involved in the constitutionalisation of political morality and cultural inheritance, law does not succeed in the moral job of constituting the ideal community and securing authentic being and humanity. Moral expectations can never be entirely accommodated by legality.

The moral ascendancy of law cannot materialise due to the functional differentiation of legal, moral and political communication. Legality does not have ultimate social capacity to codify collective identity and sanction the system of moral values in modern functionally differentiated society. Legal symbolism is possible only at a higher level of functional differentiation of the legal system. It is an example of non-trivial interdependence between law and other social systems which needs to be studied from within a system's theoretical perspective, because the symbolic communication generated by the legal system involves both legal and non-legal expectations, limitations and paradoxes. It can hardly be formulated as a problem of social order bound by a Parsonsian shared symbolic system of culture, the normative structure of which could be expressed by the system of positive laws. The legal system does not have the capacity to codify an ultimate value consensus of modern political society. It would therefore be wrong to approach the problem of legal symbolism assuming that society hands down culture and that the *longue durée* of social institutions always uncovers culture, its sedimentation and inheritance as a precondition of social integration and evolution.

Modern society is a multitude of functionally differentiated social systems without a centre defining the supreme sources for the validity and enforcement of social norms. It is a multiplicity of differences, emerging between specific social systems and drawing on the processes of horizontalization and fragmentation. There is no simple causal relationship between the systems of morality and law according to which changes in one system would necessarily result in changes in the other systems. The relationship rather is a complex of specific meanings internally communicated by these systems and externally related to one another.

Legal symbolism reflects an internal paradox of the legal system's operative autonomy and parallel search for external foundations. Although the expressive and evaluative symbolism of collective identity and moral norms does not determine legal operations, the legal system can always reconfigure its reflections of the ideality of a political society and make them an intrinsic part of legal self-reference and autopoiesis. Instead of constructing social and epistemological unity, the symbolic rationality of law thus contributes to the communicative and operative pluralism of the functionally differentiated modern society.

## References

Arnold, T.W. 1935. The symbols of government. New Haven: Yale University Press.
Aron, R. 1965. *Main currents in sociological thought 1. Montesquieu, Comte, Marx, Tocqueville, the sociologists and the revolution of 1848*. Harmondsworth: Penguin.
Augé, M. 1982. *The anthropological circle. Symbol, function, history*. Cambridge: Cambridge University Press.
Cassirer, E. 1944. *An essay on man. An introduction to a philosophy of human nature*. New Haven: Yale University Press.
Cassirer, E. 1946. *Language and myth*. New York: Harper.
Clarke, M. 1976. Durkheim's sociology of law. *British Journal of Law and Society* 3: 246–255.
Cotterrell, R. 1991. The durkheimian tradition in the sociology of law. *Law and Society Review* 25: 923–945.
Cotterrell, R. 1992. *The sociology of law. An introduction*, 2nd ed. London: Butterworths.
Cotterrell, R. 1995. *Law's community*. Oxford: Oxford University Press.
Cotterrell, R. 2006. Law and social theory. In *Law, culture and society. Legal ideas in the mirror of social theory*, ed. R. Cotterrell, 15–28. Aldershot: Ashgate.
Durkheim, E. 1987. Sociologie et sciences sociales. *Reprinted in E. Durkheim, La science sociale et l'action*, ed. J. C. Filloux, 2nd ed, 150. Paris: Presses Universitaires de France, Paris.
Durkheim, E. 2013a. *The division of labour in society*, 2nd ed. Houndmills: Palgrave.
Durkheim, E. 2013b. Law as an index of social solidarity. In *Durkheim and the law*, 2nd ed, ed. S. Lukes and A. Scull, 54–61. London: Palgrave.
Dwyer, J.P. 1990. Pathology of symbolic legislation. *Ecology Law Quarterly* 17: 233.
Elias, N. 1991. *The symbol theory*. London: Sage.
Firth, R. 1975. *Symbols. Public and private*. London: Allen & Unwin.
Garland, D. 1983. Durkheim's theory of punishment. A critique. In *The power to punish. Contemporary penality and social analysis*, ed. D. Garland and P. Young, 37–61. London: Heinemann.
Gehlen, A. 1964. *Urmensch und spaetkultur*. Frankfurt: Athenaeum Verlag.
Hart, H.L.A. 1968. Social solidarity and the enforcement of morality. *University of Chicago Law Review* 35: 1–13.
Hegel, G.W.F. 1967. *Hegel's philosophy of right* (trans: by Knox T.M.). Oxford: Oxford University Press.
James, W. 1978. *Pragmatism and the meaning of truth*. Cambridge: Harvard University Press.
Luhmann, N. 2008. *Die Moral der Gesellschaft*. Frankfurt: Suhrkamp.
Luhmann, N. 2013. *Theory of society*, vol. 2. Stanford: Stanford University Press.
Lukes, S. 1972. *Emile Durkheim. His life and work*. New York: Harper & Row.
Meinecke, F. 1972. *Historism. The rise of a new historical outlook*. London: Routledge and Kegan.
Přibáň, J. 2007. *Legal symbolism. On law, time and European identity*. Aldershot: Ashgate.
Ricoeur, P. 1985. *Time and narrative*, vol. 2. Chicago: University of Chicago Press.
Savigny, F.C. von. 1975. Of the vocation of our age for legislation and legal science. New York: Arno Press.
Timasheff, N.S. (1974; originally published 1939). *An introduction to the sociology of law*. Westport: Greenwood Press.
Tönnies, F. 1957. *Community and society*. East Lansing: Michigan State University Press.
Turkel, G. 1979. Testing Durkheim. Some theoretical considerations. *Law and Society Review* 13: 721–738.
Turner, S.P. (ed.). 1993. *Emile Durkheim. Sociologist and moralist*. London: Routledge.
Turner, V. 1967. *The forest of symbols*. Ithaca: Cornell University Press.
Tushnet, M., and L. Yackle. 1997. Symbolic statutes and real laws. The pathologies of the antiterrorism and effective death penalty act and the prison litigation reform act. *Duke Law Review* 47(1): 1.

Van der Burg, W., and S. Taekema (eds.). 2003. *The importance of ideals. Debating their relevance in law, morality, and politics*. Bruxelles: Peter Lang.
Veitch, K. 2010. Social solidarity and the power of contract. *Journal of Law and Society* 38: 189–241.
Zeegers, N., W. Witteveen, and B. van Klink (eds.). 2005. *Social and symbolic effects of legislation under the rule of law*. Lewiston: The Edwin Mellen Press.

# Part II
# Symbolic Approaches to Biolaw: Biolaw as a Symbolic Order

# Chapter 8
# The Law and the Symbolic Value of the Body

**Jonathan Herring**

## 8.1 Introduction

Legal systems have struggled to find an appropriate legal status for the body; separated bodily parts and bodily material.[1] Not long ago bodies were of limited commercial value and no one would wish to claim legal rights over them. Grave robbing to obtain bodies for trainee surgeons would be one of the few circumstances in which bodies might have some limited commercial value. However, advances in organ transplant, genetics, assisted reproduction and artistic enterprise have made some parts of bodies of considerable economic worth. Is bodily material best regarded as property or better protected by human rights or by a sui generis approach? Do we need legislation or is it better to allow the law to develop gradually as new responses emerge? A fierce debate, fierce by the standards of legal academics anyway, has emerged (Skene 2002; Wall 2011; Goold et al. 2014).

The difficulties legal systems face in classifying the body is typically conducted in terms of pragmatic considerations. Will a property approach or a statutory system have greater flexibility to respond to an area of fast moving technology? What legal structure will ensure the most comprehensive coverage? How do bodies best fit within current forms of legal classification? It seems, although these things are hard to gauge, that, based on such criteria, the property model has the most support.[2]

In this paper it will be argued that the focus on practical questions ignores a crucial aspect of the debate: the symbolic value of bodies. The legal classification of

[1] In this chapter I have drawn on ideas I have developed in Herring and Chau (2007, 2013); Chau and Herring, 'Interconnected, inhabited and insecure: why bodies should not be property' (2014) and Herring (2014).

[2] See, for example, the support of a majority for a property approach among of contributors to Goold et al. (2014).

J. Herring (✉)
Faculty of Law, Exeter College, University of Oxford, Oxford, Oxon OX1 3DP, UK
e-mail: Jon.herring@law.ox.ac.uk

 
B. van Klink et al. (eds.), *Symbolic Legislation Theory and Developments in Biolaw*, Legisprudence Library 4, DOI 10.1007/978-3-319-33365-6_8

bodily material is not simply a matter of pragmatic assessment, but says something profound about our understanding of bodies and of ourselves.

## 8.2 Symbolism and the Law

Before going further it may be helpful to explain what is meant in this chapter by symbolism and the role of symbolism in the law. At one level law can be seen as simply a set of commands and the consequences when those demands are not followed. So understood symbols may appear to play little role in the law. Indeed, calling a law "symbolic" may be seen as a form of criticism; an indication that it is a command that has no impact in practise and so is a failure (see Chap. 2 in this volume). However, I would argue that the law must be understood as more than simply involving a set of commands. It should be based upon and seek to uphold certain values. For example, many lawyers would see the preservation of human rights as being a central role of the law. Such values underpinning the law are vital because they are needed to remove ambiguities within the rules that constitute the substance of the law or determine how clashes between the rules should be resolved. The rules of the law can be symbols of, indications of, a wider set of values which are foundation to the law and the broader society. To take one example, one can tell much from the kind of values represented by a legal system and a society from its law on same-sex marriage. As with all symbols, of course, they can be misleading. One could imagine a legal system which permitted same-sex marriage but in all other respects promoted a conservative agenda. When a law on same-sex marriage is changed this is not simply a switch between one rule and another, but a reflection of a change in social attitudes and a recognition of certain values.

There is more to the role of symbolism and the law than the point summarised in the previous paragraph. For many people abstract concepts are hard to grasp and an image, a physical or mental one, is a helpful guide to understanding an issue. Even those more familiar with grasping abstract concepts will inevitably apply them to a practical situation. An example will clarify this point. When a person says the word "family" for many people, although not all, a rather cosy image of a couple with children in comfort and having fun is created. If we then ask questions about the importance of family privacy or whether the state should promote family life, these are viewed with that image in the background. The image conjured up the word in the way the law poses the issue, shapes the response to the question. As Van Klink (Chap. 2in this volume) explains: "One of the most important functions is the cognitive function: a symbolic law offers a vocabulary that affects the way in which legal and political actors perceive reality. Reality is accessed through the concepts and distinctions provided by the law." To give another example, in legal discourse in the UK, there has been a shift from talking about "child pornography" to "images of child abuse". This, perhaps, involves a recognition that the word "pornography" for many conjures up an image of something distasteful, a bit "naughty", but not seriously harmful. Whereas the phrase "images of child abuse" compels the assessor to take the material seriously.

Sometimes the legal language is used deliberately to change people's attitudes and responses. As Van Klink (Chap. 2 in this volume) put it the law sets out "open and aspirational norms that are meant to change behaviour not by means of threat but indirectly, through debate and social interaction." An example of that from UK law may be the use of "parental responsibility" as the technical legal term used to cover the rights and interests of parents in the Children Act 1989. This language was chosen to inform parents that they were to understand their role as being about responsibilities rather than rights. However, most English family lawyers would accept that the new language used has not had an impact on any particular case. Van Beers (Chap. 11 in this volume) provides another important example, that of sex, where the law's insistence that people be either male or female can close down alternative understandings of sex that people may wish to adopt for themselves form legal and social discussion. Britta Van Beers (2014) writes of the communicative aspect of the law which 'covers the ways in which the law influences 'the vocabulary by which people order their world, give meaning and attach value to it' as Van Klink writes, providing us with concepts, categories, values, and principles to structure normative discussions.'

The central theme of this chapter will be that the legal regulation of body draws on and reflects a particular picture of and understanding about the body. The legal regulation and the legal terminology used is in this sense a symbol of that image and reinforces a particular set of values concerning the body. It will be argued that not only is the picture of the body presented by the property model a false one in terms of biological reality, but also that the values reflected by the property model are ones that should not be promoted in relation to the body (see further Chap. 10, this volume)

## 8.3 Property Rights

It is understandable that in the light of the technological and social developments mentioned earlier there has been a strong movement calling for bodies to be characterised as property. Interestingly, the motivation behind the argument is not so much the nature of property itself, but the kind of legal consequences that flow from the property categorisation. With property rights come the power to control, exchange and exclude. To many people these are important interests in relation to their bodies and parts of their bodies. What we need, it is claimed, is the legal authority to determine what happens to our bodies and its products; the power to be able to sell our body parts; and the legal control over who can, or cannot, have access to our body parts and products.

Ascribing property status to bodies and bodily material appears to provide a ready response to some of the cases which have troubled courts around the world. Should men whose sperm has been stored for reproductive purposes, but which is negligently destroyed, be entitled to compensation? (Yearworth v North Bristol NHS Trust (2009)). Should a widow be entitled to use the sperm of her deceased partner? (Warren *v Care* Fertility (Northampton) Limited (2014)). Should a person who, without permission, removed body parts from a hospital to use in an artwork be guilty of an offence? (R v Kelly (1998)) Property law would give an unproblematic "yes" in answer to these questions (in contrast to the rather convoluted reasoning that the courts have had to use to resolve them). No doubt most people would intuitively think that the answer to these questions should be "yes" and hence the appeal of the property approach.

Support, then, for the property approach is partly based on these practical consequences that flow from adopting it, but it is argued here that it has symbolic significance and so reflects (explicitly or implicitly) a particular image and understanding of the body. The property model sees the body as a biological entity which is controlled, independent and static. So understood, this picture of the body presents a powerful symbol of certain moral claims such as individualised autonomy; privacy; and bodily integrity. The role of law is understood to preserve the body in its pristine controlled state. Because the body is presented in biological terms as self-sufficient, the law is seen to protect it from external interference. Because the body is presented as being immutable, the law needs to protect it from violation. I will argue in this chapter that the "biological reality" reflected in the property model is a false one. Further that we should reject the individualised rights and interests it impliedly promotes.

This chapter, however, will argue that although the property approach might have some appeal it carries with some misguided messages about the symbolic nature of the body. It will further seek to explain why that matters. The starting place must be the nature of the self.

## 8.4 The Nature of the Self

There have, for as long as there has been thought, been debates over the nature of the self. At a much simplified level the protagonists fall into two camps. On the one side there is the "individualised self" and on the other there is the "relational self". These terms require some explanation.

### *8.4.1 The Individualised Self*

Under the individualised model of the self, personhood (the attributes that make you a morally significant human) is characterised by references to matters such as intelligence; self-awareness; self-control; a sense of time; an ability to feel pain; curiosity; and ability to formulate rational ideas (Fletcher 1979: 15). These emphasise independence and the ability to be self-conscious and rational. We make sense of our lives terms of what we have done, what will do and who we are.

Under the model of the individualised self the most important values for the law to promote are autonomy; freedom and liberty (Foster 2009). People must be free to live their lives and develop their goals free from interference from others. This right of autonomy is a fundamental value, one that should only be interfered with where doing so will cause harm to others. Amel Alghrani and John Harris (2006: 192) have claimed that:

> One of the presumption of liberal democracies is that the freedom of citizens should not be interfered with unless good and sufficient justification can be produced for so doing…The presumption is that citizens should be free to make their own choices in the light of their own values, whether or not these choices and values are acceptable to the majority. Only serious danger, either to other citizens or society, is sufficient to rebut this presumption.

While we might engage on joint projects with others, it is important that people are free to disassociate themselves from others, if the relationship is no longer beneficial. Hence, for example, under the model of the individualised self a person should be free, on payment of appropriate compensation, to exit a contract; and divorce should be available with no particular difficulty.

This individualised model echoes many of the attitudes that are commonly presented about the body. "My body is mine and I can do what I want to with it" could be seen as a mantra of our age. The body is seen as something that we can use to express our identity, in artistic or political ways. Hence the law should reflect individuals' rights to control their bodies and arrange for parts of bodies and bodily material to be transferred to others, on payment if desired. This can, in part, be justified on the basis that a body is a result of a person's labour and they can control the products of their labour in a Lockean understanding of ownership (Quigley 2012). Increased emphasis on body shape, achieved through surgery, dieting or exercise is major aspect of contemporary culture. The increased popularity in tattoos and bewildering range of personal adornments reflect how people seek to control and manage their bodies to express their identity. The same is reflected in the grave distress of those who find, in a range of circumstances, their bodies to be alien to them or out of their control. Obesity, transgender issues and body dysmorphic disorder might be used as examples of that. All of this can be used to reinforce a claim that the body needs to recognise the importance of an individual's right to control their bodies; determine the use of the bodily material; and to value their bodies as they wish.

The individualised understanding of the self, and the conception of the body that results, are well captured by property law interests (Hardcastle 2007). The notion of property reinforces the concepts of control; the fact the body can be seen as the product of individual's efforts; the significance of excluding others; and the power to sell which are emphasised by the individualised understanding of our selves (Wall 2011). The body becomes a symbol of the project of a life being self-authored and a result of one's own efforts.

But now we need to look at a very different understanding of the self. One, it is suggested, is more convincing.

### 8.4.2 *The Relational Understanding of the Self*

Under the relational understanding of the self we can only make sense of our selves when we see them in relation to other people. We find our identity not in terms of our relationship with others. So if someone is asked to define themselves they do in relational terms. They may explain they a daughter; an accountant; a roman catholic; or an ice hockey fan. These all are ways in which the sense of self is found through connections with others. From our earliest moments we gain an understanding of who we are and how the world works through relationships.

Under the relational understanding the role of the law is hardly that of promoting the ability to exclude others and pursue our own goals in life. Rather the role of the law is to help maintain good relationships. We seek legal remedies that enable caring relationships to thrive, rather than rights which keep people apart (Herring 2013). That includes remedies to respond to the disadvantages which flow from relationships and for protections from abuse. It promotes a law which is designed around relationships rather than individuals (Whitney 2011). We need to recognise the fact that our bodies form parts of a complex social and biological interconnection with society, the environment and other creatures. The law needs to be based on a norm of interlocking mutually dependent relationships, rather than an individualised vision of rights (West 1997: 456). As Elizabeth Frazer and Nicola Lacey (1993: 178) argue:

The notion of the relational self, in contrast to both atomistic and inter-subjective selves, nicely captures our empirical and logical interdependence and the centrality to our identity of our relations with others and with practices and institutions, whilst retaining an idea of human uniqueness and discreteness as central to our sense of ourselves. It entails the collapse of any self/other or individual/community dichotomy without abandoning the idea of genuine agency and subjectivity.

On such a model, vulnerability and dependency are defining aspects of humanity. Caring for the bodies of others; and receiving care for our body from others, is a core human activity. Yet in caring, the division between 'you' and 'me' becomes lost. The boundaries between bodies become impossible to draw. The image of the independent body, marked by autonomy and independence, promoted by the individualised model and by the legal conception of property, seems a long way away.

In what follows these ideas will be expanded.

## 8.5 My Body Is Mine

The assumption underpinning the property approach is "my body is mine". Indeed that is perhaps at the heart of the appeal of the property approach. The idea that other people might start laying claim to parts of our body or our bodily products sounds like the story line in some kind of science fiction horror film. However, the argument that "my body is mine" is problematic (Shildrick and Mykitiuk (2005)).

To start with, it is untrue at a biological level. Crucial parts of our bodies are made up of nonhuman material. Most of our body surface is a micro-environment on which thrive many microbes, including bacteria, fungi and protozoa. I have written, with a co-author, on the role played in the body by non-human material and I will not go into the detail here again (Chau and Herring 2014). However, it is clear that the many non-human organisms play a central role to the maintenance of our bodies. We are less us than we like to think!

## 8.6 My Body Is Controlled

The notion that the body is property reinforces an understanding of the body as bounded and secured. It can be transferred to others and can be controlled. In English property law wild animals not in captivity cannot be regarded as property for most legal purposes precisely because they cannot be kept under control (Theft Act 1968, s. 4(4)). While the image of the bounded and controlled body may be a comforting image it is a false one.

The truth is that our bodies are constantly changing, dependent on others for survival and subject to environmental factors (Karpin 2005). Our bodies are profoundly leaky (Shildrick 1997). Fortunately so, because our true sense of self and identity is not found in our bounded, owned body, but in the breaking, mixing, and interaction of our bodies with others and with the wider environment (Irigaray 1993).

Our bodies continually change with cells dying and falling off, and new cells being created. By the time we die there is little of us that is biologically the same as when we were born. Whenever people are in proximity bodily material can be passed through the air. As Kenneth Gergen (2011, 23), in his important book on the nature of the self, writes:

> [T]he idea of the skin as a container seems inappropriate. The metaphor of a sieve might be more relevant, with material moving in both directions. On the one hand we could say that nothing that passes through me is distinctly mine (my body); all that I call 'my body' belongs to the larger world out of which it is but a transient conglomerate.

Whenever we meet someone they walk off with parts of our body and we walk away with parts of their body. Continuously our bodies are changing with cells dying and falling off, and new cells being created. Our intestinal lining is completely replaced about once every 2 days (Young and Hay 1995), while the average life-span of the red blood cell is about 150 days, and that of the lymphocyte about 17 days (Macallan et al. 1998). All of this makes the image of property and the rights of control over the body have more than an air of unreality. Our bodies are interacting and mixing with other bodies and the wider environment. They cannot be controlled and retained.

## 8.7 Our Bodies Are Interconnected

There are many complex ways in which bodies are interconnected and interdependent. From our beginnings, in pregnancy the fetus and pregnant woman are in deepest connect. The health and well-being of the fetus can impact on the woman's well-being, and the reverse is true. This interchange is found still after birth through breast feeding. Genetics too teach us how linked our bodies are. The difference between each body in genetic terms is minute. We share so much more than we differ. As Isabel Karpin (2005) puts it:

> The individual in the age of the gene is fundamentally connected and vulnerable.
>
> The individual in the age of the gene always contains a trace of the other; not-one but not-two.

In relationships of care, which are at the heart of our everyday lives, the bodies of carer and cared for are interdependent. For example, not only is a child dependent on their carer, but the carer becomes dependent on the child. If the child suffers an infectious childhood illness and is required to remain indoors, in effect this quarantine is imposed on the body of the carer too. If the child will not sleep, nor, in reality, will the parent. This is true not just in child-parent relationships, but in any close relationship involving caring. In a relationship involving dependence, an injury to the body of either the carer, or the person cared for, impacts significantly on the other's body. Again the picture of us owning or controlling our bodies, obscures their interdependent nature.

Dependency and care are an inevitable part of being human (Fineman 2004; Levy 2006). Eva Feder Kittay (1999, xii) writes:

My point is that this interdependence begins with dependence. It begins with the dependency of an infant, and often ends with the dependency of a very ill or frail person close to dying. The infant may develop into a person who can reciprocate, an individual upon whom another can be dependent and whose continuing needs make her interdependent with others. The frail elderly person…may herself have been involved in a series of interdependent relations. But at some point there is a dependency that is not yet, nor longer an interdependency. By excluding *this* dependency from social and political concerns, we have been able to fashion the pretence that we are *independent* – that the cooperation between persons that some insist is *inter*dependence is simply the mutual (often voluntary) cooperation between essentially independent persons.

This has profound significance for our understanding of our bodies. It shows our bodies cannot be understood as independent and controlled. As Martha Fineman (2011, 168) argues:

> The vulnerability approach recognizes that individuals are anchored at each end of their lives by dependency and the absence of capacity. Of course, between these ends, loss of capacity and dependence may also occur, temporarily for many and permanently for some as a result of disability or illness. Constant and variable throughout life, individual vulnerability encompasses not only damage that has been done in the past and speculative harms of the distant future, but also the possibility of immediate harm. We are beings who live with the ever-present possibility that our needs and circumstances will change. On an individual level, the concept of vulnerability (unlike that of liberal autonomy) captures this present potential for each of us to become dependent based upon our persistent susceptibility to misfortune and catastrophe.

Our bodies are profoundly dependant on other bodies. They are in their nature vulnerable. We typically disguise our interdependency. In a powerful article, Kate Lindemann (2003) contrasts the emphasis that is paid to the accommodations for disabled people so as to minimise the impact of their disability, with the lack of appreciation of the similar accommodations for the able bodied:

Colleagues, professional staff members, and other adults are unconscious of the numerous accommodations that society provides to make their work and life style possible. ATM's, extended hours in banks, shopping centres and medical offices, EZpass, newspaper kiosks, and elevators are all accommodations that make contemporary working life possible. There are entire industries devoted to accommodating the needs of adult working people. Fast food, office lunch delivery, day time child care, respite care, car washing, personal care attendants, interpreters, house cleaning, and yard and lawn services are all occupations that provide services that make it possible for adults to hold full time jobs.

We need an understanding of bodies which recognises their deeply interdependent nature.

## 8.8 Social Interests

Larissa Katz (2008, 275) has captured the nature of property interests well. She writes:

> First, familiar property law doctrines…carve out a position of authority for owners that is neither derived from nor subordinate to any other's. These and other rules create the institutional structure that permits the owner to function as the supreme agenda setter for the resource.

As that quote indicates the property regime gives the generator of the bodily material primary control over what happens to it. This immediately downplays the important social interests in bodily material. Information from bodily material can provide major public health benefits. We are only beginning to see the benefits of the bio-banking schemes, which contain large numbers of samples from individual's bodily material. There can be little doubt that major improvements will occur in health treatments as a result of these. Significantly the bio-banking scheme need only store a relatively small portion of human material and indeed most often merely the information gleaned from that. The individual interest in minute parts of our biological material is very limited. Indeed we drop of pieces of hair and skin all the time without a second thought. A patient undergoes surgery, with little concern with what happens to any bodily material left behind the theatre. It will be discarded, no doubt, and no one will mind. The potential benefits of research on small portions of material seem to outweigh any individual's interest in it.

Once we describe the bodily material as property that introduces a legal system which prioritises the interests of the individual over the interests of the community. It is true that property of an individual can be taken by the government in order to promote public good (e.g., where there is compulsory purchase order). But that requires a very strong interest. So taking a property approach buys into a certain ranking of social and individual interests; one that prioritising individual interests. In saying this I do not suggest that individuals have no interests in these bodies, but rather that the property model, or at least the property model that is prevalent in European legal systems, is prioritising the individual interests over societal ones.

So what precisely are the communal interests which might be included? They include the following:

- audit;
- artistic creations;
- education in medicine and allied disciplines;
- organ donation;
- public health;
- public information creations (such as the use of a picture of a person's body or organ for use in a public health campaign);
- research; and
- third parties.

The history of England's Human Tissue Act 2004 is revealing. Early drafts of the legislation gave individuals an absolute right to control what happened to any removed material forbidding any use which had not been consented to. The practical problems of doing this and the harmful public consequences which might flow allowed for a broad range of exceptions, including that human material could be used for the purposes of education and audit, without the consent of the individual donor.

There is a broader point to make here. We might imagine the well-being of our bodies to be within our control. However, they are dependent on broad social structures and health systems to promote and enable their care. Our bodies and their health (or lack of health) may in part be due to our own efforts, but also in response to societal efforts and impact. Our bodies owe to these larger systems a degree of pay back and recognition of the benefits gained from the used of other bodies.

## 8.9 An Understanding of Health

The argument that you own your own body reflects a particular understanding of health. One that is private and personal to you. Health is typically understood as a subjective matter. It is a matter of how my body is functioning. However, better understandings of health see it as a communal matter and recognise that that our bodies and our health are highly dependent upon other bodies (Foster and Herring 2014). There is great wisdom in the statement of the National Aboriginal Health Strategy Working Party (1998, quoted Boddington and U Raisanen 2009):

Aboriginal health is not just the physical well being of an individual but is the social, emotional and cultural well being of the whole community in which each individual is able to achieve their full potential thereby bringing about the total well being of their community.

The health of each of us is related to the health of those in our community. This is most obviously true in the case of infectious diseases and in the context of a national health system with rationed health care. In more subtle ways our bodily well-being is dependent on the health of a broader community.

Donna Dickenson (2013) has recently published a book entitled *Me Medicine vs. We Medicine: Reclaiming Biotechnology for the Common Good.* This is not the place to describe its significant arguments, but her book describes a shift in some political philosophies to see health as an individual concept. People are encouraged to seek a personalised medical scheme. Typically, this is provided through private health insurance. This notion that we should each seek to promote our own 'me medicine' overlooks the importance of public health initiatives. It emphasises private health-care insurance over public health provision. The language of body ownership and the emphasis on 'my body is mine', often unwittingly, plays into the hands those who wish to promote individualised health as a political ideal. It has been used powerfully by large commercial enterprises to control and manipulate donated bodily material.

## 8.10 Relational Interests

One of the problems with the individualised property model is that it can fail to recognise that bodies live in relationship. One body may, therefore, have claims over another. Indeed they may be intermingled. As Wall (2014, 20) notes:

> [W]here a parent, patient, donor or a widow, is deprived of their entitlements in bodily material, they are being deprived of the opportunity to exercise their rights as a parent, as a wife or as an embodied person. Their interest in the bodily material represents a very personal interest.

A good example of this is the organ retention scandal in the United Kingdom (Quigley 2009). The background story is a complex, but in brief it transpired that for many years doctors in a number of hospitals had been retained the organs of children, removed in surgery, without the consent or knowledge of the parents. Parents were particularly distressed in cases where they had buried their children, unaware that parts of the bodies were being stored by doctors (Sheach-Leith 2007). There are many issues raised by the scandal and I cannot go into all of these here. Most commentators accept that the doctors behaved wrongly and that the consent of parents should have been obtained. Indeed some property approach supporters have used the example of the scandal as showing how much more effective a property approach would be (Goold 2013). I would strongly question that.

First, to describe the harm done to the parents in these cases as property fails to capture the true wrong done to them. To regard the misplacement of part of one's child body as analogous to the damage to a car seems to completely misrepresent the wrong. The harm done is a relational one, not a property one. Mavis Maclean (2001, 74), who sat on the panel hearing the Inquiry at the Bristol Royal Infirmary, reported that the parents did not talk of 'parental rights to the body of their child as property' but instead as an interference in their responsibilities as parents. She notes:

The only immediate form of care which a parent can offer their child after death is to arrange the funeral. When this event is based on lack of information about the physical state of the child, this final act of care may for some families seem to be somehow be devalued and damaged.

The loss for the parents was described in terms of failing to complete the final duty of a parent (to ensure a proper burial of their child), rather than a property loss. A right of interference in respect for family life, for example, seems to capture far more effectively the loss in question.

Second, these problems are reflected when we move on to consider the remedy. Under property law the typical remedy for an interference in property rights will be the payment of compensation designed to purchase replacement property. This fails to capture the wrong at question. A statutory response to deal with cases of this kind might consider formal apologies; appropriate disposal of the body part, a memorial event or plaque could be given. These would not 'right the wrong' but enable a more imaginative response than the property model would.

## 8.11 A Unitary Picture of the Body

The argument that we should treat the body as property assumes that we have the same kinds of interests over all parts of our body and that we can use a single legal concept (property) to describe our relationship with all the parts of our bodies and in all contexts. That I suggest is misguided. We do not have the same attitude towards our eyes as we do towards excreta. We do not have the same attitude towards gametes as we do towards sweat. The image of the body as a single unity misrepresents the complexity of the different attitudes we have towards different parts of our body.

If a sui generis approach were taken towards bodies we would be able to develop a different legal response to different rights. Such a scheme can recognise the competing social and individual interests that exist in bodies and body parts and produce a sophisticated balancing of those interests, which may differ depending on the context. This point becomes more apparent when we consider the different interests that might be claimed in a body

First, there are instrumental values. There are some parts of a body, both while intact and once separated, that can be used for specific purposes. It may be it has commercial value or it can be used in the treatment of others or the self. Other parts have no use. Dried skin and waste products are rarely of interest. Indeed, not only do we have no use for them, we positively do not want to have control of them. So treating these as property may put unwanted demands on someone (Herring 2014).

Second, there are expressive values. Some parts of body represent who we are in a special way. When people see that part or when a person thinks about that part they feel a special attachment to it. Perhaps the best-known example is parenthood, where many biological parents feel a strong attachment to their children as a result of the biological connection. Another example is the unease people feel about face transplants, as compared with other types of transplants (Huxtable and Woodley 2005). This is in part due to the identifying nature of the face, unlike, say, a kidney that does not normally have expressive value. Notably organ donation systems allow people to select which organs they wish to donate, recognising, for example, that some people will feel differently about their eyes than their pancreas.

Third, there are informational values. Our bodies can revel important facts about ourselves. A piece of DNA can reveal identity, which is why it plays a crucial role in a criminal trial. It can also reveal whether one person is pre-disposed to certain medical conditions. However, note, this information is not restricted the individual. A sample of DNA can reveal if one person is related to another. It can reveal if one's children are genetically pre-disposed to a particular condition. It has value as not only an individual, but also a communal, resource.

There are other values that could be added here. However, the point of this section is that we cannot treat "the body" or "bodily material" as a single unitary concept. Our interests in, the significance of, and the communal importance various bodily parts varies greatly. This depends on the nature of the body part and the context in which the material was removed.

Consider, for example, the issues raised in *Evans v UK* (2006, discussed Lind (2006)), which concerned a couple who had donated their gametes to produce stored embryos, but then disagreed on how they should be used. The case raised complex issues over rights to reproduction, efficiency and consistency in state regulation of assisted reproduction, interests of the embryo, gender discrimination and so on. Whether the sperm and egg were property or not becomes insignificant in that case. Quite clearly the case would be completely different if the two individuals had donated hair that had been reduced to a piece of art. Even if you do support accepting property interests, this example shows that property could only ever be the first tiny step in the dealing with the legal issues raised by body parts and does not begin to deal with the host of other issues raised.

## 8.12 Me-ness

There is one aspect of the discussion we have just had which needs to be developed further. That is the idea that some parts of body, in particular contexts, can express the individual. This is a real difficulty for the property approach. As Jesse Wall (2014, 20) writes: 'The problem is, that although property is good with things, sometimes our control of things is more about us (our personality, personhood or our relationships). Our bodies can represent us to others and play a role constituting our identity.' While this me-ness[3] is especially true in relation to intact body parts, it can still apply in relation to separated body parts.

This me-ness is not captured in the property approach. In *Yearworth* some men had sperm stored frozen at a hospital as they were about to undergo medical treatment which carried a risk of infertility. The idea was that the sperm could be used in assisted reproduction if the men were indeed rendered infertile. Due to the hospital's negligence the sperm was defrosted and the sperm rendered useless. Through what is widely accepted to be opaque reasoning, the English Court of Appeal court recognised the men had a property interest in the sperm and this enabled the court to find a remedy through bailment. Supporters of the property approach would wish this case to be used to achieve a more general acceptance that we have property in our bodily material.

But does that follow? Is the damage to the sperm in this case the same as the damage when a nurse throws away a urine sample without checking with the patient first? It is true that a different kind of bodily material is involved and perhaps less loss flowing from it, but should the law classify these losses as essentially the same loss? A property harm? This completely fails to capture the nature of the wrong that has taken place. Unless we place the harm within the context of the men's plans and hopes for a family, for the precious child–parent relationship which was to come; unless we see it in the context of the special trust between a patient and a hospital committed to care for things beyond value; unless we see it in the context of the hopes and plans of the partners of the men, we cannot really understand the nature of what has happened. The property model fails, again, to capture the true significance of the bodily material in this context.

## 8.13 Dignity

Some commentators have argued that the property model fails to adequately protect the interests in dignity. Charles Foster (2014) has produced a number of examples of situations where he claims a property analysis does not capture all the interests in the body part. Two particularly stand out.

[3] I first heard this phrase in a lecture given by Jesse Wall and am grateful for his permission to use it.

1. The human-ear ashtray: Medical students steal an ear from the cadaver they are dissecting. They varnish it and use it as an ashtray. The cadaver was donated for the purposes of medical education. Liberal though he was, the donor did not regard use of an ear as an ashtray as one of those purposes.
2. The head of an unknown person: Children play football in the street. They are using, not a football, but the head of an unknown and untraceable person which a dog has retrieved from a mediaeval cemetery.

Foster argues that the property model fails to capture the wrongfulness of what is done in these cases. This is not simply a property wrong, it is harm to the special respect due to people as a result of their inherent dignity. There is not space here to fully explore the benefits and difficulties of dignity and competing understandings of it, but at the very least what supporters of dignity are identifying is the sense of value many people recognise in body parts which is beyond the physical type captured by a property analysis.

## 8.14 Developing a Non-property Model of the Body

Some critics of non-property approaches towards the body suggest that they are "merely" symbolic and offer no real solutions. That is a startling claim and is readily rebutted. The Human Tissue Act 2004 and the Human Fertilisation and Embryology Act 1990 in the UK are statutes offering sophisticated regulation of bodily material. Neither statute is based on an explicitly property model. Yet they certainly provide real solutions in the areas they cover. What can be seen in both statutes is a recognition and balancing of competing claims over the bodily material. The social, relational and personal interests are balanced and resolved in different ways in different contexts.

Adopters of the property approach towards bodily material largely close down the important debates that such statutes open up. If my bodily material is my property and only in the most extreme of circumstances might communal interests justify an interference in my bodily material then this starting point stymies any debate over how the law should deal with issues. Take for example the difficult issues of the extent to which human material removed in operations can be used for research or education; or the extent to which a person can direct who can have access to organs they wish to donate. Supporters of the property approach can offer only one response to these issues. As the material is property belonging to the person it cannot be used without their consent. Non-property approaches can offer a far more nuanced approach taking into account the competing interests that arise in such cases.

## 8.15 Conclusion

This chapter has addressed legal disagreements over whether bodies, bodily material and bodily products should be regarded as property. Into this sometimes technical argument, it has claimed that we must not overlook the importance the broader symbolic significance of bodies. In particular the chapter has argued that the property model buys into one particular understanding of bodies and selves; a highly individualised one. Under the individualised model the body is imagined as a static, independent, and controlled thing. The legal rights that flows from this are ones that promote autonomy, liberty and independence.

By contrast, it has been argued that we should understand the body in a relational way. The biological realities that bodies are vulnerable and therefore interdependent with other bodies; other biological organisms and the wider environment. This reflects an understanding of the self as a relational model. Our understandings of ourselves, the valuable things in life, and human flourishing are found in the mixing and vulnerability of bodies. We need a legal model that appreciates and promotes this more communal, mutable, interdependent nature of bodies.

## References


Alghrani, Amel, and John Harris. 2006. Reproductive liberty. Should the foundation of families be regulated? *Child and Family Law Quarterly* 18: 191–212.

Boddington, Paula, and Ulla Raisanen. 2009. Theoretical and practical issues in the definition of health. Insights from Aboriginal Australia. *Journal of Medicine and Philosophy* 34: 49–67.

Chau, P.-L., and Jonathan Herring. 2014. Interconnected, inhabited and insecure. Why bodies should not be property. *Journal of Medical Ethics* 44: 39–44.

Dickinson, Donna. 2013. *Me medicine vs. we medicine. Reclaiming biotechnology for the common good.* New York: Columbia University Press.

Feder Kittay, Eva. 1999. *Love's labour. Essays on women, equality and dependency.* Abingdon: Routledge.

Fineman, Martha. 2004. *The autonomy myth.* New York: New Press.

Fineman, Martha. 2011. The vulnerable subject. Anchoring equality in the human condition. In *Transcending the boundaries of law. Generations of feminism and legal theory*, ed. M. Fineman, 167–183. Abingdon: Routledge.

Fletcher, Joseph. 1979. *Humanhood. Essays in biomedical ethics.* Buffalo: Prometheus Books.

Foster, Charles. 2009. *Choosing life, choosing death.* Oxford: Hart.

Foster, Charles. 2014. Dignity and the use of body parts. *Journal of Medical Ethics* 40: 44–49.

Foster, Charles, and Jonathan Herring. 2014. What is health? In *Law and global health*, ed. Freeman Michael, Hawkes Sarah, and Bennett Belinda. Oxford: Oxford University Press.

Frazer, Elizabeth, and Nicola Lacey. 1993. *The politics of community.* London: Harvester Wheatsheaf.

Gergen, Kenneth. 2011. *Relational being.* Oxford: Oxford University Press.

Goold, Imogen. 2013. Property or not property? The spectrum of approaches to regulating the use of human bodily material. *Journal of Law and Medicine* 21: 299–313.

Goold, Imogen, Kate Greasley, Jonathan Herring, and Loane Skene (eds.). 2014. *Persons, parts and property.* Oxford: Hart.

Hardcastle, Rohan. 2007. *Law and the human body.* Oxford: Hart.

Herring, Jonathan. 2013. *Caring and the law*. Oxford: Hart.

Herring, Jonathan. 2014. Why we need a statute for bodily material. In *Persons, parts and property*, ed. Goold Imogen, Greasley Kate, Herring Jonathan, and Skene Loane. Hart: Oxford.

Herring, Jonathan, and P.-L. Chau. 2007. My body, your body, our bodies. *Medical Law Review* 15: 34–53.

Herring, Jonathan, and P.-L. Chau. 2013. Relational bodies. *Journal of Law and Medicine* 29: 214–225.

Huxtable, Richard, and Julie Woodley. 2005. Gaining face or losing face? Framing the debate on face transplants. *Bioethics* 19: 505–531.

Irigaray, Luce. 1993. *An ethics of sexual difference*. New York: Cornell University Press.

Karpin, Isabel. 2005. Genetics and the legal conception of self. In *Ethics of the body*, ed. Margrit Shildrick and Roxanne Mykitiuk, 2005. Cambridge, MA: MIT Press.

Katz, Larissa. 2008. Exclusion and exclusivity in property law. *University of Toronto Law Journal* 58: 275–332.

Levy, Traci. 2006. The relational self and the right to give care. *New Political Science* 28: 547–563.

Lind, Craig. 2006. Evans v United Kingdom – judgments of Solomon. Power, gender and procreation. *Child and Family Law Quarterly* 18: 576–593.

Lindemann, Kate. 2003. The ethics of receiving. *Theoretical Medicine and Bioethics* 24: 501–523.

Macallan, D., C. Fullerton, R. Neese, K. Haddock, S. Park, and M. Hellerstein. 1998. Measurement of cell proliferation by labelling of DNA with stable isotope-labelled glucose. Studies in vitro, in animals and in humans. *Proceedings of the National Academy of Sciences USA*, 95:708–718.

Maclean, Mavis. 2001. Letting go. Retention of human material after post mortem. In *Body lore and laws*, ed. Bainham Andrew, Sclater Shelley Day, and Richards Martin. Oxford: Hart.

Quigley, Muireann. 2009. Property. The future of human tissue? *Medical Law Review* 17: 457–476.

Quigley, Muireann. 2012. Property in human biomaterials—Separating persons and things? *Oxford Journal of Legal Studies* 32: 65–88.

Sheach-Leith, Valerie. 2007. Consent and nothing but consent? The organ retention scandal. *Sociology of Health and Illness* 29: 1023–1039.

Shildrick, Margrit. 1997. *Leaky bodies and boundaries. Feminism, postmodernism and (Bio)ethics*. Abingdon: Routledge.

Shildrick, Margrit, and Roxanne Mykitiuk (eds.). 2005. *Ethics of the body*. Cambridge, MA: MIT Press.

Skene, Loane. 2002. Proprietary rights in human bodies, body parts and tissue. Regulatory contexts and proposals for new laws. *Legal Studies* 22: 102–128.

Van Beers, Britta. 2014. Is Europe 'giving in to baby markets?' Reproductive tourism in europe and the gradual erosion of existing legal limits to reproductive markets. *Medical Law Review*. doi:10.1093/medlaw/fwu016.

Wall, Jesse. 2011. The legal status of body parts. A framework. *Oxford Journal of Legal Studies* 31: 783–802.

Wall, Jesse. 2014. The trespasses of property law. *Journal of Medical Ethics* 40: 19–25.

West, Robin. 1997. *Caring for justice*. New York: New York University Press.

Whitney, Shiloh. 2011. Dependency relations. Corporeal vulnerability and norms of personhood in Hobbes and Kittay. *Hypatia* 26: 544–567.

Young, Alan, and John Hay. 1995. Rapid turnover of the recirculating lymphocyte pool in vivo. *International Immunology* 7: 1607–1612.

## *Cases*

Evans v UK. 2006. 43 EHRR 21.
R v Kelly. 1998. 3 All ER 741.
Warren v Care Fertility (Northampton) Limited. 2014. EWHC 602 (Fam).
Yearworth v North Bristol NHS Trust. 2009. EWCA Civ 37.

# Chapter 9
# Revisionist Versus Broad Bioethics and Biolaw

**Herman De Dijn**

## 9.1 Bioethics and Biolaw in the Same Boat[1]

Current mainstream opinion in bioethics and biolaw, both in the Anglo-American and in the European context, demonstrates an astonishing lack of critical spirit as to the investigation of their presuppositions and basic conceptions. At the same time age old, commonly shared ethical and legal wisdom is disregarded or even discarded. If this tendency continues, certainly in the long run this will have enormous consequences. This mainstream opinion strongly influences ethical and legal practice particularly under the impulse of different kinds of bureaucracies and of market mechanisms. Although they have an air of rationality, mainstream bioethics and biolaw in fact express and participate in an ideological framework in which morality, human dignity and human rights are reinterpreted and adapted to an age of almost unhindered technological and economic progress. As can be gathered from several contributions to this volume, not only in bioethics but also in biolaw alternative views have been and are being developed that take into account the fundamental links between (bio)ethics and (bio)law on the one hand, and the deeply rooted symbolic representations of the 'life world' (*Lebenswelt*) on the other. The law is like a shared language which at the same time has its roots in and expresses the symbolic order, and to a certain degree also affects it retroactively (Van Beers, this volume, Chap. 11). Only if the symbolic basis of ethics and of law is taken into account, the deeper (symbolically determined) concerns and worries of people (e.g., with respect to the 'sanctity' of the human body) have a chance of playing the role they deserve in ethics and law, instead of being eliminated or obscured in advance

[1] This paper is a thoroughly revised version of De Dijn (2008a).

H. De Dijn (✉)
Institute of Philosophy, Catholic University of Leuven,
Kardinaal Mercierplein 2, Leuven B-3000, Belgium
e-mail: herman.dedijn@hiw.kuleuven.be



B. van Klink et al. (eds.), *Symbolic Legislation Theory and Developments in Biolaw*, Legisprudence Library 4, DOI 10.1007/978-3-319-33365-6_9

in favour of abstract, misguided and strongly pragmatically motivated categories and considerations (Herring, this volume, Chap. 8).

What are the largely unquestioned presuppositions in prevailing bioethics and biolaw? The central presupposition is that we can solve ethical and juridical problems by a combination of attending to scientific facts on the one hand, and to basic 'rational' normative principles on the other. This presupposition disregards some fundamental truths: that in real, communal life we normally and primordially do not view other human beings (and even things) from the scientific point of view; neither are human beings considered there as pure subjects with certain preferences, related to other pure subjects and to neutral objects, or as relatively complex sensitive organisms. Ethics and law are intrinsically connected with a view which is *incommensurable* with the scientific perspective. It is the view which philosophers in the previous century have named by different names: the view of common sense or common culture, the view of the 'life world', the view embedded in the sphere of our reactive attitudes to each other, etcetera.[2] It is the sort of view people typically adopt when they are spontaneously dealing with each other: a view consisting of 'thick' concepts and of descriptions which are at the same time evaluative (courage; respect for human beings; "he is a hero")[3]; a view in which realities are *symbolically* determined (children and not offspring; friends and enemies and not beneficial or harmful organisms; etc.)[4]; and in which people have meaningful reactive attitudes (Strawson 1974: 1–25) to themselves, their own body, others, different sorts of animals and objects, etc. The detached, scientific point of view cannot replace nor (completely) eliminate the life world view. But it does confront us, in ethics and law, with ever new problems related to new scientific information and new technical possibilities or realities. However, this confrontation can only arise because ethics and law are always already there in the first place. Therefore science is unable to solve the existential, moral and legal problems it itself produces in the context of the life world. From the techno-scientific point of view there simply are no ethical or legal problems; the only problems arising there are problems concerning the frontiers of our knowledge or of our technological abilities. Furthermore, there is no overarching point of view from which both the scientific and the ethical can be combined in perfect harmony (Williams 1984). So, there always will be tensions produced by the impact of the techno-scientific developments on our ethical and legal views and practices. This does not mean that we have to reject scientific and technological

[2] I borrow these terms respectively from David Hume, Ludwig Wittgenstein, Edmund Husserl, Peter Strawson, and Roger Scruton; see also the work of other Wittgensteinians like Peter Winch, Cora Daimond, Raimond Gaita.

[3] For the distinction between thick and thin moral concepts, see Williams (1984: 209–228).

[4] About symbolic meanings and law, see the work of anthropologist Geertz (1983) and about the symbolic dimensions of reproductive laws see the work of Van Beers (2015). For references about the intrinsic relation between ethics and symbolic meanings, see note 12 in Chap. 2.

progress; it means that with each new development we have to ask ourselves if it does not go against fundamental values and norms.[5]

Both ethics (or morals) and law are inescapably full part of, and at the same time peculiar expressions of ways of seeing and evaluating belonging to the life world, and presupposing its 'thick' concepts and characteristic descriptions.[6] Morality itself is inseparable from ordinary life and the implicit principles embedded in everyday conduct; therefore it is inevitably 'a broad ethic' and not a set of narrow principles, however large (see Wiggins 2006: 350).[7] Of course, ethics and law constitute different domains; they are not congruent, they only touch each other at certain points. They should not be identified and the character of the rules governing ethical behaviour may be fundamentally different from those operating in the juridical sphere.[8] Yet, I take it that 'behind' the law there were and are at least some deep ethical conceptions related to 'thick' notions such as responsibility, human dignity, etcetera, themselves incomprehensible without their link to a human life world with its system of symbolic categories and distinctions (e.g., between humans and animals, life and death, family and non-family, etc.). This is why I think that, in the context of the present discussion, (bio)ethics and (bio)law are fundamentally in the same boat. To divorce (bio)ethics and (bio)law from the symbolic representations of the life world inevitably produces the kind of ethics and law which denies or at least misunderstands the lived meaning of the human body, of human procreation, of human relations, illness and death. Maybe symbolization cannot guarantee the normative unity of modern, pluralistic society (Priban, this volume, Chap. 7), but denying or neglecting the importance of symbolization can only lead towards a bioethics and biolaw for aliens, or for individuals considered as 'elementary particles' (Houellebecq 1998).

---

[5] For a discussion of the relation between ethics and technological progress, particularly in the field of health, see De Dijn (2002: 15–34).

[6] In this contribution, I do not make a distinction between ethics and morals. However, like David Wiggins, I distinguish ethics or morals from a philosophical (meta-)reflection on, or reconstruction of morality or ethics. Philosophical reflection can of course partially and (usually) indirectly support ethics or morals, but it cannot justify them. See Wiggins (2006) and Burms and De Dijn (1982).

[7] Wiggins (2006: 350): " [A] morality does not consist of a set of moral propositions, even a very large one. Moral judgments are indeed partial expressions of the findings or demands of some particular mode of being and its associated sensibility… [M]oral judgments themselves, even when spelled out cannot even be understood as they are intended except against the background of a lived understanding that will never be fully articulated. In the absence of such a background, you have no hope of being understood exactly as you intend to be understood even if you say something as simple as 'It is wrong to say what is not true/what you don't know is true/what you know is not true…'. In the absence of such a background, which it would be an endless process, that is impossible, to spell out fully explicitly, you could not even keep in balance, as is second nature to us, the conflicting claims of a pair of fully compacted proverbs such as 'He who hesitates is lost' and 'Look before you leap'."

[8] Ethical rules (such as "Thou shall not kill") are really abbreviations intrinsically related to concrete contexts of human living (see Wiggins 2006: 350) and not strictly encoded principles such as juridical rules (about murder for example).

Since the beginning of the modern age attempts have been made to rationalize the domains of both ethics and law. This has led to revisionist (or should we rather say utopian) forms of ethics and law, divorced from the life world, from 'grass roots' ethics and its 'intuitions', and from 'common' law as determined (at least in part) by a particular history and particular traditions (for an example of a 'Kantian' revisionist bioethics and law see Beyleveld and Brownsword 2001). These revisionist forms of ethics and law became and are ideological tools propagating 'rational' interpretations of good and bad, right and wrong, as well as tools of individual and social engineering. When these tools do not work in the real (life)world – as is the case, for example, with laws against human egg trafficking and reproductive tourism – there is even a strong temptation to give up the moral stance altogether and to take a morally neutral, purely 'pragmatist' attitude (for a critique of this attitude see Van Beers 2015). No wonder this is happening, since the whole revisionist mentality already in itself betrays a fundamental lack of insight into what ethics and the law mean.

More and more we seem to have manoeuvred ourselves into a new ethical and juridical predicament, characteristic of a more pervasive 'malaise of (late) modernity'. On the one hand, it is clear that humans are 'symbolic animals', who cannot consistently treat each other as pure subjects, profit-maximizing individuals, or highly sensitive organisms; therefore traditional forms of ethics and law survive at least in part. But on the other hand, traditional ethical and legal customs are eroded by the onslaught of revisionist ethics and law; and this particularly in relation to the biological aspects of human life. This process is exacerbated because these revisionist forms are now promoted by huge bureaucracies and (national and international) political institutions, who use them in function both of social regulation, and of the control of technological developments. Add to this that markets and new technologies could not care less whether new developments go against traditional sensitivities or not. On the contrary, they actively contribute to the liquefaction or hybridization of fundamental symbolic categories (such as life and death, male and female, man and animal), and to the transgression of the boundaries between them (De Dijn 2014a; Van Beers, this volume, Chap. 11). What markets and biotechnology today expect from bioethics and biolaw, if they expect anything at all, and do not simply ignore them, is that they can be certain where the legal boundaries lie within which they are required to operate.[9]

In our late modern times, the challenges for bioethics and biolaw are clearly manifold, and interconnected: the tabula rasa mentality of revisionist ethics and law; the liquefaction or transgression of fundamental categories and boundaries; the instrumentalist attitude so pervasive in contemporary technological and market society with its commercialization and commodification even of the most sacred.

[9] About the impact of markets and pragmatic market thinking in the discussion of reproductive tourism, see Van Beers (2015: 121–127).

## 9.2 An Example of a Bioethical Debate

In December 2007 I participated in a conference in Antwerp on the marketing of human egg cells. The problem under investigation was that of the ethical acceptability of selling human eggs, either for IVF or for the production of embryos from which stem cells are harvested for scientific research. The differences of the positions argued for in the debate exemplify the present ethico-juridical situation and the enormous gap existing between different approaches to biomedical problems.

Josiane Van der Elst, professor of IVF embryology (Vrije Universiteit Brussel and Universiteit Gent) and head of the IVF laboratory of the University Hospital Brussels, argued in favour of exercising restraint in the use of mature eggs for research. Their use is only acceptable in well-planned experiments. In cases having to do with IVF, the sale of eggs should be allowed, but it should be a question of *fair trade* (Van der Elst 2007). Guido Pennings, professor of bioethics (Universiteit Gent) argued that "from a moral point of view" commercialisation is not the optimal solution, for it would be unfair towards the less fortunate, whereas everyone should have the same access to health care. For him donation is the only acceptable option, be it with a certain indemnification which takes into account the effort, time and discomforts of the acquisition, as happens for people who are compensated for clinical trials or drug trials (Pennings 2007; Mertes and Pennings 2007: 629–634).[10] The 'moral point of view' adopted by both Van der Elst and Pennings is the dominant view in bioethics, which is called 'principalism' because of its stress on the role of principles, particularly the principles of autonomy and no-harm, in bioethical discussion (De Dijn 2002: 17–18; Burms and De Dijn 2011: 7–13). The two views differ from the radical liberal argument for full freedom, which is based on the idea that any woman is the owner of her own eggs and can therefore freely dispose of them in the way she wants.[11] Van der Elst and Pennings also stress the importance of the principle of autonomy, but they think it has to be accommodated to other principles, such as the principle of no-harm, and of equality (equal access to care), etc.

A third position was argued for by Donna Dickenson (emeritus professor of Medical Ethics and Humanities at the University of London). Dickenson made two points (2007, 2008: ch 3).

1. Because the human body and its parts – especially those parts which serve to pass on human life – have a special dignity, their commercialization is unacceptable. Selling and even exchanging of human eggs (for instance to obtain cheaper IVF treatment), and generally making a profit based on selling eggs should be completely forbidden, also to institutions such as universities and research labs. We must enforce the traditional prohibition on absolute rights of property in the

[10] Pennings also advocated this position in an interview in the Flemish weekly *Tertio* on 5 December 2007. Pennings' pragmatist attitude is also noticeable in his discussion on reproductive tourism (Pennings 2002: 337–341; Van Hoof and Pennings 2011: 578–583).

[11] Position of prominent Anglo-American bioethicists like John Harris and Julian Savulescu.

body across the board. The fact that the selling of eggs is already taking place and will continue to happen, is not a valid counterargument. The violation of fundamental prohibitions (such as murder or rape) ought not to be a cause to abandon them.

2. However, the *donation* of eggs for IVF (and research?) can be accepted, similarly to the donation of other human tissues, provided that this does not result in one-sided altruism asked from women, since this would be tantamount to exploitation. Female donors rightly want respect for what they donate, and want to control what happens to it.

Dickenson further focused attention on a double hypocrisy hiding behind medical rhetoric aiming to convince women of the importance of the potentially life-giving and life-saving gift of eggs. Firstly, this rhetoric contrasts sharply with the pursuit of prestige and potential profit of researchers and institutions who need human eggs. Secondly, the rhetoric seems to hide something behind its appeal to the ideal of helping and healing: a kind of 'fetish' idea that life itself is and can be the object of endless manipulation.[12] Dickenson fears a generalised 'prostitutioning' not only of the female body, but of the human body *tout court*.

Although the positions argued for in the conference all looked like 'moderate' ones certainly in comparison to the strong liberal view, they also markedly differ from each other. They should be compared in light of the question what the *ethical* (or *moral*) good is in the objectification, instrumentalization and commercialization of the human body or its parts.

Dickenson's rejection of any form of commercialisation of human eggs (and of other human organs and tissues) is based on her understanding of the *intrinsic value* of the human body. This, she says, is the 'traditional' ethical viewpoint that has also been taken into account juridically both in continental civil law (the body is *une chose hors commerce*) and in the Anglo-Saxon common law tradition (the body is *res nullius*, i.e., it cannot be anyone's property). It is this understanding of a special dignity or a certain 'sacredness' of the human body that is still common among the general public. The persistent presence of this ethical understanding is not in line with typically modern ideas and practices, such as the idea that humans *own* their own bodies, or medical practices or research where human tissues, organs or bodies are looked upon as mere materials or spare parts.

The fact that the general public is largely unwilling to see the body as merely a neutral object, means or instrument, is evident from various data. Relinquishing organs or tissues of deceased family members is still problematic to a lot of people,

[12] See Scheper-Hughes (2001) Bodies for Sale – Whole or in Parts. Body and Society 7(2–3): 1–8 (quoted by Dickenson): "Global capitalism, advanced medical and biotechnologies, have incited new tastes and desires for the skin, bone, blood, organs, tissue and reproductive and genetic material of the other. What is different today is that the sacrifice is disguised as a "donation", rendered invisible by its anonymity and hidden under the medical rhetoric of life saving and giving. In all, the ultimate fetish is the idea of life itself as an object of endless manipulation." This opinion about the ultimately mercantile driving force behind stem cell research (e.g., in Great-Britain) has been confirmed by Franklin (2004: 87–102).

hence the use of subtle manoeuvres by the political and medical world to obtain a sufficient supply of organs (such as the legal arrangement of 'presumed consent', which stipulates that organs can be harvested unless the patient has explicitly objected to it). People still feel enormous outrage when stories of illegal trade in organs are brought to light, tombs are desecrated, human remains are handled with disrespect, etc. The claim that these 'traditional' ethical feelings can only be found among the unsophisticated general public, and especially with religious believers, is unsubstantiated. They also exist in the absence of religious beliefs, e.g., in immortality. However, it is a fact that all major religions combine respect towards the human body and its parts with an understanding of the special dignity of the body, even though different religions may not have exactly the same prohibitions (see Goris 2006).

It is striking that, apart from prominent bioethicists, several of the most important philosophers dealing with ethics share the view that ethics is inseparable from the human life world and its categories and values, thereby at least implicitly rejecting freestanding, revisionist ethics: Ludwig Wittgenstein, Peter Winch, Stuart Hampshire, Bernard Williams, Leszek Kolakowski, Jürgen Habermas, Francis Fukuyama, Michael Sandel, David Wiggins, Roger Scruton, Raimond Gaita. The majority of these philosophers is not religious; they position themselves both on the progressive and the conservative side. They of course express their conception of ethics in their own philosophical vocabulary: ethics is determined by symbolic meanings; it is fundamentally characterized by taboos; it inevitably is a broad ethics; etc. In my view, these different ways of expression are compatible with each other.

## 9.3 Narrow Versus Broad Ethics: The End of Taboos?

Dona Dickenson's refusal to objectify and commercialise the human body and its parts and tissues is based on a spontaneous, pre-reflective ethical understanding that is expressed in her indignation at the use of the human body as raw material or a reservoir of spare parts. This is why her viewpoint agrees with the traditional or common sense view. It is not simply the subjective belief of an individual, but a very stubborn and almost universally held understanding (see also Herring, this volume, Chap. 8). Very likely it is connected to the way in which we have evolved as a species into the kind of humans we are today (Wiggins 2006), i.e., beings who believe that their fellow humans are different from animals because they have a special dignity which extends to their bodies, even their dead bodies. If we did not believe that human bodies have a special significance, there would for instance be no reason why we cannot eat humans, or why we cannot process their remains as produce, as we do with animals (but, interestingly enough, *not* with pets) (Diamond 1996: 319–334). Eating human flesh is considered repugnant. Using human skin to make lampshades causes deep indignation. And – as Dickenson said – "there is something *shocking* in body shopping" (just as there is something shocking in pragmatism towards such transgressions). These sentiments and attitudes are the *real* basis of ethics in daily life: they are connected to our experiencing and *understanding* of a complex set of meanings and values embedded in the practices constituting the *life world*. The real world we live in is not the world of science where humans and animals differ only gradually, instead of categorically. It is a world where descriptive categories also have an *evaluative* function. Bernard Williams explains that the ethical categories of the life world are both *world-guided* and *action-guiding*: they are forced upon us by our life world ("This is a hero"; "that was a cowardly action"), but they also tell us how to act (by praise or loathing) (Williams 1984: 209–228). "Is 'this' a human being?": in that case 'this' deserves a kind of respect which we do not give to other things, it is something 'sacred'[13]; – ignoring this fact would be horrible, unbearable.

Our everyday ethical attitudes are spontaneously connected to – let us use the word – *taboos*. The word usually has a pejorative meaning, but it expresses the essence of a fundamental form of ethical understanding: certain things are simply 'out', prohibited. Whoever denies this, either from a scientific or from a pragmatic point of view (e.g., those arguing in favour of 'recycling' human remains or body parts), may be extraordinarily *rational*, but seems to be missing some self-evident reasonableness. Anyone who systematically ignores all taboos is not an exceptionally rational, liberated person, but a monster. Ethical taboos are intrinsically related to a complex set of *symbolic* meanings and values which determine the life and world of a group, society or culture.[14] For instance, in every culture there is a deep connection between the taboo of incest and the *identity* of persons as determined by their relation to their father and mother, to their siblings, and to other individuals to whom they may be related by blood. As the term 'ethics' itself indicates, it cannot be separated from the *mores*, the customs, values and norms present in a way of life, in a culture. Cultural anthropology has taught us that each culture presupposes fundamental 'symbolic' distinctions (between being alive and dead, human being and animal, man and woman, adult and child, mine and thine, honour and disgrace, and so on). Ethics has primarily to do with respecting such distinctions, and with sentiments like blame and guilt, indignation and horror in reaction to transgressions of fundamental boundaries. This is why a good deal of ethics concerns questions of life and death, marriage, sex, property, relation to animals, food, dead bodies, children vis-à-vis parents, etc. Centuries of progress or enlightenment have not succeeded in jettisoning this 'traditional' ethics and replacing it completely with some purely neutral principles and procedures.

Perhaps the time has come for the concept of taboo to be rescued from misunderstanding and to see it as intrinsically related to the prevalence of symbolic realities

[13] Notions like 'sacredness' and 'sanctity of' are used in a non-religious sense by different philosophers; see Sandel (2007: 93); Kolakowski (1990: 72); Dworkin (1993: chapter 3); Joas (2011).

[14] On the intrinsic relation between ethics or morals and symbolical meanings see Burms 1995: 95–107; Burms 2001: 13–34, 2011: 15–29; Burms and Vergauwen 1991: 101–105; Burms and Breeur 2008: 134–146; and De Dijn 2003: Chap. 1.

in the life world.[15] This is also the opinion of Leszek Kolakowski (1990: 13): "When I try … to point out the most dangerous characteristic of modernity, I tend to sum up my fear in one phrase: the disappearance of taboos … The taboo regarding respect for the bodies of the dead seems to be a candidate for extinction, and although the technique of transplanting organs has saved many lives and will doubtlessly save more, I find it difficult not to feel sympathy for people who anticipate with horror a world in which dead bodies will be no more than a store of spare parts for the living or raw material for industrial purposes; perhaps respect for the dead and for the living – and for life itself – are inseparable. Various traditional human bonds which make communal life possible … are not likely to survive without a taboo system, and it is perhaps better to believe in the validity of even apparently silly taboos than to let them all vanish. To the extent that rationality and rationalization threaten the very presence of taboos of our civilization, they corrode its ability to survive".

The narrow rationalism which is rampant in modern society irrevocably leads to attempts to erase all taboos, even the universally accepted ones which, as already stated, are most likely intrinsically connected to the development and preservation of a typically human way of life (rejection of incest and cannibalism, some or other form of respect for the dead).[16] Taboos often seem purely irrational because of their contingent and seemingly gratuitous character. Most bioethicists therefore try to replace them by principles which are strictly universally valid and rationally justified: the principle of doing no harm, the principle of autonomy or self-determination, the principle of proportionality etc. Using these as a starting point, ethicists try to recover the existing ethical sense on a rational basis, without any taboos (for a critique of this 'principalism' see Burms and De Dijn 2011: 7–13). Of course, at the outset the rationalist is prepared to compromise by making the new rules accommodating and moderate, and by attempting not to offend anyone who might still be stuck within old attitudes. In any case, the new (bio)ethics is a *revisionist* ethics, and its foundations are radically different from those of traditional ethics. One way to explain the difference between both forms of ethics is to analyse the different kinds of ethical concepts which operate in each context: *common sense* morals uses *thick concepts*; revisionist morals uses *thin concepts* (see Williams 1984: 239–269; Hampshire 1993: 82–100; Strawson 1974: 1–25). The content of 'thick' ethical concepts (such as 'grateful', 'coward', 'cruel', vengeful') cannot easily be circumscribed in detail, because they presuppose some knowledge of a concrete way of life with its specific attitudes and sentiments (such as indignation, loathing, etc.). In contrast, 'thin' concepts (such as 'right', 'useful', 'quality', 'preferred result') seem universally acceptable and perfectly applicable (even to 'aliens', as Kant would say). Another characteristic of much current bioethics is the predominant use of

[15] See for such an attempt De Dijn (2003).

[16] According to Gianbattista Vico, marriage and funeral rites (together with religion) constitute the real points of difference between man and beast; they are undermined in times of cultural decline (Offermans 2000: 332). Mencius, the Chinese philosopher (fourth century B.C.), has similar ideas about funeral rites (Mencius 1983: 105).

technical and scientific terms: talk about 'sperm', 'egg cells', 'unborn or potential human life' is replaced by talk about 'male or female gametes', 'oocyte', 'embryos', 'blastocyst', etc. This gives the language of revisionist ethics an additional air of 'objectivity'.

Strikingly, 'thick' concepts (such as 'decency', 'disgrace', 'human dignity') have been and are still used in legislation. So it has been – and indeed still is – possible to make ethical and juridical decisions based on these concepts (see Foster 2011 for a kind of Aristotelian approach). Nevertheless they are considered as problematic: (1) because of their link to a shared way of life and its common morality: is this not a purely contingent basis? is it really shared?; (2) but especially because, from a revisionist point of view, ethics should be based on strictly rationally constructible and justifiable principles and concepts. This is why many believe that not only ethics, but the law too should be adapted in a revisionist way: 'human dignity' for example should not be considered as a basic concept, it should at least be reformulated in terms of the more rational concepts and principles related to 'human rights' (Beyleveld and Brownsword 2001).

From the point of view of revisionist ethics, traditional taboo ethics is completely out of touch with our contemporary situation and even obstructs the development of rational judgments concerning the new ethical problems arising in the wake of techno-scientific progress and the emancipation of human beings. But is that really the case? On the contrary, it is the attempts of revisionist ethicists to get rid of taboos *in general* that are the real threat to civilised society and perhaps even human survival. This is how my argument can be formulated: (1) if ethics or morals are fundamentally different from science (ethics tells us *in concreto* what we can or cannot do, science does not tell us anything about values, but only gives us facts and explains how things work); (2) and if ethical judgments inevitably imply 'thick' concepts that cannot be recovered on the basis of 'thin' ones; then (3) it would be *ir*rational to want to replace taboo ethics by revisionist (rational or 'scientific') ethics. This piece of reasoning matches the arguments of Bernard Williams in his contribution, "The Scientific and the Ethical", in which he too defends ethics as a broad common sense domain.

It is therefore not the reference to taboos or the use of 'thick' ethical concepts which is irrational, but on the contrary, it is the attempt to create a revisionist ethics doing away with all this, that is wrongheaded. This revisionism is exactly what is now rife in many (most?) bioethical debates. Of course, it sounds paradoxical to say that 'striving towards rationality' can be irrational. But this is also the case in other contexts: whoever thinks purely rationally (e.g., in a utilitarian way) in friendship, acts irrationally, and does not follow the 'logic' of friendships. Ethical debate cannot be a sort of scientific or strictly rational debate, based purely on incontestable and universal principles of whatever nature.[17] Without faith in the 'thick' concepts of the life world or of *common sense*, we would simply no longer know what the meaning of good and evil is; instead we would acquire a way of thinking for which

[17] This is not to say that ethical principles are to be rejected outright; they can be used as abbreviations of more complex thoughts and considerations; see Wiggins (2006: Chap. 11).

anything might be good or evil. Imagine that someone deduces on strictly rational grounds something that is radically contrary to our moral sense (for instance that human beings do not deserve any respect, either individually or collectively), and that we would not be able to counter this deduction as it presents itself to us. As Kolakowski puts it (1990: 13): "… in the normal sense of "rationality" there are no more rational grounds for respecting human life and human personal rights than there are, say, [grounds] for forbidding the consumption of shrimps among Jews, of meat on Friday among Christians, and of wine among Muslims". Should we then give up our spontaneous moral reactions (e.g., the special respect reserved for human beings)?[18] Wouldn't this be a kind of madness? (Chesterton 1959: chapter 2)

The view that only rational considerations and neatly determinable concepts can be taken into account must lead to the repression of substantial, broad ethical views and their taboos. How could anyone ever give a 'strictly rational' explanation of why cannibalism or incest between *consenting adults* is unethical, especially if it does not affect any third party? What revisionism effectively leads to can be learned from proposals by ethics councils and lawmakers to end legal prohibitions for example on incest between siblings or from their hesitation to condemn cannibalism with mutual consent.[19] The consequence is that revisionist bioethics does not contribute to serious ethical thought, but on the contrary augments the unwieldiness of the new problems that are created by medical and biotechnological progress.[20] The well known Harvard ethicist and legal philosopher Michael Sandel affirms that present-day (Anglo-American) bioethicists are unable to comprehensively deal with the ethical problems created by the continuing evolution of genetic technology (Sandel 2007; see also Habermas 2003; De Dijn 2007). His approach too leads to the opposite of abandoning common, 'broad' ethical sensibility and its taboos. The idea that (in general) we can manage without taboos, is not only irrational, but also incredibly unlikely, says Kolakowski (1990: 13) "… it is quite improbable that taboos, which are barriers erected by instinct and not by conscious planning, could be saved, or selectively saved, by a rational technique [like in revisionist ethics]; in this area we can only rely on the uncertain hope that the social self-preservation drive will prove strong enough to react to their evaporation, and that this reaction will not come in barbarous form".

---

[18] This is effectively argued for by transhumanists; see, e.g., the work of Gilbert Hottois, a Belgian bioethicist, who holds a moderately transhumanist position; see further my review of his work (with many references): De Dijn (2000: 743–751).

[19] As to incest, see Die Welt/Topstories, 24 September 2014; "German Ethics Council in favor of lifting ban on incest with siblings" (last consulted on May 29, 2015). Reports on famous cannibalism trials (e.g., in 2003 and 2014 in Germany) can easily be found on the internet.

[20] About the ease with which fetuses diagnosed with spina bifida are almost systematically aborted on the basis of a vague category such as 'quality', see Mertens (2006: 15–22).

## 9.4 Why Nevertheless Broad (Taboo-)Ethics Is on the Defensive

One can wonder why – if taboos are so fundamental and in some sense unavoidable – ethical and philosophical positions defending them are so strongly on the defensive in the current bioethical debate. Even in Anglo-American bioethics, some ethicists (such as Donna Dickenson, Leo Kass and Michael Sandel) do not hesitate to defend 'old' taboos; but it is clear that they are on the defensive, while the revisionist position is pervasively shared.

My hypothesis is that, apart from the rationalist and pragmatist attitudes among mainstream bioethicists, a number of elements render the defence of a broad *common sense* or taboo ethics more difficult:

1. The principles and concepts of revisionist (bio)ethics seem to receive general approval from a larger audience than just philosophically trained scholars. Today, many ordinary people consider certain concepts or notions both as self-evident and as uniquely central in moral debates. This is not because these notions really are so clear and unproblematic in themselves, but because they, rather than others, express the spirit of the times. Take the concept of 'autonomy', which is closely connected to the ideas of free will and autonomous thinking. Are these concepts and their implications truly based on irrefutable rational considerations and arguments? Are people fully aware of how problematic the idea of free will is and of the fact that there is no philosophical consensus about it? Of course not. Unfortunately, because concepts like autonomy or equality take central stage, other *common sense* concepts which should have their proper role to play in ethical debates (such as "humility, responsibility, and solidarity") (see Sandel 2007: 86; Wiggins 2009: 239–269; see also Herring, this volume, Chap. 8), are being neglected or ignored. Notions such as 'human dignity' or 'respect for the human body' are sometimes even flatly discarded because they are too 'obscure', or even 'stupid'.[21]
2. Although, with Kolakowski, we can take the view that taboos cannot simply disappear, the system of taboos is under strong pressure from the rationalist and individualist mindset present in revisionist ethics, as well as in other domains of late modern culture or liquid modernity (Bauman 2000). Despite the survival of the taboo of the inviolability of the human body, we seem to be evolving towards a mentality which increasingly sees the human body as private *property*: something we simply *have*, instead of something we also *are* (Herring, this volume, Chap. 8). This mentality is related to 'the logic of possessive individualism', with its emphasis on *ownership* and the right to manipulate one's own body as seems suitable. Both body and mind are often *de facto* considered to be a complex set of *assets* and *capabilities* (or competences) which we, as *self-managing* individuals, coordinate and use to achieve the kind of success we are striving for:

[21] Some even reject outright the notion of human dignity: see Macklin (2003: 1419–1420) and Pinker (2008). For a positive account of human dignity see De Dijn (2012) and Joas (2011).

the individual as capitalist of its own body. Last, but not least, the human body is more and more considered as a means to construct the sort of identity desired in function of recognition: a means that can be shaped at will according to the models available on the market of identity markers (Elliott 2003; De Dijn 2014a).

3. A third element is the pressure of technological advances whether or not in combination with economic profit seeking. These factors influence even our most intimate relationships, and inevitably have a repercussion on our ethical thinking in these matters. Again the relationship with one's own body is a perfect case in point. Dickenson describes how biotechnology breaks down the distinction between the personal and the neutral, and between the inside and the outside of the body (strange objects, prosthetic hip implants and pacemakers are inside; organs can survive outside, in an artificial environment). The idea of the exchangeability of the individual's body has been considered ever since the start of modern times (Descartes and Locke).[22] Husserl already introduced the difference between the lived body as 'my own self' (*Leib*; *le corps-sujet* in Maurice Merleau-Ponty's terminology) and the body as something that can become foreign to us (*Körper*) – something we are confronted with when it thwarts us, fails us, no longer seems to fit us, and becomes something we would rather get rid of.[23] In certain circumstances (such as illness or accident) this distancing and objectifying attitude towards the body is, as it were, forced upon us. Time and again, treatments that were developed for healing are now used for grooming and enhancement in view of success or profit.[24] Through techno-science, we can now intervene (or allow someone else to intervene) in the lived body, perfecting it (e.g., by doping and bodybuilding) and treating it like an object, a means to an end. The combination of techno-science and *self-management* actually realizes the Cartesian project of a pure mind in a purely external body. Because of its extreme tendency towards 'excarnation', this capitalist 'spirituality' can only lead to further loss of meaning.[25]

Revisionist ethics has a huge impact on the practices of bioethics and biolaw. Revisionist bioethics and biolaw, and policies based on individualistic, rationalist, technological and pragmatic attitudes have themselves *symbolic effects* which disturb ethical sensibilities and attitudes without being able to supersede them altogether (see Van Beers 2015: 131–132).

---

[22] Lately a very popular theme in science fiction literature and movies: Hanif Kureishi (2003) *The Body*; Michel Houellebecq (2005) *La possibilité d'une île*; Roger Spottiswoode (2000) *The 6th Day*; Andrew Niccol (1997) *Gattaca*.

[23] See the issue edited by Jenny Slotman & Annemie Halsema (2007) of *Wijsgerig Perspectief* 47(2) Grenzen van het lichaam.

[24] Extreme examples of this Promethean attitude are mentioned in the work of philosopher and psychiatrist Carl Elliot (2003).

[25] I borrow the term 'excarnation' from Taylor (2007: 554, 610). I do not give the term exactly the same meaning as in Taylor where it primarily refers to the opposite of 'embodied forms of religion'. See further De Dijn (1999: 371–370), (2008b): 19–27.

## 9.5 Que Faire?

More and more people resist the ‘progress’ that has been brought about by the combination of biotechnology and rampant commercialization. But at the same time, this evolution seems almost unstoppable.[26] We are still struggling to make up our minds as to how to deal with the new developments. Of course, it would be unacceptable to prohibit technical innovations that have clear curative aims. If we went that far, we would have to prohibit modern medicine as a whole. Dickenson’s attitude seems to be the right one, even if it is not the easiest path to follow: (1) not to take part in the abandonment of taboos, but to reject the radical revisionist ethics and politics; (2) not to dismiss new developments out of hand, but to judge them on a case to case basis; (3) to judge them *not* by means of a number of abstract principles, *but* in light of the concrete moral values and attitudes that shape our lives and society. In view of the special dignity of the female body, this leads her for example to find the *commercialization* of human eggs unacceptable. This stance is in line with the rejection of the objectification and commercialization of human tissues (e.g., for the benefit of the cosmetic industry) and of women’s and children’s bodies in the sex industry.

The dismissal of revisionist ethics in itself does not automatically give us concrete answers to concrete ethical questions such as: is the *donation* of human eggs for IVF (for oneself or for others) and even for stem cell research morally acceptable? Dickenson seems to answer this question positively, provided that the donation does not conceal exploitation. However, whether to be in favour of donation or not, also depends on whether, for instance, the production of embryonic stem cells and even the use of ‘superfluous’ embryos in research is acceptable. Clearly, even among people who share the understanding that the human body possesses a special dignity, dissension is possible. Individuals who object to lifting the ban on the use of embryos for research, are often seen as the last defenders of an outmoded, if not fundamentalist position. Whatever the view one may have on this issue, the resistance of these individuals or groups should perhaps be appreciated as another brake on the galloping horse of capitalism that is affecting every nook of society. In any case, it is important to preserve a certain reticence in the face of the enthusiasm of biotechnology promising us paradise on earth.

Those who want to engage (again) in a substantial ethical and legal debate that takes real ethical intuitions originating in the life world into account, are confronted with several opposing forces which even seem to strengthen each other: the flourishing of a type of revisionist ethics and politics (with its proliferation of bioethical commissions and bureaucracies); the liquefaction of fundamental symbolic distinctions and boundaries typical of liquid modernity; and both of these so closely intertwined with the overpowering desire for control also of the symbolic realm. These forces cannot simply turn off the generally shared symbolic attitudes and desires,

[26] See about this phenomenon of compliance with and even addiction to certain medical advances, Elliott (2003).

but they nevertheless succeed to a certain degree in transforming and manipulating symbolic meanings and values in function of narcissistic and instrumentalist aims. Instead of playing a critical role in this debate, as one might expect of intellectuals, many academic bioethicists are on the side of revisionist ethics and, wittingly or not, in the service of a biotechnology that wants to be as free as possible in its race towards ever newer discoveries and technical innovations.

Whatever the difficulties to be expected, symbolic legislation theory is absolutely needed to focus again on the real value issues involved in the developments in culture and technology, e.g. in the attitudes to sex difference, or in the production of the 'monsters' of biotechnology (Van Beers, this volume, Chap. 11; see also De Dijn 2003). But there is no guarantee that simply by itself this theory will show us the ethical and legal way we can all agree upon. For example, it is not evident at all what attitude to take vis-à-vis the 'monsters' or *Fremdkörper* already there, and still to come. Is it sufficient to plead for a kind of ethical and legal accompaniment so as to be able to fit them more or less into the existing legal-symbolic order (Van Beers, this volume, Chap. 11)? What does the anthropological function of the law require: why accommodation, why not resistance (at least sometimes)? Would resistance invariably be the sign of 'a naturalization of the anthropological narrative'?

In the present cultural context, philosophy can and must offer a hermeneutical and philosophical-anthropological clarification of the type of values and perceptions, and of the type of 'objectivity' that is *part of* human life in the life world – a life world that almost inevitably today will be characterized by a diversity of 'traditions'. For example, in order to counter revisionist tendencies, philosophical support for the idea that the human body has a special dignity clearly is of the utmost importance (Herring, this volume, Chap. 8). The incomprehension with respect to the survival of this idea in a thoroughly secularized culture shows the necessity to develop a 'post-religious' understanding of the 'sacredness' of the human body. Such a way of thinking – reflexively and philosophically – could interpret the human body as a gift that gives life in its turn (Dickenson 2008; Sandel 2007: chapter 5). This view probably also presupposes an alternative idea and experience of nature as a source of life (and not simply as mere matter) and a renewed attention to the significance of 'the mystical body', the succession of generations, to which we human beings belong and which will carry on in our children after us (Kellendonk 1986; De Dijn 2014b). The mysteries of the *meaning* of life and of the existence of good and evil do not (have to) disappear in the course of techno-scientific progress; philosophers must be the custodians of these mysteries.[27] But is this enough? We not only need an hermeneutics of the life world (a life world itself in full change), we also need a critique of the late modern mind and its utopianism (De Dijn 2015).

---

[27] About good and evil, see Gaita (1991).

## References


Bauman, Zygmunt. 2000. *Liquid modernity*. Cambridge: Polity Press.

Beyleveld, Deryck, and Roger Brownsword. 2001. *Human dignity in bioethics and biolaw*. Oxford: Oxford University Press.

Burms, Arnold. 1995. Moral taboos and the narrow conception of morality. In *Does religion matter morally? A critical reappraisal of the thesis of morality's independence from religion*, ed. Musschenga Bert, 95–107. Kampen: Kok Pharos.

Burms, Arnold. 2001. Wetenschap, ethiek en symbolische betekenis. In *Moeten, Mogen, Kunnen. Ethiek en Wetenschap*, ed. Raymaekers Bart, 13–34. Leuven: Universitaire Pers Leuven.

Burms, Arnold. 2011. Natuur en symbool. In *De sacraliteit van leven en dood. Voor een brede bio-ethiek*, ed. Burms Arnold and Herman de Dijn, 15–29. Kalmthout/Zoetermeer: Pelckmans-Klement.

Burms, Arnold, and Roland Breeur. 2008. Persons and relics. *Ratio* 21(2): 134–146.

Burms, Arnold, and Herman de Dijn. 1982. Moreel objectivisme. *Algemeen Nederlands Tijdschrift voor Wijsbegeerte* 74: 207–223.

Burms, Arnold, and Herman de Dijn. 2011. Impliciete inzichten en expliciete principes. In *De sacraliteit van leven en dood. Voor een brede bio-ethiek*, ed. Burms Arnold and Herman de Dijn, 7–13. Kalmthout/Zoetermeer: Pelckmans-Klement.

Burms, Arnold, and Roger Vergauwen. 1991. Natural kinds and moral distinctions. *Philosophia Philosophical Quarterly of Israel* 21(1–2): 101–105.

Chesterton, G.K. 1959 (reprint). *Orthodoxy.* New York: Doubleday.

De Dijn, Herman. 1999. Incarnation and values. In *Incarnation*, ed. M. Olivetti, 371–379. Padova: Cedam.

De Dijn, Herman. 2000. De donkere transcendentie van Prometheus. *Tijdschrift voor Filosofie* 62(4): 743–751.

De Dijn, Herman. 2002. New technologies in healthcare. A philosophical-ethical appraisal. In *Between technology and humanity. The impact of technology on health care ethics*, ed. Gastmans Chris, 15–34. Leuven: Leuven University Press.

De Dijn, Herman. 2003. *Taboes, monsters en loterijen. Ethiek in de laatmoderne tijd.*. Kampen/Kapellen: Klement-Pelckmans.

De Dijn, Herman. 2007. Michael Sandel and the ethics of genetic engineering. *Ethical Perspectives* 14(2): 207–211.

De Dijn, Herman. 2008a. De bio-ethische discussie over de vermarkting van eicellen. De bio-ethiek als probleem. *Ethische Perspectieven* 18(1): 54–68.

De Dijn, Herman. 2008b. How and why marriage matters even in postmodern times. *Intams Review* 14(1): 19–27.

De Dijn, Herman. 2012. Nadenken over menselijke waardigheid. *Algemeen Nederlands Tijdschrift voor Wijsbegeerte* 104(4): 252–267.

De Dijn, Herman. 2014a. *Vloeibare waarden. Politiek, zorg en onderwijs in laatmoderne tijd.* Kalmthout/Zoetermeer: Pelckmans-Klement.

De Dijn, Herman. 2014b. *Hoe overleven we de vrijheid. Modernisme, postmodernisme en het mystiek lichaam. Twintig jaar later*. Kalmthout: Pelckmans.

De Dijn, Herman. 2015. Soft utopisme in de zorgsector. *Streven Cultureel maatschappelijk maandblad* 82(10): 881–895.

Diamond, Cora. 1996. Eating meat and eating people. In *The realistic spirit. Wittgenstein, philosophy, and the mind*, 3rd ed, ed. Diamond Cora, 209–228. Cambridge MA/London: MIT Press.

Dickenson, Donna. 2008. *Property in the human body. Feminist perspectives*. Cambridge: Cambridge University Press.

Dickenson, Donna. 2007. The lady vanishes. Human eggs as commodities. Position advocated at the congress '*Van stamcelonderzoek tot eiceldonatie. Een ethische reflectie over de vermarkting van het vrouwelijke lichaam*'. Antwerpen: Universiteit Antwerpen.

Dworkin, Ronald. 1993. *Life's dominion. An argument about abortion, euthanasia, and individual freedom*. New York: Alfred A Knopf.
Elliot, Carl. 2003. *Better than well. American medicine meets the American dream*. New York/London: Norton.
Elst, Josiane van der. 2007. The oocyte gold rush. A plea for fair trade. Position advocated at the congress *'Van stamacelonderzoek tot eiceldonatie. Een ethische reflectie over de vermarkting van het vrouwelijke lichaam'*. Antwerpen: Universiteit Antwerpen.
Foster, Charles. 2011. *Human dignity in bioethics and law*. Oxford/Portland (Oregon): Hart Publishing.
Franklin, Sarah. 2004. Stamcelonderzoek en voortplantingstechnologie in Engeland. In *Factor XXX. Vrouwen, eicellen en genen*, ed. Huijer Marli and Horstman Klasien, 87–102. Amsterdam: Boom.
Gaita, Raimond. 1991. *Good and evil. An absolute conception*. London: Macmillan.
Geertz, Clifford. 1983. Local knowledge. Fact and law in comparative perspective. In *Local knowledge. Further essays in interpretive anthropology*, ed. Geertz Clifford, 167–234. New York: Basis Books.
Goris, Harm. 2006. *Bodiless and human dignity. An intercultural approach – Lieblichkeit und Menschenwürde. Interkulturelle Zugänge*. Berlin: LIT-Verlag.
Habermas, Jûrgen. 2003. *The future of human nature*. Cambridge: Polity Press.
Hampshire, Stuart. 1993. Morality and pessimism. In *Morality and conflict*, ed. Stuart Hampshire, 82–100. Oxford: Blackwell.
Houellebecq, Michel. 1998. *Les particules élémentaires*. Paris: Flammarion.
Houellebecq, Michel. 2005. *La possibilité d'une île*. Paris: Fayard.
Joas, Hans. 2011. *Die Sakralität der Person. Eine neue Genealogie der Menschenrechte*. Berlin: Suhrkamp.
Kellendonk, F. 1986. *Mystiek lichaam. Een geschiedenis*. Amsterdam: Meulenhof.
Kolakowski, Leszek. 1990. *Modernity on endless trial*. Chicago/London: The University of Chicago Press.
Macklin, R. 2003. Dignity is a useless concept. *British Medical Journal* 327: 1419–1420.
Mencius (Reprint 1983) *Mencius,* transl. D.C. Lau. Harmondsworth: Penguin.
Mertens, P. 2006. Je zal niet doden. In *Zorg om het beginnende leven. Prenatale diagnostiek, foetale therapie en zwangerschapsafbreking in ethisch perspectief*, ed. Bodifée Gérard et al., 15–22. Antwerpen: UCSIA-The Dialogue Series 9.
Mertes, Heidi, and Guido Pennings. 2007. Oocyte donation for stemcell research. *Human Reproduction* 22(3): 629–634.
Offermans, C. (ed.). 2000. *Het licht der rede. De verlichting in brieven, essays en verhalen*. Amsterdam/Antwerpen: Contact.
Pennings, Guido. 2002. Reproductive tourism as moral pluralism in motion. *Journal of Medical Ethics* 28(3): 337–341.
Pennings, Guido. 2007. Medisch-wetenschappelijk onderzoek als ethisch kader voor eiceldonatie. Position advocated at the congress '*Van stamcelonderzoek tot eiceldonatie. Een ethische reflectie over de vermarkting van het vrouwelijke lichaam'*. Antwerpen: Universiteit Antwerpen.
Pinker, S. 2008. The stupidity of dignity. Conservative bioethics latest, most dangerous ploy. *The New Republic*.
Sandel, Michael. 2007. *The case against perfection. Ethics in the age of genetic engineering*. Cambridge MA/London: The Belknap Press of Harvard University Press.
Scheper-Hughes, Nancy. 2001. Bodies for sale –Whole or in parts. *Body and Society* 7(2–3): 1–8.
Strawson, Peter. 1974. Freedom and resentment. In *Freedom and resentment and other essays*, ed. Peter Strawson, 1–25. London: Methuen.
Taylor, Charles. 2007. *A secular age*. Cambridge MA: Harvard University Press.
van Beers, Britta. 2015. Is Europe 'giving in to baby markets?' Reproductive tourism in Europe and the gradual erosion of existing legal limits to reproductive markets. *Medical Law Review* 2015: 103–135.

van Hoof, Wannes, and Guido Pennings. 2011. Extraterritoriality for cross-border reproductive care. Should states act against citizens travelling abroad for illegal infertility treatment. *Reproductive Biomedicine Online* 23: 578–583.

Wiggins, David. 2006. *Ethics. Twelve lectures on the philosophy of morality*. London: Penguin.

Wiggins, David. 2009. Solidarity and the root of the ethical. *Tijdschrift voor Filosofie* 71: 239–269.

Williams, Bernard. 1984. The scientific and the ethical. In *Objectivity and cultural divergence*, ed. Stuart Brown, 209–228. Cambridge: Cambridge University Press.

# Chapter 10
# Bioeconomy, Moral Friction and Symbolic Law

**Klaus Hoeyer**

## 10.1 Introduction

Biotechnological innovation as a source of economic growth is high on the political agenda in Europe and elsewhere (Organisation for Economic Cooperation and Development, OECD 2005, 2011). Decades of work have sought to facilitate translational research and patenting of European research (Commission of the European Communities 2005; Foray 2004), and during the Portuguese presidency of the European Union (EU) in 2000, member states agreed on the so-called *Lisbon Strategy* to stimulate economic growth through the bioeconomy. A key feature was innovation in the medical and pharmaceutical fields. In March, 2012, the Danish presidency for the European Union (EU) hosted a high-profile conference behind closed doors in Copenhagen also on the theme of 'bioeconomy' proving the tenacity of the political agenda. In the area of health, such dreams of economic progress typically depend on expedient access to bodies and bodily material. In Denmark, it is a key element of the governmental Plan of Growth to facilitate such access by way of enhancing industrial exploitation of national biobanks and health registries as well as by strengthening collaborations between public and private health researchers ('Handlingsplan for styrkede rammer for offentligt/privat samarbejde om klinisk forskning', 2014).

In parallel with initiatives to stimulate market access to bodies and healthcare data, various legal initiatives (e.g. Council of Europe 1997; UNESCO 2005) have been put in place to safeguard the body against economic exploitation. Such initiatives can be related to a fear of what is often called 'commodification' of the human body and bioinformation (Rose 2005). Ostensibly, such initiatives are enacted to protect bodies against markets. Sometimes they take the form of conventions, such

K. Hoeyer (✉)
Department of Public Health, Centre for Medical Science and Technology Studies, University of Copenhagen, Øster Farimagsgade 5, Room 10.0.09, Copenhagen DK-1014, Denmark
e-mail: klho@sund.ku.dk



B. van Klink et al. (eds.), *Symbolic Legislation Theory and Developments in Biolaw*, Legisprudence Library 4, DOI 10.1007/978-3-319-33365-6_10

as the Oviedo Convention, and serve as soft law, and sometimes the declarations feature in preambles to EU Directives governing other issues such as safety (The European Parliament and The Council of the European Union 2004).

Typically the two sets of political ambitions remain unrelated in public policy debates, but a growing social science critique has suggested that granted the constant expansion of economic interests in bodies, we should understand the legal initiatives that claim to 'protect' bodies more as a veil covering up the means of exploitation than as actual protection against it (Scheper-Hughes 2002; Sharp 2007). From a legal perspective, conversely, it has been suggested that we should accept that bodies simply have become commodities in a market place, and to adjust our legal landscape accordingly: "The fact is that there *is* a market, and so to say that there is not is to perpetuate a fiction" (Mason and Laurie 2001: 715). Though drawing different moral conclusions, both positions imply that the treaties meant to safeguard the body have no immediate impact.

If these observers are right, the treaties should be seen as merely symbolic in the sense of superficial and ineffective texts without legislative effect. With this chapter I suggest that they are not. Rather, the symbolism involved is productive and it has real effects. However, the effects are different from those one might have expected if one is reading the treaties out of context from the competing policy ambitions propagated by the same authorities through other agencies. The 'symbolic' soft law intentions do influence everyday practices, but we will not understand how unless we acknowledge the interplay between competing governmental ambitions as they develop on the ground. Furthermore, I believe we should apply a principle of symmetry in the way we approach the two sets of legislation. The measures to enact economic growth are marked by symbolic traits too, and in many ways they operate through similar mechanisms. These mechanisms relate to what social scientists have called *performative effects* (Butler 1993; Callon 1998).

It is important not to associate performativity with 'theatre' in the sense of 'just a play'. From the perspective of performativity theory, performative effects are just as real as anything else. Performativity theory is about acknowledging that human agency relies on conceptions of how things ought to be. Everybody is constantly trying to enact their world according to the images they uphold. Both sets of governmental ambitions described above become real through people's enactment of them; they become part of practice because people are trying to 'perform' them. If we acknowledge this premise, we cannot make any clear distinction between symbolic and material work, but must acknowledge the symbolic as real and the material as engrained with symbolism.

This chapter is focused on exploring the performative effects in practice as people engaged in tissue procurement deal with the competing governmental ambitions. Following a theoretical discussion of the role of law in defining body parts as exchange objects, I describe the performative effects of governmental friction at the level of organizations procuring and distributing tissue in the European Union. I focus on how competing understandings of the body as an economic resource and as subject of moral integrity interact in practices of tissue procurement, and on how body parts are made available through what I will term 'practices of division' and 'practices of production'. By focusing on practices, I point to the way in which concrete material work is simultaneously symbolic in its effect (see also Mohr 2014).

## 10.2 Body and Thing: Symbolic Law, Friction and Performativity

The bioeconomic ambition of growth necessitates contemplation of a prominent legal conundrum: the legal status of the body in property regimes. Post-slavery legal systems rest on a distinction between persons and things: the former can be owners and the latter owned, but the two are not supposed to be mixed (Skegg 1975). When building a business on practices in which human body parts contribute to the generation of commercial profit, it becomes important to clarify the status of these body parts as either 'part of persons' or 'thing' (Harrison 2002; Herring and Chau 2007, see also Chap. 8, this volume). The definitions of what can and cannot be subject to property (as well the concept of property as such) have changed over time and are contingent on broader political movements (Gold 1996). Finding that the body is in fact increasingly enrolled in processes of capitalization, some legal scholars have suggested simply redefining bodies as property (Fabre 2006; Goodwin 2006). Others suggest that there is no way of avoiding ambiguity and that new legal options therefore need to develop alternative ways of thinking of the body (Hyde 1997), for example by inventing a kind of third place between persons as owners and things, or commodities, as plain property (Fox 2000). Irrespective of one's position on this topic, the bioeconomic drive towards commercialization relies in one way or another on clarification of commercial entitlements.

The classificatory predicament is nicely summed up by Lenk and Beier who suggest that "'property' is the decisive concept for controlling an entity" (Lenk and Beier 2012: 348) while simultaneously noticing that:

> …according to most jurisdictions, the human body constitutes a significant exclusion from this general rule, a fact that is also mirrored in a number of important normative national and international documents. This exclusion is commonly known as the no-property rule (Ibid.).

Lenk and Beier outline the treaties and legal documents designating the body as beyond trade and then remark:

> It is a peculiarity of the field of human tissue and body material that, besides the common official normative framework (i.e. the no-property rule), different approaches to tolerating the commodification of human tissue and body material can be observed in practice (p. 349).

The tolerated practices are in fact fully legal, they observe, and stimulated through attempts of promoting commercialization (see also Pirnay et al. 2013). The task we face is not just to clarify the legal status of the body, but how to understand the co-existence of two competing and conflicting legal frameworks, or in Lenk and Beier's words: "How can this obvious divergence ….be explained?" (p. 350). This chapter is a step towards an answer to their question.

One option is to consider one of the two types of law less 'real' than the other. Social science critique of the commercialization of the body has chosen this option by viewing 'ethics rules' as merely symbolic window dressing; a veil of economic exploitation (Dickenson 2007; Scheper-Hughes 2002; Sharp 2007). When conventions and treaties talk about a "body beyond commercialization" and make repeated calls for educational campaigns to generate willingness to donate body parts altruistically, these scholars see it as a cover-up for the actual underlying economic forces. Pirnay and colleagues take a similar position, when they note that treaties aimed at safeguarding the dignity of the body fail to counter what they describe as the EU's 'business oriented' legislation for tissue and cell therapies (Pirnay et al. 2013). They want to strengthen the protection of the body because "human cells and tissues (one's own or somebody else's) should not be degraded to tradable goods" (p. 543). The current protection of the body is, it appears, merely a 'symbolic' gesture, or what van Klink (Chap. 2 in this volume) identifies as symbolism in a negative sense where the symbolism serves the instrumental goal of furthering business interests.

In legal thinking, the notion of a symbolic law was introduced with Aubert's study of a law which was said to govern the working conditions of Norwegian housemaids, but which apparently had no effect. Aubert proposed the following conclusion: "Law may on occasion move in the sphere of symbolism and magic rather than in the everyday sphere of practical solutions to practical problems" (Aubert 1966: 115). Later work has questioned the analogy to magic. Van der Burg and Brom suggest that law brings about interaction through which norms take shape, and as such even law which is not used to regulate conduct is part of shaping conduct. Symbolism can work on positive terms by propagating norms which, in time, influence practice. Though providing few sanctions, symbols are still seen as having effects. Van der Burg and Brom argue that we should not see such laws as either reflecting pre-existing norms or as defining new norms for the constituency. Rather, they are part of generating norms and this works through symbolism as much as other regulatory means. Therefore, they note,

> someone who looks carefully at reality will understand that [legal symbols] are not so much ineffective as effective in a different way. They are effective in using discussion and persuasion as a means rather than external sanctions and in appealing to morally responsible behaviour of citizens rather than to merely strategic action of subjects (Van der Burg and Brom 2000: 68).

The mechanism here alluded to resembles what in the social sciences is typically thought of as performativity theory. As discussed above, performativity theory suggests that symbols produce action as people enact their world through them: what people think about the world is part of shaping what the world becomes. People do not simply reproduce symbols, and it is sometimes unexpected interpretations of symbols that make them part of everyday practices. Through people's agency the symbols themselves may materialize in new ways and acquire meaning and tangibility. They are performed in action. In Judith Butler's seminal work on performativity it is stressed that performative effects are ontological; it is how things become real, not just some sort of ephemeral symbolism unrelated to 'real things' (Butler 1993). The idea is that the world is in a constant process of becoming, and therefore words such as 'bodies' or 'markets' are not names of pre-existing universal entities; rather, what we get to know as 'bodies' and 'markets' emerge through the agency of people engaged in making them real (Hoeyer 2013).

Importantly, symbols are never free-standing or solitary. They operate in densely populated spaces with many competing agendas. Hence, people perform symbolism through interaction with several competing systems of meaning and in reflection of all the complexity of their material, economic and political options. I suggest that the co-existence of *competing* symbols is essential to the performative effects of 'symbolic law'. I thereby also invite us to approach regulatory initiatives such as the growth plan and regulatory facilitation of industry access as symbolic. They provide modes of logic according to which people enact their everyday practices, but they cannot fully determine what people do. We need to embrace the competing governmental ambitions and investigate what emerges through the resulting friction (Wadmann and Hoeyer 2014). We cannot elevate one set of legal initiatives to 'real' law and reduce another to 'symbolic' as if they did not operate in the same world. In the following, I will discuss the practical implications of having two ostensibly contradictory policy ambitions (to stimulate a bioeconomy based on bodily products and to keep the body out of the market) based on studies of the everyday practices of tissue exchange in Europe.

## 10.3 Methodological Approach

This paper builds on several years of fieldwork in and around the agencies regulating and procuring organs and tissue in Europe. Through committee work in Danish and European associations for organ and tissue procurement I have come close to the informal discussions about ethics and economics in relation to human biological material, and I have supplemented these insights with interviews with members of the European Parliament who have been engaged in developing the complex regulatory framework for inter-European exchange of blood, cells, tissue and organs as well as members of Danish regulatory agencies translating the EU rules into national law. Along with this work, I have conducted several case studies, e.g. an interview study including representatives from all 17 Danish bone banks (Hoeyer 2010), and also in Denmark, I have with colleagues interviewed health professionals working with organ procurement as well as potential organ, tissue and dissection donors (Hoeyer and Jensen 2013; Hoeyer et al. 2015). I have also travelled to several major European tissue banks to understand the impact of the European regulation in different constituencies.

In this chapter, I draw mostly on material from three very different European tissue banks. Tissue banking is a small world and two of the three were very strict on the need for absolute confidentiality. In the following I have therefore covered their identities by giving them new names; avoiding reference to the countries in which they operate; and, in one instance, by attributing a different type of tissue to its portfolio. The three banks represent different organizational forms: *Quality Tissue Trust* is a public multi-tissue type procurement facility which harvests material from bodies originating in the national jurisdiction while delivering tissue grafts globally. *International Eye Bank* is a privately established mono-tissue type procurement facility which was established by health professionals from several countries, using their own money for start-up capital. It receives tissue from donors in several countries and it ships tissue world-wide though mostly in Europe. It is embedded in a public facility and its price setting is carried out by the national authorities in the country where it is located. Finally, *VitalGrafts* is multi-tissue type procurement, research and distribution bank, which harvests cadavers nationally, but ships grafts internationally. It is affiliated with public health authorities. I have interviewed managers, doctors and one secretary from these facilities. In some instances, I was not allowed to record the interview and the direct quotes from these interviews are quite short and limited to what I could make a note of while interviewing. The urge to avoid electronic recording is in fact indicative of the type of environment in which these tissue bankers operate, and as such it illustrates the governmental frictions these biobankers need to handle on a daily basis, as I will now show.

## 10.4 Ethics and Economics: "A Very Strange Sector"

It is about 10 min into a conversation with the manager of VitalGrafts that I reach a point where I think it is appropriate to ask if I may record the interview. He looks hesitant and then says that it would be better if we could keep it 'informal'. He explains: "It's a very strange sector! And it's a small world with a high emotional level." I ask him what he means by 'strange', and he replies that as tissue bankers they "have to balance ethical with economic aspects". When visiting the International Eye Bank, it took an hour before we had reached levels of confidence sufficient for the manager to allow recording of the remaining one and half hours of our conversation. He said that journalists would often want to write about tissue recovery just to provoke a sense of horror and it was important to protect their bank from negative publicity. It is a strange sector indeed, if we consider how practitioners in the field wish to protect themselves against publicity and yet simultaneously agree on a need for educational campaigns and increased awareness.

What is this 'strangeness' about? At the end of each interview, I asked every tissue banker "What are in your opinion the major ethical issues in tissue banking?" They all pointed to the balancing of 'ethics' and 'economy' mentioned by the manager of VitalGrafts above. Two staff members from VitalGrafts interviewed together responded in chorus to my question: "commercialization!" Then they began elaborating on a need for public guidelines on legitimate purposes (should they, for example, deliver material for cosmetic enhancement?); how should they balance requests from the research industry with requests from hospitals ("many people want to do research with this tissue, and it is difficult to see if it's all covered by 'research into transplant purposes'"); and, finally, "The ethical issue is how to respect donors' intentions in terms of tissue usage and ensure transparency". The responsible doctor in International Eye Bank responded to the same question saying:

> In my opinion, tissue banking must be non-profit, absolutely, the, the… in different senses we may not *sell* the tissues, we may *cover* our costs and the tissue banks must remain very, very strictly eh.. controlled, in the sense of who would use the tissues, for which, which purpose the tissues can be used.

Again he points to a balancing of ethics, donor control and economic concerns as the major ethical issues, and while doing so, he illustrates the difficulties in finding a language adequately expressing both these concerns and the nature of the transactions that do take place. Respect for donor wishes also came across in an interview with a doctor from Quality Tissue Trust, operating in a country which has a presumed consent rule for organ recovery, but not for tissue recovery. She thought for a while about the ethical issues and after mentioning concerns about some procurement agencies being too focused on procuring a lot of bodily material for economic reasons she then decided that it was not economics as such but *consent* that was the major issue. I asked why consent was important and she replied:

> …because I think that the world of transplantation is not perfect. It is not one hundred per cent perfect. And it is important that the consent has been done without pressure. And if the consent is well done, the security and the safety of the tissue and organ is well, or better.

Note that as she shifts from concerns about economic incentives for procuring too much to concerns about consent, she builds a bridge to another main concern, namely safety. If the informed consent is not in place, proper investigation of the medical history is impossible. In practice, economics, medical safety and ethics belong to interrelated registers, and though subject to different legal initiatives, tissue bankers need to address all of them concurrently. For people working with cadaveric tissue recovery, the concerns about economy, ethics and safety all converge.

If ethics, safety and economics converge at the practical level of tissue banking, it is interesting to note also how policymakers seem to slip from topic to topic. One of the first things which struck me when moving around the EU offices was the ways in which discussion about directives said to govern *safety* inevitably glided into comments about ethics. Even specific questions about technical standards could somehow lead to reflections on informed consent and 'voluntary and unpaid donations'. Of course, this reflects also the fear of the gaze of the outsider, the interviewer, but this fear is part of the everyday context for their work.

In one particular instance, a legislator identified this moving back and forth between economic considerations and ethics talk as a clear mistake. I had noted that the directives implied increased expenditure and I therefore (naïvely) asked a member of parliament whether they had discussed the economic implications of the directive when preparing the organ directive. He quickly corrected me: "No, we're looking more to principles, for example safety, that it is unpaid and voluntary, traceability etc. The cost is not my job." Cost calculations belong under national jurisdiction. When I then proceeded with questions about principles having to do with rules for aftercare which could limit organ trade (i.e., European citizens purchasing organs abroad and opting for aftercare back home), he limited the domain of the European Parliament even more: "No, we're interested in safety and quality only." Notice, how 'unpaid and voluntary' was now left out in this formulation though part of the directive and heavily discussed in Parliamentary groups. After a pause, however, he then added that "we should never accept smuggled organs." He then embarked on a discussion of ethics irrespective of his former insistence that the directive was about safety only. I found this intriguing, not least because I had just asked his secretary what had been most surprising to her in the process of preparing the directive and she said it was all the meetings with Chinese representatives. It was her impression they had strong interests in the wording of, for example, human rights issues, and I therefore had the impression in advance that it would be something he would have wanted to bring up. And yet, when I asked the Member of the European Parliament (MEP) if he had ever met with anybody from China he said no. Some political aspects of the intermingling of ethics, economic interest and directives are apparently too sensitive to be discussed with outsiders. Clearly the intermingling of concerns found at the ground-level of tissue bankers are reflected also in policymaking circles, though in different ways.

The EU directives enacted to ensure a 'safe and stable supply' of human tissue influence the economic considerations in various indirect ways too. Initial drafts suggested exclusion of industrial actors from procurement of tissue, but one MEP I interviewed claimed that industrial lobbying had succeeded in convincing policymakers that industry had a greater familiarity with Good Manufacturing Practices than the small tissue banks in the public sector, and it was therefore impossible to suggest exclusion of industrial actors. Representatives from the public tissue banks that I interviewed found it difficult to finance the new standards, and one tissue banker remarked that nobody knew how the industrial actors did it, but 'nobody was checking'. As a consequence, some tissue bankers found that the EU Tissue and Cells Directive (from 2004) privileged international (mostly American) for-profit organizations over local not-for-profit tissue banks. There are, however, no aggregate financial data available to document such a transition of market shares.

In some countries, governments furthermore seek to stimulate competition between local tissue banks in line with New Public Management (NPM). VitalGrafts explained that they were now subject to a process of tendering where they should compete on price for grafts to gain access to cadavers from which the grafts could be produced: "We have to fight over the bodies", as one of them noted during a meeting. Note, however, that it is not a normal commodity market; here competition features as demand for optimization of processes, and bodies as such do not acquire a price. Regulation aimed at stimulating competition serves symbolically in that tissue bankers then need to find ways of enacting this regulation in practice. The prices for grafts are also said to 'recover' expenses, and do not follow supply/demand principles (see Hoeyer 2009). The demands for detailed description of services, the tendering process, and the changeable safety standards were experienced as frustrating: "The government thinks it steps back and allows competition, but actually it turns on every knob and decides on every detail," the manager said with his tongue actually in his cheek while holding out his hands in the air as if turning on two imaginary knobs.

The point is that it is very difficult handling the friction between contradictory governmental policies. They feel pressured to perform as businesses and yet they are constantly told that it is very important that they are not seen as trading in body parts; they must be 'ethical'. We cannot reduce one set of governmental actions to 'symbolic law' and claim that the other determines what happens in reality, on the ground. Both serve as symbols in varying degrees, and it is the co-existence of two competing policy ambitions that produce the predicament faced by the tissue bankers. The legislation aimed at stimulating competition and market growth does not create a straightforward commodity market; and the treaties on non-commodification do not protect the body from entering a process of competition, tendering and financial interests. However, they both shape the practices on the ground. They have performative effects. Furthermore, ethics and market thinking are both ingrained in the EU directives on 'safety'. Before I turn to the practices through which tissue bankers enact bodies in ways which are compatible with *both* governmental ambitions, I wish to give a brief glimpse of the tensions between the organizations described above in relation to the notions of profit and non-profit.

International Eye Bank and Quality Tissue Trust were both subject to a governmental agency setting the 'recovery prices' they could charge. They were pressured financially because the fees were adjusted irregularly and not always in response to increasing costs. Also, fees were too low for generating savings for essential investments such as new clean rooms, as a normal business model would otherwise advise. Nevertheless, the two organizations were successful in harvesting sufficient material and were both sending grafts abroad based on availability and medical need. VitalGrafts, conversely, did not get the cadavers they needed, but they were free to set their own prices. Hence, they would sometimes request material from, for example, International Eye Bank and ship it on to their contact network after adding a so-called 'handling fee'. International Eye Bank was frustrated with this practice and described it as illegitimate profit-making. In an interview, the manager of International Eye Bank called VitalGrafts "a commercial organization", which produced leaflets "that they distribute everywhere and they say 'we send the cornea' and so on, but they don't have them, VitalGrafts don't have the cornea." I was surprised and asked "They don't have them?" and he said:

> No, VitalGrafts is asking for cornea almost every day from here and they increase the prices, for I don't know eh 30–40 percent and they sell them to Germany or to Denmark or, I don't know, to the other countries. VitalGrafts is a commercial organization.

Knowing that VitalGrafts is a statuary non-profit organization I could not help inserting "They are non-profit…" but he interrupted me saying "They, [laughing] they say they are non-profit of course, but they are commercial, a commercial organization." I therefore asked: "But why are you then sending cornea to them?" His replied:

> It's a very difficult issue you know, it depends on the emergency (…) I have to give them the cornea because they need them for the patients (…) I'm not very happy to, I prefer that [the treating doctor] calls us directly and says 'I need a cornea can you, do you have available tissues' and so on.

It is important to note that he does ship the cornea, and does so for the sake of the patients. This illustrates that it is not just an instance of two competing businesses. The organizations are dedicated to treating illness and disease and money-making is a matter of organizational survival needed to facilitate patient survival, not an objective in its own right. The sector is strange, from a business perspective, in that they compete not just for market shares, but for ethical reputation and both income and reputation is secondary to patient interests. When I asked VitalGrafts about the price-setting principles and accusations of profit-making they said: "price is somehow secondary (…) it's a finder's fee. We're not brokering tissue; we're facilitating [contacts]". He paused for a while and then continued that for a surgeon faced with a sick child "the price elasticity is such that it doesn't matter. It's the life of the child that matters." Certainly, market thinking is present when organizations are governed as businesses in an NPM regime in which nobody can be sure of their own survival. Incentives to optimize business models have performative effects, and yet the contradictory demands for non-commodification have their performative effects too: he ended his explanation with the clear confirmation "We're non-profit."

I have now shown how the two governmental inclinations, bioeconomic market thinking and ethics rules of non-commodification, generate a sense of moral friction with performative effects at the grass roots level of tissue procurement. We cannot reasonably claim that one set of legal work is 'merely symbolic' and subordinated to instrumentally serve the other. In fact, both governmental ambitions are being played out in the socio-economic conditions under which tissue is procured. Tissue procurement involves handling the transition of bodies from being 'persons' to becoming 'therapeutic tools' in a very concrete way (Hogle 1999). It takes work to make a tissue graft. We see in this work the practical ways through which the legal tension between person and thing gains ontological presence in shaping exchangeable units. In the following two sections I describe how this transition is achieved (though never irreversibly) through what I will call *practices of division* and (in the following section) *practices of production*.

## 10.5 Practices of Division

In considering practices of division, I aim to discuss the separation of body and person, i.e. of materiality and of the emotional meaning associated with a person who has an identity. As Herring (Chap. 8 in this volume) notes, the body is full of symbolic value, and to make the body available for exchange, there is a need to divide body parts from the meanings associated with the persons from whom they are harvested. Bodies as wholes are not ontologically given: the notion of a body as a person depends on a perceived 'imaginary' whole, as Zwart (2014) has argued.

Still, some kind of symbolic division between body and person needs to take place to make the material body available as a medical tool without affecting the sense of respect for the body as representative of a person. How is this division performed?

One of the first things we should notice is that not all body parts are equal and therefore they do not demand the same kind of treatment. As pointed out by also the Nuffield Council, the legal system already distinguishes between more worthy and less worthy forms of human biological material being subject to different commercial restrictions (Nuffield Council on Bioethics 2011). Blood, for instance, is subject to tendering processes; and donors of breast milk and gametes can be compensated, unlike tissue donors. Tissue grafts, on the other hand, figure on the internet with price tags (calculated as recovery costs), unlike organs, which are fiercely guarded as non-commodities. The logic seems to be that the bigger the part the closer to personhood, and thereby the less obvious is the commercial aspect of its exchange. One of the practices of division is thus to distinguish between parts and wholes. A division into parts can help separate body from person.

Next point is to note how notions of *ethics* are essential to the practices of division. Remember how the doctor from Quality Tissue Trust thought that consent was a central concern. Informed consent also has significant implications for preparing tissue for global exchange. By installing informed consent procedures, governments create an obligatory point of passage for the division of person and body. The person, or the person's relatives, can bequeath the body to the procurement agency and in that process the body is supposed to shift meaning. It should stop being a person; if we refer now to 'person' it should rather be to the person who was the agent involved in handing over the material remains, while the remains themselves now become available as 'things': therapeutic tools.

Informed consent processes are essential for the practices of division and the tissue market in other ways too. Consent is supposed to sort out liability issues and protect the receiving organization against complaints for illicit procurements (Halpern 2004). The consent process also serves to transfer disposition rights and in some jurisdictions property rights. It has been discussed in great detail in the literature following the famous case of John Moore who opted for property rights to cells from his spleen, but was granted only the right to an informed consent, and I will not expand upon it here (Harris 1996; Hoeyer 2007; Landecker 1999; Rabinow 1992; see also Chap. 13, this volume).

In some cases, consent requirements seem to follow a form of reasoning similar to that which I describe above in relation to the relative fear of commercialization dependent on the size of the body part. In Denmark it is striking how relatives are allowed to donate the organs of their deceased, but not whole bodies for anatomical training and dissection. The whole body can be donated only with the informed consent of the donor prior to death. Smaller tissue samples, in contrast, are regularly exempted from consent. Still, in most instances there is an insistence on gaining some form of consent (if not from the donor, then from a research ethics committee). The body is not treated as an ordinary natural resource or a plain commodity. The treaties for the protection of the body bring about a special moral status, which in turn has performative effects: on the ways in which exchanges take place, on the measures for transferring rights of disposition, and on the ways in which economic aspects are handled. Elsewhere, I have in detail described the means by which a piece of bone acquires a price without forming anything which can meaningfully be termed ‘a market’ (Hoeyer 2009), and the point here is again that no-property rules and anti-commodification legislation is perhaps symbolic in one sense, but then these symbols exert significant agency. In practice, they structure everyday practices in important ways. It is no coincidence that the tissue bankers think of ‘commercialization’ as the primary ‘ethical’ issue; the two are different sides of the same coin: ethics practices are performative effects of the friction between competing governmental ambitions to divide bodies into exchangeable units while ensuring the dignity of the whole they once formed.

## 10.6 Practices of Production

The practices of division tie in with practices of production: once the tissue is separated from the person through the consent process, it can be enhanced into therapeutic tools. It is via this work it becomes an object of utility value rather than part of a subject associated with values of dignity and innate human worth. It takes work to shift the order of worth (Stark 2009). It might be helpful to consider an analogy in a classical discussion in anthropology – the distinction between the raw and the cooked: cooking is a way to make material objects shift domains, connotations and status (Lévi-Strauss 1969). The enhancement of the plain material to grafts implies a form of de personalization through which the body shifts domain, connotations and status. Preparation of grafts can therefore be thought of as a form of ‘cooking’. It is a material practice, yes, but it is intertwined with symbolic implications. The symbolism is co-produced with the material form of production. In this way, it is not only the anti-commodification legislation that can be described as ‘symbolic’: the very process of handling body parts at a material level involves symbolic work too.

This work performed in the tissue banks serves as a form of ‘boundary work’ (Gieryn 1983). Gieryn talks about boundary work as the work done to create separation where there previously was no clear line. Some would see a lack of boundary as a dangerous form of mixing of what ought to be kept separate (see also Douglas 1995). Gieryn suggests that the separate ‘entities’ are the result of work rather than pre-existing entities ‘saved’ from pollution. It is through boundary work we learn to see grafts as totally different from persons, as if they had always belonged to two unrelated and different registers.

The work of production in the tissue banks is totally practical, and being carried out in sterile environments, demands great care and an exceptionally high level of competence. It is through this work that cadavers turn into tendons, heart valves, cornea and pieces of bone. The body of a person is procured, tested, rinsed, cut in the right forms and packaged in ways facilitating storage and transport. Along with handling the tissue, the banks must process the paper trail accompanying the tissue and keep track of grafts they have already shipped. The tissue banks install and utilise systems aimed at confidentiality by anonymizing the origin of the tissue while allowing traceability. Such audit apparatus not only entrenches the separation of tissue from person, but involves tedious systematic low-tech work comparable to that at an assembly line in a chain of production.

It takes sophisticated accounting systems and production facilities to deliver these services, and such facilities come at a high financial cost. This in turn necessitates the establishment of sources of income. In NPM systems of governance, income is supposed to reflect production to stimulate increased efficiency. Practices of production are structured accordingly and the need to ensure financial stability is never off the agenda. The accounting systems therefore deliver proof of not only safety and consent but also of productivity. Partly as a consequence of the demands of increased efficiency and partly as a consequence of attempts to meet increasing demands for more grafts, the amount of tissue taken from each corpse is gradually increasing. The three banks I have described provided me with no statistics, yet we know from the literature that not only are increasing numbers of grafts produced, but more and more are taken from each individual donor. When The Musculoskeletal Transplant Foundation was founded in 1987 they procured 1500 grafts from approximately 100 donors. That is 15 pieces per cadaver. By 2002, they procured 297,644 grafts from 4468 donors, which equals 66 per cadaver (Gocke 2005). Such expansion involves changing senses of appropriateness. In 1987 the manager of Virginia Tissue Bank stated that "We don't try to make the largest number of deposits, we try to take what is needed" (Kirn 1987: 304). Yet the sense of 'need' changes over time, and it is driven not so much by 'supply and demand' in an economic sense, but by the pride tissue banks take in facilitating treatment.

The sense of pride also came across in my interviews. VitalGrafts explained their ambition to me as a matter of making better products: "Although we're non-profit, we're seeking to become more advanced. We have new products under way." They clearly experienced an ongoing competition with both financial and medical implications. This competition was also threatening their survival if they were not at the forefront of developing the practices of production. A doctor working with enhancement of heart valves proudly explained:

> In young patients, newborns to five years, the failure of the valves is much higher than in other patients, and due to that there are many initiatives of engineering of the valves and arteries. If you are able to remove the cells, without damaging the matrix of the valves, this tissue will not be recognized as a foreign, [it is called] de-cellularization, and it will be re-cellularized by the recipient in few weeks. After a few months it will be integrated in the body as its own tissue. So there are some projects on this issue actually, and eh.. eh, this is *probably* now considered as an advanced product (original emphasis).

The latter comment refers to the EU directives governing the types of product. A 'plain' heart valve belongs under the EUTCD while 'enhanced' or 'engineered' products fall under product regulations (see also Faulkner 2012). The logic of elements of production making a tissue into something more fit for market regulation is in this sense written into the regulatory framework. It does not erase the ambiguities, however, as the little word 'probably' illustrates.

## 10.7 Conclusion

I have described how tissue procurement is neither ‘gift exchange’ with no economic importance nor ‘market economy’ driven by so-called market forces. Indeed, as I argue also elsewhere, we need to move beyond this dichotomy to understand the intricate intermingling of governmental ambitions on the ground (Hoeyer 2013). The demands imposed on tissue banks dealing with the mixture of symbols as they are asked to comply with both ideals create an enduring sense of tension. Governmental measures aimed at ensuring growth, efficiency and competition co-exist with demands for ensuring that the material procured and exchanged is treated differently from plain commodities.

The two conflicting sets of governance frameworks described here both have performative effects as tissue bankers try to model their daily work according to the competing images of what is true, morally right and socially sustainable. It would be mistaken to elevate one set of laws to a higher ontological status and downgrade the other to ‘symbolic’, as if ‘symbols’ were less real, since both forms of legislation work through symbolism and strategic appeals. Both operate through people trying to make them ‘real’ through their daily work with bodily boundary objects.

I have illustrated the work environment and work practices that are produced through the competing legal frameworks. To make bodies available for exchange, the tissue banks have to separate the person from the body, and the body from the graft, while continuously reminding themselves of the enduring relation to personhood which makes it appear as ‘beyond trade’. It generates a morally tense atmosphere full of friction and puts the individual tissue banker at a constant risk of blame (Wadmann and Hoeyer 2014). If not sufficiently business oriented, they might be seen as unprofessional and inefficient. If too business-oriented, they might be seen as unethical and motivated by the wrong incentives. It is no coincidence if they worry about publicity.

To provide a different work environment we need to appreciate the constitutive social effects of legal symbols, even when the laws as such do not seem to work as intended. They are part of moulding the conditions under which people work; or, rather, it is by performing their symbolism that people generate modes of work in reflection of dissonance between incompatible orders of worth (Stark 2009). Symbols acquire meaning in action and this action takes place in contexts already densely populated with competing symbols and inclinations. Rather than writing off some laws as ‘merely symbolic’, such as treaties said to protect the body from economic exploitation, we should understand what kind of tension they reflect and what kind social inclination they produce in the practices they are said to govern. Rather than claiming they are ineffective, we should explore their unintended effects. Then we might be in a better position to understand what symbolic law amounts to in practice.

## References

Aubert, V. 1966. Some social functions of legislation. *Acta Sociologica January* 10(1/2): 98–120.

Butler, J. 1993. *Bodies that matter. On the discursive limits of "sex"*. New York: Routledge.

Callon, M. 1998. The embeddedness of economic markets in economics. In *The laws of the markets*, 1–57. Oxford: Blackwell Publishers.

Commission of the European Communities. 2005. *Report from the Commission to the Council and the European Parliament. Development and implications of patent law in the field of biotechnology and genetic engineering*, 1–6. Brussels: The European Parliament.

Council of Europe. 1997. *Convention for the protection of human rights and dignity of the human being with regard to the application of biology and medicine. Convention of human rights and biomedicine*.

Dickenson, D. 2007. *Property in the body. Feminist perspectives*. New York: Cambridge University Press.

Douglas, M. 1995[1966]. *Purity and danger. An analysis of the concepts of pollution and taboo*. London: Routledge.

The European Parliament, and The Council of the European Union. 2004. *EU Tissue and Cells Directive (EUTCD) Directive 2004/23/EC of the European Parliament and the Council of 31 March 2004 on setting standards of quality and safety for the donation, procurement, testing, processing, preservation, storage and distribution of human tissues and cells*.

Fabre, C. 2006. *Whose body is it anyway?* Oxford: Clarendon Press.

Faulkner, A. 2012. Commensuration and proliferation. Similarity and divergence in law's shaping of medical technology. *Law, Innovation and Technology* 4(2): 165–184.

Foray, D. 2004. The patent system and the dynamics of innovation in Europe. *Science and Public Policy* 31(6): 449–456.

Fox, M. 2000. Pre-persons, commodities or cyborgs. The legal construction and representation of the embryo. *Health Care Analysis* 8: 171–188.

Gieryn, T.F. 1983. Boundary-work and the demarcation of science from non-science. Strains and interests in professional ideologies of scientists. *American Sociological Review* 48(6): 781–795.

Gocke, D.J. 2005. Tissue donor selection and safety. *Clinical Orthopaedics and Related Research* 435: 17–21.

Gold, E.R. 1996. *Body parts. Property rights and the ownership of human biological materials*. Washington, DC: Georgetwon University Press.

Goodwin, M. 2006. *Black markets. The supply and demand of body parts*. New York: Cambridge University Press.

Halpern, S.A. 2004. *Lesser harms*. Chicago: The University of Chicago Press.

Harris, J.T. 1996. Who owns my body. *Oxford Journal of Legal Studies* 16(1): 55–84.

Harrison, C.H. 2002. Neither *Moore* nor the market. Alternative models for compensating contributors of human tissue. *American Journal of Law & Medicine* 28: 77–105.

Herring, J., and P.-L. Chau. 2007. My body, your body, our bodies. *Medical Law Review* 15: 34–61.

Hoeyer, K. 2007. Person, patent and property. A critique of the commodification hypothesis. *BioSocieties* 2(3): 327–348.

Hoeyer, K. 2009. Tradable body parts? How bone and recycled prosthetic devices acquire a price without forming a 'market'. *BioSocieties* 4(2–3): 239–256.

Hoeyer, K. 2010. An anthropological analysis of European Union (EU) health governance as biopolitics. The case of EU tissues and cells directive. *Social Science and Medicine* 70: 1867–1873.

Hoeyer, K. 2013. *Exchanging human bodily material. Rethinking bodies and markets*. Dordrecht: Springer.

Hoeyer, K., A.M.B. Jensen., and M. Olejaz. 2015 Transplantation as an abstract good – Practicing deliberate ignorance in deceased organ donation in Denmark. *Sociology of Health and Illness* 37(4): 578–593.

Hoeyer, K., and A.M.B. Jensen. 2013. Transgressive ethics. Professional work ethics as a perspective on 'aggressive organ harvesting'. *Social Studies of Science* 43(4): 599–619.

Hogle, L.F. 1999. *Recovering the Nation's body*. New Brunswick/London: Rutgers University Press.

Hyde, A. 1997. *Bodies of law*. Princeton: Princeton University Press.

Kirn, T.F. 1987. How does tissue banking work? Virginia bank, while not typical, may offer some insights. *JAMA* 258(3): 304–305.

Landecker, H. 1999. Between beneficence and chattel. The human biological in law and science. *Science in Context* 12(1): 203–225.

Lenk, C., and K. Beier. 2012. Is commercialisation of human tissue and body material forbidden in the countries of the European Union? *Journal of Medical Ethics* 38(6): 347–350.

Lévi-Strauss, C. 1969. *The raw and the cooked. An introduction to the science of mythology, vol. 1*, 1–387. Trans. J & W Weightman. New York: Harper & Row.

Mason, J.K., and G.T. Laurie. 2001. Consent or property? Dealing with the body and its parts in the shadow of Bristol and alder hey. *The Modern Law Review* 64(5): 710–729.

Mohr, S. 2014. Containing sperm – Managing legitimacy. Lust, disgust, and hybridity at danish sperm banks. *Journal of Contemporary Ethnography* 45(3): 1–24.

Nuffield Council on Bioethics. 2011. *Human bodies. Donation for medicine and research*, 1–254. London: Nuffield Council on Bioethics.

Organisation for Economic Cooperation and Development. 2005. *Proposal for a major project on the bioeconomy in 2030. A policy agenda*. Paris: OECD.

Organisation for Economic Cooperation and Development. 2011. *The bioeconomy to 2030. Designing a policy agenda*.

Pirnay, J.P., A. Vanderkelen, D. De Vos, J.P. Draye, T. Rose, C. Ceulemans, N. Ectors, I. Huys, S. Jennes, and G. Verbeken. 2013. Business oriented EU human cell and tissue product legislation will adversely impact Member State's health care systems. *Cell and Tissue Banking* 14: 525–560.

Rabinow, P. 1992. Severing the ties. Fragmetation and dignity in late modernity. *Knowledge and Society: The Anthropology of Science and Technology* 9: 169–187.

Rose, C.M. 2005. Afterword. Whither commodification. In *Rethinking commodification. Cases and readings in law and culture*, ed. M. Ertman and J. Williams. New York: New York University Press.

Scheper-Hughes, N. 2002. *Commodifying bodies*. London: Sage.

Sharp, L.A. 2007. *Bodies, commodities, and biotechnologies. Death, mourning, and scientific desire in the realm of human organ transfer*. New York: Columbia University Press.

Skegg, P.D.G. 1975. Human corpses, medical specimens and the law of property. *Anglo-American Law Review* 4: 412–424.

Stark, D. 2009. *The sense of dissonance*. Princeton/Oxford: Princeton University Press.

United Nations Educational, Scientific and Cultural Organization. 2005. *Universal draft declaration on bioethics and human rights*. Paris: UNESCO.

Van der Burg, W., and F.W.A. Brom. 2000. Legislation on ethical issues. Towards an interactive paradigm. *Ethical Theory and Moral Practice* 3(1): 57–75.

Wadmann, S., and K, Hoeyer. 2014. Beyond the 'therapeutic misconception'. Research, care and moral friction. *BioSocieties* 9(1): 3–23.

Zwart, H. 2014. The donor organ as an 'object *a*'. A Lacanian perspective on organ donation and transplantation medicine. *Medicine, Health Care and Philosophy* 17(4): 559–571.

# Chapter 11
# From Winged Lions to Frozen Embryos, Neomorts and Human-Animal Cybrids: The Functions of Law in the Symbolic Mediation of Biomedical Hybrids

**Britta van Beers**

> *"The trouble concerns the fact that the 'truths' of the modern scientific world view, though they can be demonstrated in mathematical formulas and proved technologically, will no longer lend themselves to normal expression in speech and thought. The moment these truths are spoken of conceptually and coherently, the resulting statements will be 'not perhaps as meaningless as a "triangular circle," but much more so than a "winged lion"' (Erwin Schrödinger). We do not yet know whether this situation is final. But it could be that we, who are earth-bound creatures and have begun to act as though we were dwellers of the universe, will forever be unable to understand, that is, to think and speak about the things which nevertheless we are able to do."*
>
> (Hannah Arendt, *The Human Condition*)

## 11.1 Introduction

In the opening pages of *The Human Condition* Hannah Arendt makes the intriguing observation that "the wish to escape the human condition" is becoming apparent as an underlying motive of certain scientific developments. She points out that space technology can be regarded as a "first step toward escape from men's imprisonment to the earth" (Arendt 1998: 1) and the automation of labor as the abandonment of one of the vital aspects of the human condition.

B. van Beers (✉)
Department of Legal Theory and Legal History, Vrije Universiteit Amsterdam,
De Boelelaan 1105, 1081 HV Amsterdam, the Netherlands
e-mail: b.c.van.beers@vu.nl



B. van Klink et al. (eds.), *Symbolic Legislation Theory and Developments in Biolaw*, Legisprudence Library 4, DOI 10.1007/978-3-319-33365-6_11

Interestingly, Arendt detects a similar form of what she calls "earth alienation" (Arendt 1998: 264) in scientific endeavors "toward making life also artificial". As examples of this tendency she mentions "the attempt to create life in the test tube", "the desire to mix frozen germ plasm from people of demonstrated ability under the microscope to produce superior human beings", and "the hope to extend man's life-span far beyond the 100-year limit" (Arendt 1998: 3).

As her influential book dates back to 1958, Arendt could not have foreseen that the scientific attempts, desires and hopes that she refers to would one day result in the spectacular achievements of biomedicine. At the time of her writing, biomedical knowledge was still in its infancy. Moreover, most of the terms that Arendt uses are not current anymore. Nevertheless, each of the three medical technologies that she mentions have in the meantime become serious fields of research. They are nowadays known as: in vitro fertilisation (IVF), germline genetic modification and life extension technologies.

Although IVF has been available for over 25 years, and although both germline genetic modification and life extension have not lived up to their promises yet, these technologies do not cease to give rise to urgent legal, ethical and political questions. For instance, IVF has set into motion the development of a vast range of technological possibilities for artificial reproduction, better known as assisted reproductive technologies (ARTs). ARTs are nowadays accompanied by a worldwide fertility industry, which to an increasing extent enables prospective parents to assemble and customise their offspring, even from behind their computer screens.[1]

As to germline genetic modification, in the United Kingdom the legal ban on mitochondrial replacement, also known as 'three parent in vitro fertilisation', has been lifted in early 2015. This technology aims to prevent the transmission of mitochondrial diseases to the baby by using the egg cell from a third party. This donated egg cell is enucleated and filled with the nucleus of the prospective mother's egg cell. The effect is that the resulting baby's mitochondrial DNA originates from the egg donor. Therefore, genetically, the baby has two mothers and one father. Because mitochondria replacement concerns genetic changes that are passed down to future generations, it constitutes a form of germline genetic modification. As such it is currently prohibited in most countries and by several international conventions in this field.[2]

A similar procedure is at the root of another controversial technology that equally led to heated debates in British politics. In this case enucleated egg cells of animals are filled with human DNA. The results are called human-animal cybrids. In 2008 the British Human Fertilisation and Embryology Authority (HFEA) granted the first licences to create these cybrids.

---

[1] In a previous article I have analysed the symbolic functions of law in light of global reproductive markets and their accompanying streams of reproductive tourism (see Van Beers 2015).

[2] See for instance Article 13 Convention on Human Rights and Biomedicine (Council of Europe): "An intervention seeking to modify the human genome may only be undertaken for preventive, diagnostic or therapeutic purposes and only if its aim is not to introduce any modification in the genome of any descendants."

Life extension science has also elicited strong public reactions. For example, in 2011 Aubrey de Grey, a well-known scientist in the field of regenerative medicine, made headlines by predicting that the first persons to become 150 years or older had already been born. Lastly, the more general attempt at "making life also artificial" can be recognised in the human enhancement movement, and more specifically in the rise of synthetic biology, both of which are currently hotly debated.

This artificialisation of human life has led to the creation of previously unthinkable 'hybrids' of human origin. The frozen embryos of IVF, the 'Google babies'[3] of global reproductive markets, the three parent babies of mitochondria replacement, the human-animal cybrids of stem cell research, the artificially grown human tissues of regenerative medicine and the brain-machine-interfaces of human enhancement: each of these technological creations blurs the hitherto static lines between persons one the one hand, and things, machines, animals and commodities on the other (also see Chaps. 9 and 12 in this volume). The materialisation of these human hybrids illustrates in various ways, what Habermas calls, the 'dedifferentiation, through biotechnology, of deep-rooted categorical distinctions which we have as yet, in the description we give of ourselves, assumed to be invariant' (Habermas 2003: 42). Similarly, both in politics and academia the struggle to integrate these novel biomedical entities in existing legal-ethical frameworks is becoming visible.

In the quotation preceeding this paper, Arendt expresses her concern that the creations of emerging sciences and technologies are so much beyond our lived experiences and current frames of reference, that we may be fundamentally unable to understand or discuss them, even if we are capable of producing them. If we would nevertheless try to put these biomedical realities into words, the resulting concepts are likely to be even more beyond our comprehension than 'winged lions', as Arendt quotes quantum theorist Erwin Schrödinger.

Such a lack of vocabulary has major consequences for the regulation of these developments, as Arendt also briefly points out. If we were indeed speechless upon our encounter with these *Fremdkörper* of the symbolic order, that would also mean that these hybrids escape from legal-political deliberation. The result of this legal-political vacuum would be an endless, unreflected and aimless "proliferation of hybrids", to use Latour's expression (Latour 1997: 7).

Now, almost 60 years after the first publication of *The Human Condition*, we can see that Arendt's predictions have fortunately not entirely come true. We turn out not to be completely speechless in our understanding of novel hybrids and have taken up the challenge to integrate the novel biomedical entities in our symbolic orders. As will be argued in this paper, the symbolic order of law has played a special role in this process. Nevertheless, this integration has proven to be a demanding process, as each of these revolutionary developments appear to be questioning the founding frameworks, categories and distinctions of our moral experience (Sandel 2007: 9; Dworkin 2000: 443–445).

[3] See Zippi Brand Frank's documentary with the same title on the phenomenon of worldwide reproductive markets, in which sperm, egg cells and surrogate mothers are all commercially available (*Google Baby*, see <http://www.imdb.com/title/tt1490675/>).

This chapter aims to throw light on several dimensions of law's normative symbolisation of biomedical hybrids. In a way, lawyers have become as inventive and prolific as biomedical scientists themselves in the invention of new and hybrid legal constructions. For instance, a common strategy in law to come to a classification of hybrid objects of human origin is to create equally hybrid legal regimes in which property and personal rights are blended in novel ways (see Chaps. 8, 10, and 13 in this volume). Nevertheless, the question remains how this process of legal symbolisation is to take place and is already taking place. How can and should law respond to the shifting and hybrid realities of biomedical practices? After all, these human hybrid creations equally surpass the foundational categories and distinctions of law. To answer these questions, this chapter offers an analysis of the changing understanding of the biogenetic facts of life in the light of recent biomedical developments. The main reason for this changing understanding in law is that legal regulation of biomedical technologies brings with it an infiltration of law into the biological aspects of life. In the next section, this expansion of the legal domain will be explained and examined against the background of philosopher of science Bruno Latour's 'politics of nature'.

## 11.2 Law as a Prime Symbolic Mediator of Biomedical Realities

Although Arendt expresses her doubts on the possibility to even speak about the novel creations of science, she also insists on its absolute necessity for political action in this field. As she writes, the question whether we wish to use our new scientific and technical knowledge in a particular direction is "a political question of the first order" (Arendt 1998: 3) that should not be left to scientists themselves. With this assertion, Arendt goes beyond the conventional division of labour between politics and science. If scientists are considered to be neutral and detached intermediaries who lay bare the cold facts of nature, as in the stereotypical idea of the scientific process, then politicians have indeed no business to meddle with their activities. Contrary to Arendt's suggestion, their task would then be confined to decisionmaking on questions of a strictly social nature.

This conventional 'separation of powers' between science and politics, if it ever was convincing, has lost its credibility in light of recent technological developments. Artificial satellites and test tube babies, to use two of the examples that Arendt mentions, can by no means be called the detached 'objects' or cold facts of a scientific discovery. Rather, they are the novel creations of engaged and impassioned 'engineers' of nature. Such a perspective of science opens up the possibility to detect the choices and decisions involved in the scientific process. Moreover, it allows one to see the political aspects of scientific procedures. One of its consequences is an expansion of the political domain.

This new division of labour between politics and science has become common knowledge within more recent disciplines such as science and technology studies (STS). In fact, the type of politics that Arendt alludes to could be called a 'politics of nature', after the title of one of the books by prominent STS scholar Bruno Latour (Latour 2004). What has received less attention in academic circles, is what this expansion of the political domain implies for the functioning of law. For instance, Latour does not provide us with much reflection on the role of law in this new division of labour in his classic work *Nous n'avons jamais été modernes*, nor in its follow-up *Politiques de la nature*, and not even in his etnographic account of his stay at the French Council of State (*La fabrique du droit*). Nevertheless, as will be argued hereafter, Latour's work does provide several clues to come to a more elaborate account of the role of law in a politics of nature. First, the outlines of this new function of law will be explored.

The Roman maxim *ubi societas, ibi ius* is still a widely acknowledged truth among lawyers. Without a doubt, wherever there is a society, there is also law. However, now that the scope of human action has reached the building blocks of life, especially in the field of biotechnology, it could be argued that the maxim *ubi natura, ibi ius* is becoming equally valid. In fact, in one of the first comprehensive studies of the relation between law and the biosciences this tendency is referred to as a "colonisation of the natural by the just" (Buchanan et al. 2000: 82). This infiltration of law in the previously inaccessible domain of nature, including human nature, brings with it a fundamentally new relation between law and biological reality.

Since biotechnology has rendered many biological aspects and boundaries of life fluid and malleable, the law can no longer reflect these as a natural given or a static reality. Instead, law is becoming an active mediator in the biotechnological creation of novel entities and in the determination of previously static boundaries. An example is the legal boundary between humans and animals. Whereas this boundary used to be tacitly assumed in, for example, the traditional legal qualification of animals as legal objects, or in legal standards for animal welfare, the boundary itself is now increasingly becoming the object of legal-political decisionmaking in biomedical issues. This boundary's transformation from unreflected legal premise to overt legal construction is illustrated by statutes which determine the legal conditions under which human-animal hybrids may be created, or by legal bans on the creation of chimeras. In this process the lines between man and animal are no longer merely factual, but also acquire a certain normativity.

The result is that the human-animal divide, that had already been exposed by the animal rights movement as a largely social-cultural construct, has become even less 'natural' and self-evident, also in law. In the next sections, the shifting legal boundaries between life and death and male and female will be studied in legal approaches to brain death and transsexuality. In all of these cases, the boundaries between the foundational categories of law are gradually losing their naturalist aura as a result of biomedical developments.

More importantly, these cases indicate that no matter where political communities will decide to draw the lines in issues such as the creation of human-animal hybrids, the use of human organs for donation or the use of human embryonic stem cells for research, they will inevitably rely on law's categories, distinctions and principles to phrase and express the final outcome of these deliberations.

It is interesting to note at this point that also Latour uses mainly legal terms and metaphors to describe the conditions under which hybrids are created in his "politics of nature". In this vein, he refers to the conventional dualistic metaphysical system, in which the separation of powers between science and politics is anchored, as "the Constitution" (Latour 1997: 26):

> De même que la constitution des juristes définit les droits et les devoirs des citoyens et de l'État, le fonctionnement de la justice et les passations de pouvoir, de même cette Consititution [...] définit les humains et les non-humains, leurs propriétés et leurs relations, leurs compétences et leurs groupements.

In fact, Latour's work can be read as a plea for a radical 'constitutional' reform, that is, a new, "experimental-metaphysical" framework (Latour 2004, 97) in which technological hybrids can be accommodated and represented without distortion. As Latour argues in his classic work *Nous n'avons jamais été modernes*, biotechnology and other recent technologies have served to expose how technosciences have always gone beyond the constitutional boundaries, by creating hybrids of nature and culture, science and politics, objects and subjects, things and persons and also facts and norms. Until recently, societies managed to accommodate these 'quasi-objects' within one of the sides of this great divide through a process of concealment that Latour calls 'purification'. However, the radical nature of recent technologies has caused the Consitution to collapse, revealing a hybrid network reality in which "les faits socialisés et les humains devenus monde naturels" can finally be truly recognised.

If political orders would take this collapse seriously, and accept the inherently political aspects of the scientific process, they would from then on subject the production of hybrids to a procedure of collective decisionmaking, thereby constituting a new 'separation of powers' (instead of the old division of labour between politics and science). According to the new Constitution we will then have only two categories of hybrids: those that have been created 'dans les formes' (according to the correct legal procedure), and those that have not (Latour 2004, 249):

> Au lieu de distinguer, comme l'exigeait la tradition, le fait et le droit, [le collectif] exige des faits qu'il deviennent légitimes; il distingue dorénavant les amalgames de faits et de droits mal formés des associations d'humain et de non-humains obtenus dans les formes.

Although Latour's use of legal metaphores can be regarded as merely a way of speaking, just like his idea of a "parliament of things" (Latour 1997, 194) is not meant to be taken literally, his anthropology of science is able to open up new perspectives on the role of law in the regulation of new technologies. Indeed, biomedical regulation shows that legal norms and natural facts have indeed become highly entangled. To use the earlier example, the legal provisions on the creation of human-animal hybrids and the British debates on human-animal cybrids, show that the boundary between humans and animals is overtly evolving into an imbroglio of biological facts and legal norms, a hybrid of *Sein* and *Sollen*.

## 11.3 The Inscription of Life into Law

So far I have argued that the infiltration of law in the previously inaccessible domain of nature brings with it a fundamentally new relation between law and biological reality. In fact, biomedical developments have revealed an increasing entanglement between legal norms and biological facts. These observations raise the vital question to which extent, if at all, and in which manner the biological aspects of life can be inscribed in the language of law.

In this section I distinguish between three understandings of the relation between law and the biological aspects of life, which can each be recognised in legal-political debates: a neonaturalistic (Sect. 11.3.2), artificialistic (Sect. 11.3.3) and legal-symbolic (Sect. 11.3.4) approach. According to the neonaturalistic stance, the most basic biological events life can be tacitly assumed and mirrored in law as basic, given facts of life. Moreover, for a correct legal understanding of biological reality, legal systems can rely on intervention by scientists. From an artificialistic perspective, however, the neonaturalistic approach is problematic, since law is at heart a human construction. As such, the legal realm is radically cut of from biological reality. The legal-symbolic position brings both approaches together through the claim that the law, through its artificial categories and constructions, offers multitudinous ways to symbolise the biological aspects of life.

To make these approaches, as well as their criticisms, more tangible, I will, throughout the rest of this chapter, draw from recent discussions on the legal-symbolic mediation of two biological boundaries: the boundary between life and death and the sex difference. Both of these 'natural' limits have become subject to renegotiation and thus reveal, albeit in different ways, law's changing role in these matters. Before I come to an analysis of the three perspectives on the relation between law and biological reality, I will first briefly introduce recent discussions on both of these boundaries.

### *11.3.1 The Renegotiation of the Boundaries of Death and the Sex Difference*

Since long, death has been determined by verifying whether the body has no pulse, is cold to the touch and without breath. However, as is commonly known, the introduction of postmortal organ donation necessitated the creation of new legal definitions of death. This can be illustrated by the practice of *controlled donation after circulatory death*. In this context, death is no longer approached as a purely biological event, waiting to be observed and determined by a medical professional (Shah and Miller 2010: 552–556). Instead, it becomes part of a complex medical process, meticulously orchestrated to serve the purposes of organ donation. According to its accompanying medical protocol, life-sustaining treatment or ventilator support will be withdrawn when further medical treatment is deemed to be pointless. After circulatory functions have consequently ceased, surgeons will have to wait a certain time to ensure that the patient's death is irreversible before starting organ retrieval. However, surgeons also need to procure the organs as soon as possible for the transplantation to be successful. Although medical protocols have solved this tension by requiring an interval between the cessation of circulatory functions and the procurement of organs, in practice physicians turn out to have some discretionary power. The most striking illustration is heart procurement from non-heart beating patients, seemingly a contradictio in terminis. In fact, "some medical interventions that doctors perform to preserve organs can either hasten death or even accidentally reanimate the donor", raising questions as to the irreversibility of circulatory death (Shah and Miller 2010: 552). This suggests that under certain circumstances the procurement process itself can be the event causing irreversibility.

A second, better-known example is *brain death*, which defines death as the irreversible cessation of all brain functions. Contrary to circulatory death, brain death facilitates 'heartbeating' donation. Without the concept of brain death, the removal of organs donation would amount to homicide. Moreover, this redefinition enables surgeons to remove organs already before clinical death has set in without violating their professional ethical guidelines. According to the so-called dead donor rule, which is the central guideline in the ethics of organ transplantation, organs can only be transplanted from persons who have been declared dead.

In this chapter I will mainly focus on brain death. Brain death has been controversial from the start. The most fundamental critique has been that this redefinition of death is not based on bare biological or scientific facts, but on choices inspired by certain hidden moral, political and cultural presumptions about what constitutes life. For philosopher Hans Jonas for instance, brain death "is a curious revenant of the old soul-body dualism" here appearing as "a new dualism of brain and body" (Jonas 1974, 139). More recent critics, however, choose a line of attack which points in a different direction, proposing to abandon the dead donor rule altogether, as we shall see.

In similar ways, the legal boundary between the sexes is currently questioned and revised in many legal systems. Ever since intersexuality[4] and transsexuality[5] appeared on the medical agenda, the traditional understanding of the sex difference has become contested, also on a legal level. Claims by transsexual and intersex

---

[4] Intersex persons are born with biological characteristics that do not fit squarely in the traditional understanding of male and female.

[5] Transsexual persons are born with clear traditionally male or female characteristics, but nevertheless strongly identify with the other sex. Often, though not always, this identification is accompanied by a longing for a sex reassignment.

persons have led to amendments of the traditional legal construction of gender. A clear indication is that in 2003 the European Court of Human Rights ruled that the right to private life entails the right of post-operative transsexuals to have their new sex identities recognised in official documents and birth registries.[6] Moreover, in a growing number of countries[7] a third gender has recently been introduced in passports and birth registration forms.

### *11.3.2 The Neonaturalistic Approach*

How are the biological boundaries of death and the sex difference conceived and reconstructed in law? According to more traditional approaches, the law merely reflects or mirrors these boundaries as given facts. However, it is clear that this position has become untenable now that biomedical technologies have rendered these biological facts and boundaries unstable. Nonetheless, it is also evident that the artificialisation of human life has not progressed to such an extent as to make the category of the natural or the biological completely redundant. The complexities of contemporary biomedical practice have only made the biological dimensions of life much more difficult to grasp. From that perspective, the main problem is a lack of expertise.

Therefore, a first possible way to come to a legal understanding of biomedical realities is to rely on the medical explanation of these realities by professionals involved in these practices. An example is the tendency to let the medical profession itself or so-called expert commissions decide on questions of a fundamentally political nature, such as the use of PGD or postmenopausal pregnancies.[8] Another example is the legal-dogmatic view that the embryo's increasing legal protection is a reflection of its biological growth as understood in medical discourse.

In this chapter I refer to this approach as neonaturalistic. The term is used to distinguish this approach from the naturalistic human rights tradition. Contrary to human rights discourse, the perspective that is discussed in this section approaches the category of the 'natural' as something which can be detected and explained by the natural sciences.

This neonaturalistic stance can also be detected in discussions on brain death and transsexuality. From this perspective, brain death can be understood as the legal recognition of an alternative biological event constituting the end of legal

[6] *Christine Goodwin v United Kingdom* [GC], no 28957/95 (Grand Chamber, 11 July 2002).

[7] For instance in Australia, Germany and New-Zealand. In Thailand the third gender may even receive recognition within the Constitution, as the Constitution Drafting Committee will include references to the third gender in its new draft of the Constitution (see http://edition.cnn.com/2015/01/16/world/third-gender-thailand/).

[8] For example, the Dutch government delegates the regulation of assisted reproduction to a large extent to the medical profession itself. For instance, rules on postmenopausal pregnancies can only be found in professional guidelines, established by the Dutch association of gynaecologists and obstetricians.

subjectivity: the cessation of all brain functions. Similarly, the legal creation of a third gender can be said to mirror a third category that actually exists in nature: intersexuality. Indeed, in cases of intersexuality, individuals may have biological characteristics of both sexes, or neither sexes.

Nevertheless, a closer look reveals how the neonaturalistic approach obscures the political decisionmaking involved in the legal-medical construction of death and sex. For instance, brain death involves a much more complex situation than merely ascertaining a certain biological state of affairs. The determination of brain death requires to a large extent also certain more utilitarian decisions from the surgeon. If determination of death is not considered to be expedient in certain contexts, or conflicts `with more somatic definitions, such as in the case of a pregnant brain dead patient (Shah and Miller 2010: 549), then the patient may not be declared dead. Even clearer is controlled circulatory death, in which death is facilitated by withdrawing medical treatment in such a way as to allow organ transplantation at an early stage (see previous section). Another example is the end-of-life practice of combining continuous deep sedation combined with withdrawal of treatment. This procedure may be characterised as a "complex *mise-en-scène*" resulting in a "mimicry or simulation" of natural death (Raus et al. 2012: 332). As a result, new definitions of death are not entirely a reflection of biological facts, but also to an increasing extent the product of certain decisions.

A similar line of reasoning applies to the neonaturalistic approach to the legal sex difference. Intersexuality may be an example of the existence of a third sex in nature. But this leaves the phenomenon of transsexuality undiscussed. It is clear that the phenomenon of transsexuality fundamentally differs from intersexuality: transsexuality is the result of an established longing to belong to the other sex, even if clear biological traits of one of the sexes are present. Accordingly, many of those who claim to belong to a third gender are not intersex persons, but do not feel at home in either sex for other reasons than the presence or absence of certain biological traits. Admittedly, it seems that also in cases of transsexuality certain biological or physical differences continue to function as the basis for the legal sex difference. For instance, in many legal systems one cannot be assigned a new legal sex without certain physical adjustments and operations such as sex reassignment surgery and sterilisation.[9]

Yet a closer look reveals that that the resulting physical characteristics are at heart a physical *mimicry* through plastic surgery of traditional, biological sexual traits. Under these circumstances, one's legal sex is no longer the reflection of a certain physical state of affairs. Rather, it is the other way around: the law requires physical changes in order to assign a new legal sex.

[9] For instance, according to the organisation Transgender Europe in 2014 21 countries in Europe still required by law that transgender people undergo sterilisation before their gender identity is recognised (see http://www.tgeu.org/sites/default/files/Trans_Map_Index_2014.pdf). Nevertheless, a trend towards legal reform in this area can be detected throughout Europe, with Denmark, Germany and the Netherlands recently abandoning surgery requirements (see Saner 2014).

More generally, the specific legal constructions and bureaucratic requirements constituting the legal recognition of brain death and sex changes make it impossible to see the legal boundary between life and death and male and female as merely mirroring 'given' biological facts. Instead, biomedical science has rendered these biological boundaries manipulable and fluid. In short, as the discussions on brain death and transsexuality clearly demonstrate, in the age of biomedical technology, biological boundaries are increasingly the creation of legal-political decisionmaking.

### 11.3.3 The Artificialistic Approach

In the previous subsection I have argued that attempts to come to a legal understanding of biomedical realities by getting back to the underlying complex of biological facts, is doomed to fail. One of the major problems with this approach is that the law is at heart a human construction, an artefact of the human imagination, even when the law deals with the biological aspects of life. After all, the legal system relies on its own representations of the material world.

For example, as is often taught to first year students, the legal understanding of persons, animals, time, place and truth should not be confused with the lay person's understanding of these concepts. In the artificial world of law, corporations are persons, animals are legal objects, subjects are not bound by time and place (e.g., the concept of legal representation) and legal truth can be attained by openly lying (legal fictions). Hart refers to this gap between legal and non-legal terms as "the great anomaly of legal language – our inability to define its crucial words in terms of ordinary factual counterparts" (Hart 1983: 25).

Contrary to the neonaturalistic position, an artificialistic approach takes law's autonomous, distinctive and sometimes even fictional vision of reality as its starting point. In essence, artificialism can be described as a nominalistic position on the function and significance of legal language. The artificialistic lawyer believes, in the manner of Humpty Dumpty[10] in *Through the Looking Glass*, that when the law uses a word, it means just what the law chooses it to mean – neither more nor less. This boundless ability of law to create and bend its own concepts is an illustration of "the separateness and completeness of [...] the legal plane" (Lawson 1957: 913).

The artificialistic position also pervades discussions on transsexuality and brain death. From the artificialistic perspective, legal references to the sex difference and the boundary between life and death are either aberrations of law that should be relinquished as soon as possible or legal fictions that should be recognised for their artificiality. For instance, several legal and ethical scholars have proposed to view the recent definitions of death as legal fictions (Shah and Miller 2010). They argue that these definitions of death are so far removed from our ordinary, biological understanding of death, that it is better to be open about the artificiality of this legal construction, by calling it what it is: a fiction of biolaw.

---

[10] "When I use a word," Humpty Dumpty said, in rather a scornful tone, "it means just what I choose it to mean – neither more nor less."

Some take it one step further and propose to abandon the criterion of death for organ donation altogether. In their view, the dead donor rule has become a relic of the past which only serves to obscure the complex *mise-en-scene* which the contemporary practice of organ transplantation, in the quest to preserve as many potential transplant organs as possible, has actually become. Moreover, to them "it is not obvious why certain living patients, such as those who are near death but on life support, should not be allowed to donate their organs, if doing so would benefit others and be consistent with their own interests" (Truog et al. 2013: 1287).

In similar vein, some scholars argue that the sex difference has had its best time in law. They wonder why the law should hold on to certain natural traits of law's person, when it is clear that personality in law is nothing but a construction. As Naffine argues (Naffine 2004, 632): "Although the prevailing legal view is that nature is there to take care of sex difference, law nevertheless enforces sexual nature, from the moment of our birth, sometimes with an insistence that seems to border on cruelty." Interestingly, these artificialistic ideals on legal sexing now seem to be on the verge of becoming a legal reality. In family law, the sex difference is gradually disappearing, with for instance same-sex marriage becoming an option, and with the legal distinction between fatherhood and motherhood slowly dissolving as a consequence of the rise of LGBT parenting rights.

All of these scholars presuppose in artificialistic vein that the law has no function in the representation of these fluid, biological boundaries. Instead, according to them, law's inherently artificial nature corresponds perfectly with the increasing artificialisation of the biological aspects of life. Corresponding with the seemingly boundless possibilities of medical biotechnology, the artificial world of law is limited only by the limits of lawyers' imagination. French legal historian Yan Thomas argues that law should therefore be regarded as a "technique de dénaturalisation" (Thomas 1998a: 106) or an art, as in the Roman expression *ars aequi et boni*, exactly because of the inherent artificiality of its categories and concepts (Thomas 1998a, 104–105).

In this radicalised vision of the gap between *Sein* and *Sollen*, law's categories and terms solely serve a legal-technical function. They are, in Thomas' words, "seulement des lieux ou se projettent des normes" (Thomas 1998b: 14). In other words, legal concepts are like empty moulds that can be filled with anything that the legal order deems useful at that particular moment in time. They are stripped bare of any connotations to the outside world, despite the seeming familiarity of the terms used. Similarly, Hohfeld emphasises "the importance of differentiating purely legal relations from the physical and mental facts that call such relations into being" (Hohfeld 1923: 27).

Nonetheless, as was argued in section 11.2, biomedical developments have brought about a certain colonisation of the natural by the just. It seems indeed very 'artificial', not to say contrived, to maintain that law is cut off from biological reality, at a moment in time in which law and biology have never been more intertwined. More importantly, many debates on bioethical matters revolve around the question which limits should be posed to the artificialisation of the human condition and at what point biomedical developments may start to compromise human dignity. An artificialistic position would disqualify these core concerns beforehand.

It is clear that in the era of biomedical technology lawmakers and judges have taken up the task to come to an understanding of the biological aspects of life, as evidenced by legislation and conventions on issues such as the legal status of embryos, patents on life or organ donation. In the regulation of biomedical hybrids, lawyers are creating hybrid constructions in which norms and facts, law and nature, are blended, in ways which remind us of Latour's politics of nature.

### *11.3.4 The Legal-Symbolic Approach*

Although certain neonaturalistic and artificialistic tendencies can indeed be recognised in law, both approaches show important shortcomings when coming to an understanding of biolaw's involvement with the biological aspects of life. Within a neonaturalistic approach the legal-political dimensions of biomedical practices are concealed. The other way around, within an artificialistic approach the gradual blending between the legal and the biological cannot be conceived because of "the separateness and completeness of the legal plane". One of the central arguments in this chapter is that a third way is offered by a symbolic analysis of biomedical law.

As explained earlier, biomedical hybrids seem to defy existing categories and distinctions, such as the distinctions between persons and things, the human and the mechanical, male and female and life and death. From this perspective, many of the anxieties and controversies surrounding biotechnology can be traced back to the difficulties to represent biomedical hybrids through the central distinctions from the symbolic order (Chap. 9 in this volume; Smits 2006). To be able to come to grips with biomedical developments on a social-cultural level, and thus come to a cultural "domestication of new technologies" (Smits 2006), reflection on the symbolic order and its categorical distinctions is necessary.

According to a legal-symbolic approach, legal frameworks also contribute to the social-cultural process of symbolisation of biomedical hybrids. To what extent, for instance can human biological materials be regarded as objects? In case of so-called 'three parent babies', what are the ties between the second mother and the child? To what extent can human egg cells be treated as tradable commodities? The law offers, to an increasing extent, answers to these semi-metaphysical questions.

Qualifying the legal system as a symbolic order allows one to detect the ways in which the law brings about a certain symbolic mediation of the biological facts and boundaries of life. In this manner, the legal-symbolic approach brings the naturalistic and artificialistic approach together. From neonaturalism the legal-symbolic position borrows the ambition to come to a fuller understanding of the mutual involvement of the legal and the biological. From legal artificialism, the legal-symbolic position adopts the view that "law is part of a distinctive manner of imagining the real" (Geertz 1983: 184).

Interestingly, the language of law is of increasing importance in the collective symbolisation of novel biotechnological entities. Without a doubt, other systems of meaning profoundly affect the legal process of symbolisation, such as ethical, medical, religious or economic discourse. Nevertheless, when the symbolic orders of these systems collide, as is mostly the case in bioethical matters, law has to balance and mediate between these contesting systems of value and meaning. As a result, law has acquired a certain autonomy in the process of symbolisation.

French legal scholar Marcela Iacub captures both the autonomy of law in the regulation of biomedical developments and the resulting confusion well, when she states that in the age of biotechnology, only the law seems to be able tell us what a human body or the beginning of life is (Iacub 2002: 17):

> Ce qu'est un 'corps' aujourd'hui, étant donné les nouvelles conditions techniques qui sont les nôtres, seul le droit peut nous le dire. Dans se sens, la question de la juridicisation des décisions portant sur la vie (la naissance, la santé, etc.) prend une tout autre teneur. Le droit est devenu, dans ces domaines, grâce à ses procédures de construction du monde et donc à sa place dans la production de significations communes, la seule convention ayant force suffisante pour s'imposer à tous, convention parée des attributs et bénéficiant du pouvoir de la violence légitime.

An illustration of the special symbolic role of law in the construction of an *imaginaire social* for biomedical entities, is the legal symbolisation of the human embryo. As sociologist Luc Boltanski writes in his work *La condition foetale*, the human foetus until recently suffered from a certain *underrepresentation* ('sous-représentation') in the symbolic order. This absence seems quite remarkable. After all, as Boltanski explains, "in the field of social relations […] not only actual human beings are present, but, in certain cirumstances, also the dead, animals and plants, supernatural beings, future human beings, etc." (Boltanski 2004. 37). An illustration is the general absence of rituals as a last respect to foetuses in cases of miscarriage or abortion. According to Boltanski, this tradional "quasi-absence" of the foetus is mainly due to the seemingly irreconcilable moral imperatives surrounding abortion: on the one hand humans are believed to be unique and irreplaceable individuals, on the other hand, abortion shows that humans in an early stage can be disposable after all.

Biomedical technology has replaced this situation of secrecy and taboos with a more utilitarian approach to the creation and destruction of human life. For example, technologies such as prenatal diagnosis and IVF have brought about, what could be called, an 'exteriorisation' of the pregnancy. The visualisation of embryos through ultrasounds and the possibilities to preserve embryos outside the female body, have made the embryo more than ever part of our lifeworld and even part of our social interactions.[11] This made it necessary to develop a vocabulary for our interactions with this 'new' entity. Since the status and symbolisation of the human embryo has been fraught with ethical, religious and medical controversies from the

[11] Habermas refers to this process as an 'anticipatory socialisation of unborn life'. As he writes, the parents do not only talk about the child that is growing in utero, in a certain way they also already communicate with it (Habermas 2003, 35).

beginning of these developments, this vocabulary has, to a large extent been developed in law. In that process, lawmakers have not only adopted existing representations of the embryo, but have also called into existence new and multiple categories and distinctions. Current legal systems distinguish between embryos in vitro that are left over, embryos in vitro that are destined to be implanted, embryos that would be able to survive outside the womb, embryos in vitro that are past the 14 day limit, etc.

Interestingly, the frames, categories and constructions that law creates to provide these biomedical entities with a status, also seem to affect our own perception of reality and our basic frames of reference. Under these circumstances, legal language can be said to also have a constitutive instead of a merely interpretive function (Glendon 1987: 9). The question is what this constitutive function, which points to a certain autonomy of law in processes of symbolisation, exactly entails. In the next section I analyse the role of law in the collective efforts to integrate new hybrids into the symbolic order in more detail.

## 11.4 Three Symbolic Functions of Law in Biomedical Regulation

How does the law contribute to the symbolisation of new biomedical hybrids? To answer that question, the notion of the symbolic needs further explanation. What is clear from the outset is that a characterisation of the law as a symbolic order goes beyond the following, still common view of law: that the law is solely about effectuating direct and visible changes in the behaviour of its subjects. From a symbolic perspective, the law also appears as a source of narratives, vocabularies and frameworks to make sense of and give meaning to the world around us. In other words, law is part of the "meaningful stories" necessary for speech and action (Arendt 1998: 236). As such, a symbolic approach to law is premised on a broad understanding of the notion of obligation, one that surpasses the level of mere commands. As Paul Ricoeur poignantly phrases in his *Reflections on the Just* (Ricoeur 2007: 84):

> [...] the adjective 'symbolic' [encompasses] within a single emblematic notion the different ways in which language can give figure to obligation: as an imperative, to be sure, an injunction, but also as counsel, advice, shared customs, founding narratives, the edifying lives of heroes of the moral life, the praise of moral sentiments, including – besides respect – admiration, veneration, guilt, shame, pity, solicitude, compassion, and so on.

Indeed, the law's imaginative use of language and its elaborate systems of multiple and inter-related terms, qualifications, constructions and categories, seems to underline the view that the law is perhaps "the most prodigious laboratory of the collective imaginary" (Edelman 2007: 132).

To gain more insight into law's symbolic dimensions, I will highlight and analyse three aspects of the symbolic dimension of law that are relevant to the process of legal-symbolic representation of biomedical hybrids: law's expressive, communicative, and anthropological function.

### 11.4.1 Law's Expressive Function

The legal approaches to death and the sex difference offer striking illustrations of the extent to which the law is indeed about more than effectuating clear and visible changes in the behaviour of its legal subjects. After all, brain death and transsexuality reveal other functionalities of law: the law *assigns* a sex and *redefines* what constitutes death. In both cases, the law confers a certain status to its subjects: the legal subject is either dead or alive, male or female.

In its designation and attribution of statuses, the law exercises an expressive function, that is, "the function of law in 'making statements' as opposed to controlling behavior directly" (Sunstein 1995: 2024). More generally, lawmakers rely on this expressive function to state the main aspirations of the legal order, to formulate which core values or legal principles are at stake in certain developments or to articulate the status of certain entities.

It could be said that biomedical developments have revitalised to a certain extent law's expressive function. Illustrations of the expressive function of biolaw can be found in the recitals and articles of biomedical conventions, declarations, and directives, which appeal to constitutive values such as to human dignity (Lembcke 2013) or "respect for the human being both as an individual and as a member of the human species".[12] Additionally, these international documents mobilise certain representations of human biological life, such as the symbolisation of the human genome as "the heritage of mankind".[13] More generally, human dignity, as one of the central values of biolaw, can be understood as a legal-symbolic representation of what it means to be human (van Beers et al. 2014: 9–15; also see Chaps. 9 and 12 in this volume).

The main reason why law's expressive function resurges in regulation of biomedical developments seems to be that, as Michael Sandel writes, these technologies raise "questions largely lost from view in the modern world – questions about the moral status of nature, and about the proper stance of human beings toward the given world" (Sandel 2007: 9). Indeed, where "the movement of the progressive societies has hitherto been a movement from Status to Contract" (Maine 1908: 151), biomedical developments seem to have caused a movement in the other direction: they have put status questions back on the legal agenda. Sandel admits that these questions are challenging, and barely able to be answered within the vocabularies currently available to us. Nevertheless, according to him "our new powers of biotechnology" have made them unavoidable (Sandel 2007: 10).

Numerous examples can be offered of the status questions with which medical biotechnology confronts us. Should human genetic sequences be regarded as objects of patent law (see Chap. 13 in this volume)? Do embryos in frozen embryo disputes qualify as objects of property law? Does compensation of 1000 euro for donation of egg cells turn the latter into a commodity? Do commercial surrogacy contracts

---

[12] Convention on Human Rights and Biomedicine (Council of Europe), recital.

[13] Article 1 Universal Declaration on the Human Genome and Human Rights (UNESCO).

commodify both the surrogate and the commissioned child? What is the legal relation of egg cell and sperm donors to their genetic children? And what is the legal status of a child conceived abroad through donated gametes and commercial surrogacy, both concerning the child's relation toward its parents and the child's citizen rights? Therefore, it could be said that biomedical law "exposes the uncomfortable but inescapable place of status distinctions in even the most progressive legal systems" (Fagundes 2001: 1766).

### *11.4.2 Law's Communicative Function*

Law's categories and constructions often also set the stage for discussions on biomedical inventions. Moreover, they may even affect the general frames of thought that are required to come to an understanding of novel biomedical hybrids. Such effects of the law exemplify law's communicative function. As Van Klink writes, the communicative role of law refers to the ways in which legal concepts influence 'the vocabulary by which people order their world, give meaning and attach value to it', (Van Klink 2005: 127) as well as the way in which they structure public debates (Van der Burg 2001: 47–52). In his chapter to this volume, Van Klink also refers to this function as an *epistemic* one.

The communicative function of law can also be recognised in discussions on brain death and transsexuality. Firstly, although intervention by specialists and bureaucrats is legally required to determine brain death in case of organ donation or gender identity in case of gender dysforia, both concepts have managed to infiltrate common frames of reference. Notwithstanding the continuing influence of more traditional accounts of death and gender,[14] the new concepts seem to have been generally accepted as possible exceptions to the conventional boundaries between life and death and male and female.

Moreover, both concepts have already informed and will inevitably further shape public debates that touch upon the definition of death or the sexes. For instance, new concepts of death may influence debates on the possibility of organ donation after euthanasia,[15] or on reforms of the national system of organ donation. Similarly, the recognition of transsexuality and the blurring of the categories of male and female can potentially also influence discussions on possible revisions of the legal arrangements surrounding kinship and marriage, such as the possibility to legally descend from two women[16] or same-sex marriage.

[14] For example, rejection of the concept of brain death is regarded as one of the three main arguments from Dutch family members of the deceased against organ donation (see Den Hartogh 2003: 60).

[15] In Belgium and the Netherlands, the combination of euthanasia and postmortal organ donation is currently emerging.

[16] For instance, in the Netherlands it has recently become possible to legally descend from two women (see Pessers 2013).

Lastly, when confronted with the warm, pulsating and breathing bodies that are used for organ transplantation, many people will struggle to find the right words to describe the status of these human entities. Linguistic difficulties of a different nature occur as a result of inter- and transsexuality.[17] Should we call a transwoman who has been able to father a child the father or a mother to the child? Or, how should we refer to someone who feels neither male or female, and has physical characteristics of both sexes? In such cases, when traditional concepts and categories in common language seem to fall short, the communicative function of law can be of special importance in the development of a vocabulary to make sense of biomedical developments. More importantly, because those involved may ultimately bring such cases before both the legislature and the judiciary, the law is often *de facto* called upon to provide an answer to these semi-metaphysical questions, even if the law is perhaps not the most likely or best equipped candidate to do so. The decisions from the European Court of Human Rights on status questions related to transsexuality,[18] involuntary abortion,[19] IVF[20] and organ donation[21] may serve as clear examples of this tendency.

Lawyers may of course try to evade the metaphysical question altogether, and confine themselves to the case at hand. However, in many legal cases a line will necessarily have to be drawn somewhere, which will often also imply a form of symbolisation. Therefore, a more apt strategy seems to be to deal with the symbolic aspects hands on. For these purposes, lawyers may for instance stretch the meaning of existing definitions (the transwoman who fathered a child may be called a mother) or may come up with new categories (the sexless person can register as a person of the third gender).

Whichever approach is chosen, the resulting legal redefinition or legal construction is also likely to influence the public's understanding. Of course, the attention that brain death and transsexuality have received in popular media have equally ensured that virtually nobody in society remains ignorant of these phenomena. Nevertheless, official legal recognition will serve as an accelerator in this process. For instance, recognition of a third gender in passports and birth registration will probably further the acceptance in society of a grey zone between male and female. However, as the continuing resistance against the concept of brain death among certain parts of the population illustrates,[22] law is not omnipotent. As a result, the symbolic power of legal language remains of necessity limited.

---

[17] As Naffine correctly writes about the legal sex difference: "There is no third term, apart from 'it', but this would drive the person into property which is why men and women are never described this way. 'It' debases" (Naffine 2004: 637).

[18] *Christine Goodwin v United Kingdom* [GC], no 28957/95 (Grand Chamber, 11 July 2002).

[19] *Vo v. France* [GC], no 53924/00 (Grand Chamber, 8 July 2004).

[20] *Evans v. United Kingdom* [GC], no 6339/05 (Grand Chamber, 10 April 2007); *S.H. v. Austria* no 57813/00 [GC] (Grand Chamber, 3 November 2011).

[21] *Elberte v. Latvia*, no 61243/08 (Fourth Section, 13 January 2015).

[22] See Den Hartogh (2003): 60.

### 11.4.3 Law's Anthropological Function

The relation between the legal and the biological aspects of life is at the heart of the third symbolic function of law in the symbolisation of biomedical hybrids: law's anthropological function, as it has been called by a group of French legal scholars (Supiot 2007; Labrusse-Riou 2007; Edelman 1999).

This function concerns law's role in the symbolic mediation of the bare, biological facts of life (*zoè*) into the meaningful events of the lives that we lead (*bios*).[23] The most basic biological aspects of our lives, such as birth, death, reproduction, sexuality, kinship and, more generally, our corporality, have since the first civilisations been surrounded by symbolic values and rituals. The fact that we are barely able to spreak about a corpse as an object of property, that we consider the human body to be more than live meat and that we generally refuse to reduce parenthood to a genetic relation (not merely a *genitor* but a *parens*, see Pessers 2013): all of these facts illustrate the high extent to which these biological events are subject to symbolic mediation.

With Ernst Cassirer it can be said that we, as members of the species *animal symbolicum*, cannot interact with the physical world except by symbolising it (Cassirer 1972: 26). From this perspective, we also live in two worlds: the biological and the symbolical universe. In this vein, Boltanski speaks of a double birth of the child: *l'engendrement par la chair* (through the flesh) and *par la parole* (through speech). The transformation of physical facts into meaningful elements of the human experience, which enables our birth in the symbolic universe, also constitutes our identities as persons (see Chap. 12 in this volume). The primary example is the way in which each child upon birth receives a family name, a gender, and a place in the larger family story. More generally, the child's symbolic birth is illustrated by its introduction in the world of language.

This anthropological perspective is able to reveal how the law equally assists in this second birth. As French legal scholar Alain Supiot writes in his book *Homo Juridicus*, which is entirely dedicated to the study of the anthropological function of law (Supiot 2007, 37):

> A legal system does not fulfil its anthropological function unless it guarantees that every newcomer on this earth finds a world that pre-exists them and guarantees their identity over time, while also providing them the opportunity to transform this world and leave their personal mark on it.

The anthropological function of law is especially visible witin family law and the law of persons. For instance, the law requires that everyone upon birth receives a family name and gender. Other clear examples are the legal arrangements surrounding the institutions of marriage, kinship and parenthood. These examples from family law illustrate each in their own way how the law contributes to the cultural-symbolic mediation of the sex difference.

---

[23] See for further reflection on the distinction between life as *zoè* and life as *bios*: Arendt (1998: 96–97); Agamben (1998: 9–14).

Similarly, the biological fact of death is surrounded by symbolic values and rituals, also on a legal level. Although biological death generally coincides with legal death (the legal subject can no longer be a party to contracts or inherit etc.), the right to privacy, physical integrity and respect for human dignity have postmortal effects, which result in a special legal status special for the corpse.[24] An illustration is the fact that corpse and grave desacration are criminal offenses in most legal systems.

From this perspective, the law offers a larger anthropological and institutional narrative which permeates, constitutes and connects the stories we tell about ourselves and each other. In that sense, the legal order, through the exercise of its anthropological function, leaves certain marks on the legal subject.

Notwithstanding its importance, appeals to the anthropological function of law are often controversial. For example, according to Italian philosopher Giorgio Agamben, brain death is a legal construction constituting the human body as a form of bare life over which sovereign power can be exercised (Agamben 1998: 94). More specifically, to him brain death illustrates how the constitution of bare life marks "a threshold in which law constantly passes over into fact and fact into law, and in which the two planes become indistinguishable" (Agamben 1998: 97). He argues that under such circumstances, the law seems to become a pure extension of supposed scientific facts and is thus stripped of its fundamental symbolic, mediating function. Likewise, recognition of transsexuality is sometimes conceived as a threat to existing legal arrangements based on the sex difference, such as marriage and kinship, and is therefore said to be at odds with law's anthropological function (Salas 1994: 128).

In both examples, the symbolic function of law is used as an argument against new concepts of death and the sex difference. Nevertheless, the anthropological of law is not necessarily oriented towards preserving the existing symbolic order. After all, when confronted with phenomena such as organ donation and transsexuality, the law does not have to hold on to existing anthropological arrangements. Instead, it can come to a renewed symbolic mediation of the boundaries of death and the sex difference. Just as same-sex marriage, for example, can be interpreted as a resymbolisation of the institution of marriage rather than its abolition, recognition of transsexuality does not have to lead to an abolition of the legal sex difference altogether.

Similarly, interpreting the recognition of brain death entirely as a victory of utilitarian values over symbolic ones, is too simplistic. Agamben's reading of the practice of organ donation misses important aspects of the special legal-symbolic status which the law provides to the brain dead patient. To Agamben, the *neomort*, as he refers to the braindead patient,[25] is a biomedical version of the ancient Roman figure *homo sacer*: the outlaw who despite his or her exclusion is still included within the legal order as naked life. However, in the neomort traces of the legal subject can still be found. For instance, a form of informed consent, whether presumed or explicit, is necessary for postmortal organ donation in virtually all legal systems.

---

[24] See most recently *Elberte v. Latvia*, no 61243/08 (Fourth Section, 13 January 2015).

[25] With a term borrowed from Willard Gaylin (see Gaylin 1974).

Therefore, both examples only reveal the contested nature of the specific interpretation that is given to law's anthropological function. Most of those who argue for recognition of brain death or transsexuality, do not contest the function itself (they do not call for the abolition of the dead donor rule or birth registries), but aim for amendments in its elaboration. Nevertheless, one of the risks of the anthropological function of law is that those who are in charge of its execution and elaboration, ignore the fact that legal-anthropologial narratives and categories may change over time as a consequence of certain social, cultural or technological developments. If this is ignored, then the anthropological narrative is naturalised and reified.

## 11.5 Conclusion: The Fragility of the Legal-Symbolic Order

The central argument in this chapter is that law is becoming a prime mediator in the symbolisation process of novel biomedical hybrids. From that perspective, an analysis of the symbolic functions of law is indispensable to come to a proper understanding of the relation between law and biological reality in the age of medical biotechnology.

Nevertheless, understanding the legal system as a symbolic order also implies recognising a certain fragility of that system. According to Ricoeur, this fragility follows from:

> the difficulty that everyone has in inscribing his or her action and behavior into a symbolic order, and in the impossibility a number of our contemporaries have in comprehending the meaning and necessity of this inscription, principally those whom our sociopolitical order excludes (Ricoeur 2007: 85).

The primary source of this fragility can already be recognised at an etymological level. The earliest meaning of *symbolon* ('thrown together'[26]) illustrates how the symbolic order of law is something which is necessarily shared (Ricoeur 2007: 88). From a symbolic approach, the law resembles a shared language. However, the necessarily collective nature of the symbolic order of law also means that the process of legal symbolisation is contested from the start. After all, who decides how novel biomedical hybrids should be represented in the language of law (see Chap. 7 in this volume)?

Nevertheless, as traditional concepts and categories in common language seem to fall short when coming to an understanding of novel biomedical hybrids, we have come to rely on law's symbolic order in the development of a new framework and vocabulary. As a result, the symbolic functions of law in these matters have been able to become of great importance.

This brings me back to the opening pages of Arendt's *The Human Condition.* Why did Arendt commence her fundamental study on the human condition with

---

[26] *Syn-* is for 'together' and *bol* is derived from *ballein* which is 'to throw'.

reflections on certain technological developments? How do these technologies relate to the human?

One possible answer is that the human is directly involved in some of these technologies. After all, in technologies that aim to make human life artificial, the human is mechanised, objectified, modified, sequenced etc. However, this answer cannot satisfy as it defines the human in biological terms, as a set of natural properties. Such a 'material' approach to man is at odds with Arendt's philosophy of "the common world", the sphere of speech and action.

Alternatively, the connection could be human dignity. Arendt would then follow along the lines of contemporary international conventions that warn that the misuse of biology and medicine may lead to acts endangering human dignity. However, this approach would conflict with Arendt's refusal to define human nature and her resistance against essentialist approaches.

The most probable answer can be found by focusing on Arendt's notion of modern day 'earth alienation' in scientific contexts (Arendt 1998: 264). To her, contemporary scientists, in their quest for an Archimedean point, the search for a view from nowhere so to speak, "seem to move in a world where speech has lost its power" (Arendt 1998: 4). Without the means to symbolise these technological creations, they remain alien to the human world of speech and action. It is in that sense, that the symbolic functions of biolaw are also the humanising functions of biolaw.

## References

Agamben, Giorgio. 1998. *Homo Sacer. Sovereign power and bare life*. Stanford: Stanford University Press.

Arendt, Hannah. 1998. *The human condition*. Chicago: University of Chicago Press.

Beers, Britta van, Luigi Corrias, and Wouter Werner (eds.). 2014. *Humanity across international law and biolaw*. Cambridge: Cambridge University Press.

Beers, Britta van. 2015. Is Europe 'giving in to baby markets'? Reproductive tourism in Europe and the gradual erosion of existing legal limits to reproductive markets. *Medical Law Review* 2015: 103–134.

Boltanski, Luc. 2004. *La Condition Foetale. Une Sociologie de l'Engendrement et de l'Avortement*. Paris: Gallimard.

Buchanan, A., D. Brock, N. Daniels, and D. Wikler. 2000. *From chance to choice. Genetics and justice*. Cambridge: Cambridge University Press.

Burg, Wibren van der. 2001. The expressive and communicative functions of law, especially with regard to moral issues. *Law and Philosophy* 20: 47–52.

Cassirer, Ernst. 1972. *An essay on man. An introduction to a philosophy of culture*. New Haven: Yale University Press.

Dworkin, Ronald. 2000. Playing God. Genes, clones, and luck. In: *Sovereign virtue. The theory and practice of equality*, ed. R. Dworkin, 427–452. Cambridge/London: Harvard University Press.

Edelman, Bernard. 1999. *La Personne en Danger*. Paris: Presses Universitaires de France.

Edelman, Bernard. 2007. *Quand les juristes inventent le réel. La fabulation juridique*. Paris: Hermann Éditeurs.

Fagundes, D. 2001. What we talk about when we talk about persons. The language of a legal fiction. *Harvard Law Review* 114: 1745.

Gaylin, W. 1974. Harvesting the dead. *Harper's*, September 23..

Geertz, Clifford. 1983. *Local knowledge. Further essays in interpretive anthropology*. New York: Basic Books.
Glendon, Mary Ann. 1987. *Abortion and divorce in western law. American failures, European challenges*. Cambridge: Harvard University Press.
Habermas, Jürgen. 2003. *The future of human nature*. Cambridge: Polity Press.
Hart, H.L.A. 1983. *Essays in jurisprudence and philosophy*. Oxford: Clarendon.
den Hartogh, G. 2003. *Gift of bijdrage? Over morele aspecten van orgaandonatie*. The Hague: Rathenau Instituut.
Hohfeld, W.N. 1923. *Fundamental legal conceptions applied in legal reasoning and other legal essays*. New Haven: Yale University Press.
Iacub, Marcela. 2002. *Le crime était presque sexuel*. Paris: Flammarion.
Jonas, Hans. 1974. Against the stream. Comments on the definition and redefinition of death. In *Philosophical essays. From ancient creed to technological man*, ed. H. Jonas, 132–140. Englewood Cliffs: Prentice-Hall.
Klink, Bart va. 2005. An effective-historical view on the symbolic working of law. In *Social and symbolic effects of legislation under the rule of law*, ed. N. Zeegers, W. Witteveen, and B. van Klink. Lewiston: Edwin Mellen Press.
Labrusse-Riou, C. 2007. *Écrits de Bioéthique*. Paris: Presses Universitaires de France/Quadrige.
Latour, Bruno. 1997. *Nous n'avons jamais été modernes. Essai d'anthropologie symétrique*. Paris: La Découverte.
Latour, Bruno. 2004. *Politiques de la nature*. Paris: La Découverte.
Lawson, F.H. 1957. The creative use of legal concepts. *New York University Law Review* 32: 909.
Lembcke, Oliver W. 2013. Human dignity – A constituent and constituted legal principle. Some perspectives of a German discourse. *Archiv für Rechts- und Sozialphilosophie*, Beiheft 129 (Human Dignity as the Foundation of Law), 137:207–231.
Maine, H.S. 1908. *Ancient law. Its connection with the early history of society and its relation to modern ideas*. London: John Murray.
Naffine, Ngaire. 2004. Our legal lives as men, women and persons. *Legal Studies* 21: 621.
Pessers, Dorien. 2013. De Terugkeer van de Bastaard. *Nederlands Juristenblad* 88: 2189.
Ricoeur, Paul. 2007. *Reflections on the just*. Chicago: University of Chicago Press.
Raus, K., S. Sterckx, and F. Mortier. 2012. Continuous deep sedation at the end of life and the natural death hypothesis. *Bioethics* 26(6): 329–336.
Salas, D. 1994. *Sujet de chair et sujet de droit. La justice face au transsexualisme*. Paris: PUF.
Sandel, Michael. 2007. *The case against perfection. Ethics in the age of genetic engineering*. Cambridge: Belknap/Harvard University Press.
Saner, Emine. 2014. Europe's terrible trans rights record. Will Denmark's new law spark change? *The Guardian 1 September*. http://www.theguardian.com/society/shortcuts/2014/sep/01/europe-terrible-trans-rights-record-denmark-new-law. Accessed 24 Feb 2015.
Shah, S.K. and F.D. Miller. 2010. Can we handle the truth? Legal fictions in the determination of death. *American Journal of Law & Medicine* 36: 540–585.
Smits, M. 2006. Taming Monsters. The cultural domestication of new technologies. *Technology in Society* 8: 489–504.
Sunstein, C.R. 1995. On the expressive function of law. *University of Pennsylvania Law Review* 144: 2024.
Supiot, Alain. 2007. *Homo Juridicus. On the anthropological function of law*. New York: Verso.
Thomas, Y. 1998a. Le sujet de droit, la personne et la nature. Sur la critique contemporaine du sujet de droit. *Le Débat* 100: 106.
Thomas, Y. 1998b. Introduction du dossier 'Le droit et la nature des choses'. *Enquête* 7: 14.
Truog, Robert D., Franklin G. Miller, and Scott D. Halpern. 2013. The dead-donor rule and the future of organ donation. *The New England Journal of Medicine* 369: 1287–1289.

# Chapter 12
# The Symbolic Meaning of Legal Subjectivity

**Dorien Pessers**

## 12.1 Introduction

Why and when are people willing to obey the law? Researchers into legal psychology are quite sure in their response to this recurrent question regarding the legitimacy of the law. They maintain time and again that citizens are willing to obey the law and judicial decisions – even when their obedience is at odds with their own interests – if the law has been established by democratic procedure, applied by the judge in a properly reasoned judgement, after a fair trial in which the parties have been treated with dignity and respect. Procedural justice would appear to be the main source of the legitimacy of the law and judicial decisions (Tyler 2000, 2006). These studies seem to confirm the famous theorem of Luhmann, who in *Legitimation durch Verfahren* stated that in pluralistic societies where unanimous consensus about values is difficult to achieve, only procedures can generate legitimacy (Luhmann 1969). Legal procedures – as such – bring stability, regardless of the conflicts `of values to which they relate. In a legal procedure all parties involved are heard, and after their interests are balanced by the administration or the judge, the final decision will be made and accepted by the parties. In this stabilising effect the legitimacy of the law is – according to Luhmann – to be found.

Is this explanation of the foundation of the legitimacy of the law convincing? Does, indeed, procedural justice respond to such a deeply-rooted legal conviction that citizens are willing to obey the law and, for that reason, are prepared to set aside their personal view on substantial justice? Or is substantial justice, especially in pluralist and multicultural societies, so difficult to define that procedural justice is necessarily accepted as second best choice? Does, indeed, legal practice generate legitimacy by itself?

D. Pessers (✉)
Faculty of Law, Vrije Universiteit Amsterdam, Amsterdam, the Netherlands

Faculty of Law, University of Amsterdam, Amsterdam, the Netherlands
e-mail: d.w.j.m.pessers@vu.nl

 
B. van Klink et al. (eds.), *Symbolic Legislation Theory and Developments in Biolaw*, Legisprudence Library 4, DOI 10.1007/978-3-319-33365-6_12

The explanation of the legitimacy of the law by its procedures seems to overlook a crucial condition, which is the symbolic function – even the symbolic core function – of the law. When many people appear to adhere to the law, an ingenious process of disciplining has preceded that consent. A process repeating itself moreover in every new generation: the insertion of the new-borns into the community of legal subjects. This symbolic insertion is as old as law itself, but most ambitiously expressed in the famous first article of de *Universal Declaration of Human Rights*: "All human beings are born free and equal in dignity and rights." At first sight a remarkable statement, implying that human beings are born three times. First in a biological sense; then in a social sense, as a member of a family and social environment; and finally in a legal sense, i.e. as a member of the legal community. Needless to say, this last birth is a miraculous one. Not a human being is born, but an ideal human being: free, equal, dignified and in the possession of human rights. Of course, this "birth" has little to do with real life. On the contrary, on a global scale the opposite tends to be case. Nevertheless, the legal birth of an ideal man is of great symbolic importance. French legal scholars speak of the *homme rêvé* of the human rights as a *fiction protective*. The utopic dream of the legal community in which all humans are included as legal subjects, represents a "counterfactual anticipation": man is born as an *homme situé*, but the fact that he from the outset is received into a legal community that anticipates another reality, referring to a life in freedom, equality and dignity, offers him the protection of dialectic, critical legal principles.[1] This, at least, is the idea behind the *Universal Declaration of Human Rights*. As such, it illustrates the positive meaning of symbolic legislation, as explained by Van Klink in this volume (Chap. 2).

We cannot deny that since the *Declaration* was adopted (1948), this idea has produced real and historic effects. Principles of law, fundamental political and social rights and the extensive administrative law that has been developed to realise these rights, have undoubtedly raised other citizens than would have been the case without these endeavours. Legal subjectivity has become an identity that during the second half of the last century has repeatedly proven its emancipatory potential. People not only submit themselves to the law, they also appeal to the law when they feel violated in their dignity, freedom or equality. It is this identity-forming capacity of the law – as a precondition for obedience to the law – that legal psychologists are overlooking in their search for the foundation of the legitimacy of the law.

One might call the symbolic transformation of human beings into legal subjects an "anthropotechnology". An unusual term, explained later on in this article. For now it suffices to conceive it as a device that, besides controlling people's behaviour by rules and sanctions, tries to influence people's very identity. That is why – at least from an anthropological perspective – Fullers famous description of the purpose of the law as "subjecting human conduct to the governance of rules" is not complete. The law not only aims to subject human conduct, but – if not in the first place – to

[1] See for a comprehensive survey of the counterfactual dimensions of the law: Van Beers (2009, pp. 573–589).

I have made gratefully use of this excellent dissertation in my research for this article.

subject man himself. Therefore, the law is not only a technique for social engineering, but for human engineering as well.

How does this transformation of humans of flesh and blood into legal subjects – the symbolic core activity of the law – proceed? The answer needs a detour which, in appearance alone, distracts us from the theme.

## 12.2 Law and Human Desire

More than any other human faculty, the world is driven by human imagination. Man's imagination makes him long for another world, a world in which he hopes to find redemption from his human condition. In the midst of chaos the *homme situé* longs for order; in the midst of injustice for justice; in the midst of war for peace; and in the midst of dependence and helplessness for freedom and autonomy. In short, man longs for a counterfactual world in which he will find salvation. His desire can stagnate into dreams and daydreams, illusions and projections, but his desire might also function as an incentive to action. At the same time – and history shows this time and again – this desire can become boundless, fanatic and even destructive. So, human desire requires a free and inviting space on the one hand, and moral orientation and direction on the other.

This channelling of unfocused desires proceeds by what is called "symbolic orders". Man is an *animal symbolicum*. Once upon a time he succeeded in leaving his universe of basic instincts and impulses behind. He became capable to order the chaos of the reality in which he was imbedded and did so, primarily, by developing a language. Language enabled him to give meaning and cohesion to his physical and social surroundings and to his own place within it. Those meanings functioned as orientations in a world that would be inconceivable without such guidance. Therefore, symbolizing, in the sense of providing a guiding system of meanings, is the beginning of any culture. In the course of time the symbolic order became increasingly complex, and differentiated into separate and institutionalized symbolic systems, such as kinship, religion, art, law, politics and trade. These symbolic systems function as supra-individual "*Sinngebilde*" in Weberian terms, or as the institutionalized expressions of the "*conscience collective*" as conceived by Durkheim. After our first, biological, birth we are born for the second time into these socializing symbolic systems. (See Chap. 9 in this volume).

Within the symbolic order as a whole, religion and law occupy a special place. These institutions provoke *and* structure the desire for salvation and redemption from the human condition. Religion promises a paradise in heaven. This reference to metaphysics satisfies the human desire for transcendence. At the same time, religion subjects man to the Laws of God and to dogmatism, in order to channel a boundless desire. The law, for its part, promises a paradise on earth where justice will prevail, according to the principles of retributive and distributive justice, rendering each his due. Also this earthly paradise has transcendent dimensions, although one might question if these stem from the historically close connection between law and religion, or from the fundamental openness of humankind to transcendence, to what surpasses man, to the invisible and the counterfactual. In any event, also in modern secular societies "the Almighty has proved remarkably difficult to dispose of", as Eagleton (2014, p. ix) observes. People continue to believe, be it not in God anymore, but in man and his human rights, as a *réligion civile* which will bring a better world closer. People may lose their religious belief, but they keep their faith in an ultimate redemption from the human condition. Their human rights are declared "*sacrés et inaliénables*". Constitutions are conceived as "sacred texts". The first principle of modern law, human dignity, is rooted in the Christian tradition, but has maintained its transcendent meaning in our secular society. Moreover, on a philosophical level there is still the debate about the question whether Kelsen's secular *Grundnorm* is, in last resort, not a shared presupposition that refers to a transcendent standard.[2] On a pragmatic level Gauchet (2006, p. 115) concludes: "This detour via the transcendence is justified by the expected results it provides in the immanence. (..) The counterfactual world is placed at the service of this world".

Whatever the origin of the transcendent dimensions of the law may be, just like religion, the law submits people to rules and dogmatism in order to channel and structure human desire. That is why modern law occasionally is defined as a "normative architecture" (Supiot 2005). An important difference however between religion and law is that religious dogmatism tends to fundamentalism, but legal dogmatism instead implies an anti-dogmatism, due to its open concepts and principles that are vague and even contradictory in their references to a world where justice, ultimately, will prevail.

Cotterrell calls this anti-dogmatism of the law "its mystificatory brilliance". He writes: "Law proclaims symbols so vague or all-embracing that most members of society can accept and support them in some interpretation. Law holds up its mutually contradictory ideals like a beacon around which otherwise divided elements in society rally." Cotterrell (1992, p. 103) continues with a quote from Thurman Arnold: "And therein lies the greatness of the law. It preserves the appearance of unity while tolerating and enforcing ideals which run in all sorts of opposing directions. It fulfils its functions best when it represents the maximum of competing symbols". In other words, the symbolic system of the law is, at the level of its principles, so vague and self-contradictory – not the least because of the transcendent character of those principles – that the law, running in any direction, still manages to keep the image of unity alive. Apparently the law successfully prevents human desire – ignited by the promises of law itself – becoming delirious, unlimited or destructive. The "greatness of the law" therefore lies in its success as an "anthropotechnology". As said earlier an unusual term in legal literature. But one that expresses the essence of the symbolic transformation of people of flesh and blood into legal subjects.

The term "anthropotechnology" came into vogue after the notorious and much debated lecture *Rules for the Human Zoo* (*Regeln für den Menschenpark*) by the German philosopher Peter Sloterdijk in 1999. In it, he argues that classical humanism

[2] An instructive article on law and transcendence: Van Roermund (2015).

as a tool for civilization is not able to cope with the arrival of genetic science which, by means of liberal eugenics, makes radical change of the human species conceivable. Humanism – according to Sloterdijk – regarded people as "animals that live under influence", i.e. under influence of their urges and instincts, their fears and aggression, their desires, fantasies and delusions. The credo of humanism was that people should be exposed to the right kind of influences that would inhibit and tame their innate bestiality. Sloterdijk describes the European literate culture as a civilization operation *par excellence*. Education and discipline is a matter of reading, learning, interpretation, knowledge and wisdom. "Ever since philosophy began as a literary genre, it has recruited adherents by writing in an infectious way about love and friendship. Not only about love of wisdom: it is also an attempt to move others to his love". That written philosophy "has been re-inscribed like a chain letter through the generations, and despite all the errors of reproduction – indeed, perhaps because of such errors – it has recruited its copyists and interpreters into the rank of brotherhood." Through their writings, the philosophers made unknown but like-minded friends and established a community of literacy which could spread civilization. The claim of humanism was that reading the right books calms the inner beast. Reading is healing. However, according to Sloterdijk, in our post-metaphysical era with its mass culture which is predominantly a culture of images, pictures and screens instead of texts, and with the domination of the market and consumerism, this claim has lost its persuasiveness. The humanistic faith in civilization by literacy, has given place to faith in redemption and civilization by technology. In particular, biotechnology. Why still discipline man in a laborious process of educating and taming, if man can just be bred? Why change the phenotype of man if the genotype can be optimized by genetic selection of embryos and manipulation of the genetic constitution? Sloterdijk warns that this new kind of anthropotechnology does not belong to science fiction anymore, but has become a reality in which worldwide billions are invested.

Neither Sloterdijk, nor the many commentators on his influential lecture, mentioned the law as a historical anthropotechnology. This omission is all the more remarkable because the law – just like philosophy – has a centuries-old, uninterrupted literate culture, which also – and even earlier (Dworkin) – can be described as a "chain letter". In addition, the legal literate culture is also characterized by the aim to inhibit and tame man. Since the Enlightenment the law is conceived as a central part of the humanist civilization project. And do not write legal academics likewise "on an infectious way about love and friendship"? The law aims to create a brotherhood of strangers, in which solidarity is understood as the "social face of love". In the words of Unger (1976, p. 206) the ideal of the law is: "maintaining a system of social relations in which men are bound to act compassionately, if not at least as if they had compassion for each other". And is it not applicable to law as well that genetic science threatens to disrupt the image of man in law, and to undermine law's function of human engineering? Consequently, all of this adds up to the claim that it might make sense to study the law from an anthropotechnological perspective.

We have finished now our detour and will return to the question about the way the transformation of humans into legal subjects – the symbolic core function of the law - proceeds.

## 12.3 Law as Anthropotechnology

Where Fuller, as a legal philosopher, asked attention for the "inner morality of law" as a major formal factor in the submission of human conduct to the rule of law, legal anthropology focusses on the informal, anti-dogmatic factors and, in particular, on the dynamic relationship between dogmatism and anti-dogmatism; between the law as a rational system and the law as a system with outspoken transcendent dimensions. Whether it is classic natural law, Christian natural law or modern natural law, these schools of thought refer all to another world in which the *homme situé* will find redemption and his ultimate destination as an *homme rêvé*, i.e. as a man who finally lives in dignity, freedom and equality. However, he will only obtain that dignity after his boundless desires are subjected to moderation, proportion and rules. In this way, dogmatism and anti-dogmatism, law as order and law as justice are tied together. The horizontal and the vertical, the visible and the invisible are brought into a dynamic and dialectical connection. If the law manages to achieve a dynamic balance, the risk is restricted that people end up in an imaginary order in which reality, experiences and desires can get any and, therefore also, a destructive meaning.

An important means to this end is setting boundaries. The idea of the installation of boundaries is the oldest idea of the law, but not until Roman law was this idea developed into a sophisticated classification system (Thomas 2011). Primordial boundaries are particularly important from the point of view of a disciplinary anthropotechnology, such as the boundary between person and commodity (*persona* and *res*); between being and having (*suum* and *proprium*); between private and public (*ius privatum* and *ius publicum*); between states and tribes (*ius civile* and *ius gentium*). These borders installed separate spheres of life and justice, all based on the three major principles of the law: *honeste vivere*, *neminem laedere* and *suum cuique tribuere* (to live honourably, to harm no one, and to give everyone his due). Principles that have – over the centuries and in an endless chain letter – resulted in an advanced philosophy of law. In democratic constitutional states these principles also have found expression in positive law.

In particular, the boundary between a personal, private sphere and a public sphere – the *summa divisio* of the rule of law – has brought civilization. The domestication of humans Sloterdijk spoke about, starts as soon as they find a dwelling, a private space of their own, in which they can seclude themselves from the threatening outside world. In Dutch language dwelling *(wonen)* is etymologically derived from "being brought to peace". Only when in the privacy of one's own home the fear for the outside world has faded away – and by consequence the aggression against others – man can live and die in peace, instead of in violence. Article 8 of the European Convention on Human Rights, protecting home and hearth, private and family life, is the expression of this original insight into the pre-political conditions for peace, and therefore perhaps the most important human right.

The characteristic of a boundary is that it separates *and* binds both parties at the same time. Their mutual promise to respect the boundaries set between them, brings them in a moral relationship which can be enforced by the law. In this way, an inter-subjectivity emerges, mediated by the law, as the main effect of the symbolic transformation of "the animal under influence" into a subject of law.

This symbolic transformation is the core of the Enlightenment version of the conceptual model of the social contract. The man of flesh and blood who demonstrated his cruellest sides in the religious wars of the sixteenth and seventeenth centuries had to be tamed, and this time once and for all. He was supposed to supress his aggressive nature and his sensitivity for irrational and obscure influences. He had after all received from God – the metaphysics remains – the gift of reason. This gift enabled him to reach reasonable agreements with his former enemies, agreements that would be in the interests of both parties, if only because there would be, at least, a period of peace. Thanks to which man could find his true destination: living in happiness. For the French Enlightenment philosophers the social contract should expressly allow for *le bonheur*, understood as a life in freedom and peace in which man could realize his desires and potential. Happiness, however, presupposes submission. Submission to the conditions of the peace agreement, i.e. to the general interest. A strong maintenance of the supra-individual, general interest is after all the best guarantee for the protection of the individual interest. This submission to the law makes man a subject of law, in the sense of a *subiectum*. But at the same time a holder of law in the sense of a *subiectus*, due to his obtained freedom and equality. The submission to the law means a liberation and salvation from the evil that people use to bring upon each other. However, legal subjectivity also means, by definition, inter-subjectivity. From both sides, parties promise to bind themselves to the social contract and thus to the interests of others. The human being is "tamed" by inserting him from his birth in a reciprocal relationship of rights and duties towards others. Only from these reciprocal relations man may start – according to the American *Declaration of Independence* – his "pursuit of happiness".

The inter-subjective character of legal subjectivity, therefore, has a threefold significance: inter-subjectivity appears to be the only way of surviving and maintaining a lasting existence of humankind; it is also a moral imperative; and it corresponds to the transcendent belief that the destination of man is located in the reciprocal recognition of his human dignity. Inter-subjectivity is thus both an ontological, ethical and teleological category. The layered nature of the law is nowhere better expressed than in this category.

Looking back, we cannot deny that the anthropotechnical transformation of people of flesh and blood into legal subjects has been successful to a great extent. From the idea of the social contract, democratic constitutional states came into being. They reached their bloom period in the welfare states of the second half of the twentieth century. People proved to practise a comprehensive solidarity. In short, the law – as a literate culture *par excellence* – has demonstrated an impressive civilization operation with remarkable innovative capacity. The basic concepts of the law, originally developed by Roman jurists, were found to possess a miraculous elasticity, as did the principles of law. We live in a radically different world than the Romans did, but still their concepts underpin modern law, like property, possession, contract, obligation, tort, unjust enrichment, good faith, to mention only a few. In fact, some private law concepts appeared to have a political charge when they were employed in the medieval covenants between princes and nobles, or between princes and the citizens of powerful cities. In the eighteenth century they even produced revolutionary effects when applied to the social contract between the state and its citizens. The historical continuity, the systematic processing, the innovative strength and the modern constitutionalization of the law and legal principles had as – anthropotechnological – consequence that Europeans derive their identity – as *bourgeois* and *citoyen* – from their legal subjectivity.

## 12.4 The End of the Law as an Anthropotechnology?

Of course, the history of civilization by law is not just a success story. As a relatively autonomous system, the law has also legitimised and even legalised barbarism, like the fascist barbarism of the previous century and the brutal excesses of capitalism in this century. Obviously, the law can also be used as an instrument for the untamed desires of despots or of global financial players. As to the counterfactual, transcendent dimensions of the law, too often they have lost – and still lose - their critical character, turning into pure ideology. Even in democratic states, governed by the rule of law, law as justice is under permanent pressure of law as order. In addition, "the greatness of the law" which Arnold spoke of, has a built-in weakness. The elusive and self-contradictory character of the principles of law might guarantee its dialectical force and might allow for a dynamic interpretation "in the light of the present day conditions", but the same dynamics might cause a loss of intersubjectivity and, as a consequence, a loss of the legitimacy of the law.

A clear example of the ambiguous nature of the law – as soon as it is practiced – is the proliferation of "claim rights". Claim rights regard the desire for individual self-determination. Citizens no longer accept an objective interpretation of fundamental rights and principles of law in all cases, but demand room for a personal, subjective interpretation. Why is the desire for a gentle death – even if there is no hopeless suffering – considered illegal? Is self-determination in life and death not the core of individual human dignity? Why is dwarf Manuel Wackenheim not allowed to let himself be tossed and thrown around in a pub? It is the only employment that suits him. Being deprived of the right to employment is the same – as Wackenheim stated in court – as being deprived of his right to dignity. Why is baby Kelly not allowed to ask for financial compensation because of the fact that she was born heavily handicapped and therefore convicted to a degrading life? Why may women not act on payment as a surrogate mother if that is their only way out from inhuman poverty? Why do sadomasochists lack the right to fustigate each other, if such an experience – although by law regarded as degrading – gives them optimal sexual pleasure? These examples, derived from case law, show that citizens no longer, as a matter of course, accept a supra-individual interpretation of the transcendent values of the law. On the contrary, they remove the mask of the inter-subjective legal subject, and demand recognition as human beings of flesh and blood who require complete self-determination, as an expression of their individual dignity. They refuse their symbolic transformation into legal subjects, perceiving this transformation as a restriction of their freedom and happiness. In other words, they withdraw from the inter-subjectivity of the law and of what is considered as the supra-individual, general interest. But, does this deprivation not also deprive them – in the same act – from the legal, normative infrastructure upon which their existence depends?[3]

The proliferation of claim rights revitalizes the centuries-old debate about the relationship between man and community and between legal subject and legal community. On the one hand, this relationship should be open to innovation in order to maintain the dynamic and emancipatory potential of the law. On the other hand, law should remain faithful to its promise to ensure the dignity of the human family. When the guarantee of collective human dignity is threatened – for example if man would be degraded to an object – the law should refuse to give in. After all, the legitimacy of the law depends on the extent to which it is able to guarantee human dignity.

This enduring debate about a legitimized room for a subjective interpretation of human dignity, has gained a new and spectacular perspective by claim rights in the field of medical biotechnology. Sloterdijk was one of the first – at least one of the most eloquent – philosophers who considered the consequences of this new branch of technology. With the combination of genetics, information technology, nanotechnology and neuroscience human enhancement becomes a real possibility. That is to say, medical science is not only able to make people better, but also to make "better" people. Better in a biological, psychological, and cognitive sense. Many biotechnological devices are already available or will be in the foreseeable future. Genetic selection and modification of embryos not only prevents the birth of disabled children, but might also promote the birth of talented children. Germline therapy might eliminate once and for all hereditary diseases. Through stem cell therapy, organs could be renewed. Nanorobots will be able to detect and cure diseases and make significant life extension imaginable. The brain can be connected to computers and thereby enhance our cognitive capacities enormously. In fact, robots can be humanized and humans can be robotized.

In short, medical biotechnology has ignited a new desire for redemption. Redemption from the fateful human condition, from suffering, diseases and imperfection. Even immortality no longer seems beyond the reach of human endeavour. High priests of the philosophy of technology, like Savalescu, Hottois and Bostrom are delighted and claim a turning point in the civilization process. The dignity of man, that ultimate promise of the law, no longer needs to be enforced by humanist anthropotechnologies. Civilization of man does not require external influences any

[3] See for a thoroughgoing discussion on this question: Van Beers (2009).

longer because man can be internally, genetically manipulated. Why make painstaking efforts to change the phenotype of man, if the genotype can be directly conditioned? Why not replace social and human engineering by bio-engineering? The *homme situé* will finally become an *homme rêvé*, not due to discipline by law and other humanist anthropotechnologies, but due to biotechnological modification.

However, these philosophers understand very well that the post-human world they are welcoming, evokes a risky imaginary order, in which basic categories lose their validity as moral guidelines. If the distinctions between person and commodity, mind and body, man and machine, natural and artificial, perfection and imperfection get lost, a disorienting chaos will emerge again, the very chaos that was suspended by the symbolic order. (See Chap. 11 in this volume). Desymbolization is therefore called a "*risque anthropologique*": when the symbolic tissue is destroyed, the social fabric also threatens to dissolve (Vacquin 2003). The international trade in organs and the reproductive tourism in which poor women sell their eggs and their wombs to rich infertile couples or single parents, has already resulted in this desymbolization and produced unabashed dehumanisation (Scheper-Hughes 2002; Van Beers 2014).

The post-humanists, therefore, appeal to the lawyers to continue their civilizing tasks in the age of biotechnology and to forcefully maintain human rights, democracy and the rule of law. However, that appeal is accompanied by a remarkable condition: the law should first be stripped of its metaphysical and transcendent dimensions. Hottois is particularly firm in this claim. He even speaks about the symbolic violence with which the law – dogmatically referring to the transcendent principle of human dignity – tries to prevent the inevitable development of a post-human species. After all, the law itself should no longer provoke the human desire for salvation and dignity, nor structure that desire in a normative way. The salvation from suffering and imperfection will soon become a matter of technology, and technology has – according to Hottois (1999) – nothing to do with metaphysics and references to another, counterfactual world. So the legal task that remains is one of symbolization of the new and hybrid forms of life, as well as regulation and fair distribution of the products of biotechnology. In particular, freedom of choice is important. People should get the right to "morphological freedom", i.e. the right to choose their own form of life: either naturally human, or technologically modified post-human.

It may be clear that this overt plea for a liberal eugenics and an anti-metaphysical law concerns not only a German philosopher such as Sloterdijk. How might the principle of human dignity still retain its dialectical and critical potential if its supra-individual, transcendent dimensions are stripped, and its interpretation is left to the individual self-determinator? The collective, historic experiences of humankind are stored in this principle. Experiences that have led to the practical wisdom that some limits should never be transgressed, if one wishes to keep social misery, exploitation and tyranny at a distance. That practical wisdom entails, among other things, the notion that the metaphysical dimensions of the law should be upheld as far and as long as critical principles can be derived from them. (See Chap. 9 on the role of taboos, in this volume).

In the context of medical biotechnology and human dignity, the metaphysical dimensions of the legal protection of the human body are of great importance. In the European constitutions the human body is declared "inviolable" or sacrosanct. An impressive example of this symbolic meaning of the human body, was tragically demonstrated in 2014, when the remains of the bodies of the victims of the MH17-air crash in East-Ukraine were transported back to the Netherlands. In the scientific-reductionist perspective of Hottois a dead body is no more than a remnant that quickly decays. It is not a person, at the most a commodity. Yet, it is a proof of practical wisdom to treat the human corpse nevertheless with respect, as if it was still imbued with the *anima* of the deceased. The respectful ceremonial, organised by the government to welcome the body remains at the airport, made a deep impression on the Dutch people, and acknowledged the desire for a transcendent meaning of the dead bodies. A desire that was all the more understandable because the corpses had lain down for weeks, scattered over the crash site, in the burning sun. A situation that was generally felt as a desecration of the victim's bodies.

Nevertheless, in the long run the profanation of the body seems inevitable. Medical biotechnology has become a multibillion dollar industry with explosive stock values. Authorities, considering biotechnology as a promising sector for innovation, promote and facilitate this industry with permits, grants, biobanks and patents. Despite a comprehensive system of treaties, directives, laws and medical-ethics committees being harnessed to protect people from the risks of medical biotechnology, domestic and international law hardly seems in control over the spectacular developments or the phantasmal desires which these developments unleash. (See Chap. 10 in this volume). Moreover, the astonishing progress of medical biotechnology not only generates the desire for final salvation from the fateful human condition, it actually creates an international market of supply and demand, eliciting claim rights of the consumers. A free market which is not mediated by effective law, however, does not create inter-subjectivity, nor does it relate human beings to the general interest. In other words: such a market is not able to structure man's limitless desires in a normative way.

If medical biotechnology and its industrialization fail to become rigorously regulated, the symbolic core function of the law will eventually disappear. Why should every new generation still symbolically be born into a transcendent legal community, promising ultimate salvation and human dignity, if that salvation can be obtained in the here-and-now, albeit probably at the cost of dignity? In any event, the symbolic meaning of legal subjectivity as the foundation of the legitimacy of the law will profoundly change.

## References

Bostrom, Nick. 2005. In defense of posthuman dignity. *Bioethics* 19(3): 202–214.

Cotterrell, Roger. 1992. *The sociology of law. An introduction*. London: Butterworths.

Eagleton, Terry. 2014. *Culture and the death of god*. New Haven: Yale University Press.

Gauchet, Marcel. 1998. *Religie in de democratie. Het traject van de laicité*. Nijmegen: Sun. French edition: Gauchet, Marcel. 1998. *La religion dans la démocratie. Parcours de la laicité*. Paris: Gallimard.

Gauchet, Marcel. 2006. *Religie in de democratie*, 115. Nijmegen: Sun.

Hottois, Gilbert. 1999. *Essais de philosophie bioéthique et biopolitique*. Paris: Vrin.

Luhmann, Niklas. 1969. *Legitimation durch Verfahren*. Frankfurt am Main: Luchterhand.

Scheper-Hughes, Nancy. 2002. The ends of the body. Commodity fetishism and the global traffic in organs. *Sais Review* 22(1): 61–80.

Sloterdijk, Peter. 1999. *Regeln für den Menschenpark*. Frankfurt am Main: Suhrkamp Verlag.

Supiot, Alain. 2005. *Homo Juridicus. Essai sur la function anthropologique du droit*. Paris: Éditions du Seuil.

Thomas, Yan. 2011. *Les opérations du droit*. Paris: Gallimard.

Tyler, Tom R. 2000. Social justice. Outcome and procedure. *Int J Psychol* 35(2): 117–125.

Tyler, Tom R. 2006. *Why people obey the law. Procedural justice, legitimacy and compliance*. Princeton: Princeton University Press.

Unger, Roberto Mangebeira. 1976. *Law in modern society*. New York: The Free Press.

Vacquin, Monette. 2003. Réflexion sur la notion de "risque anthropologique". In *Cahiers de recherche sociologique*, 175–195. Montréal: Université de Québec.

Van Beers, Britta C. 2009. *Persoon en lichaam in het recht. Menselijke waardigheid enzelfbeschikking in het tijdperk van de medische biotechnologie*. Den Haag: Boom Juridische uitgevers.

Van Beers, Britta. 2014. Hoe stoppen we de vruchtbaarheidstoerist? Trouw, Letter & Geest, December 13.

Van Roermund, Bert. 2015. Kelsen, secular religion, and the problem of transcendence. *Netherlands Journal of Legal Philosophy* 44: 100–115.

# Part III
# Legislative Strategies: Regulating Biomedical Developments from a Symbolic Perspective

# Chapter 13
# The Natural, the Informational, the Claimable? Human Body Material in US and European Patent Law

**Sigrid Sterckx and Julian Cockbain**

> *A discovery may be brilliant and useful, and not patentable. No matter through what long, solitary vigils, or by what importunate efforts, the secret may have been wrung from the bosom of Nature, or to what useful purpose it may be applied. Something more is necessary.*
>
> (Morton v. New York Eye Infirmary 1862, emphasis added).

## 13.1 Introduction: Patent Law as Symbolic Legislation

Patents provide the proprietor with a short-term (usually about 20 year) monopoly on the subject matter covered by the patent claims. Patents are granted by or on behalf of individual nations, and the requirements of patent law, though to some extent harmonized, do differ from state to state. Within Europe, since 1979, the dominant law has been the European Patent Convention (EPC) (European Patent Office 2013), under which the European Patent Office (EPO) can grant patents that can take effect in up to 40 countries. In the US, the patent law can be found in Title 35 of the US Code (35 USC) United States Patent and Trademark Office (2014a).

In this chapter we are concerned with the language used in patent law and in legal proceedings regarding patent cases relating to parts of the human body, and the effects of such language. We will analyse various key cases, starting with cases from US and, later on in the chapter, some important European cases. We should note at the outset that, whereas the EPC explicitly mentions parts of the human body, US patent law does not. However, in the US we encounter a judicially or court-derived doctrine regarding products, laws and phenomena of nature and, as we will see, the questions that arise are the same in both jurisdictions.

S. Sterckx (✉)
Department of Philosophy and Moral Sciences, Ghent University, Gent, Belgium
e-mail: Sigrid.sterckx@ugent.be

J. Cockbain
Bioethics Institute Ghent, Ghent University, European Patent Attorney, Ghent, Belgium
e-mail: julian.cockbane@btopenworld.com



B. van Klink et al. (eds.), *Symbolic Legislation Theory and Developments in Biolaw*, Legisprudence Library 4, DOI 10.1007/978-3-319-33365-6_13

Elsewhere in this volume, Bart van Klink sets out two different versions of 'symbolic legislation': the first a 'negative' version which essentially only pays lip service to social and ethical concerns, and the second a 'positive' version' which seeks to engage and drive society towards a desirable goal. *Patent law provides instances of both of these versions of symbolic legislation*, and in its operation we see a range of actors engaged in *emasculating* or *empowering* 'aspirational' passages in the law. These actors include the politicians and bureaucrats involved in drafting and amending the laws, the patent office examiners and appeal board members involved in applying the law, patentees, patent applicants, opponents and alleged infringers with direct and pressing concerns regarding particular cases, industry and patent attorney associations interested in driving interpretation in a particular direction, judges, NGOs, academics, and even individual lay members of society. Indeed, numerous groups with conflicting or incompatible interests vigorously compete with each other in this area of law, which thus clearly meets the conflict `of interests criterion discussed by van Klink as one of the pragmatic criteria for identifying a law as symbolic.

As discussed in greater detail below, the question of whether materials extracted from human bodies, i.e. naturally occurring materials simply removed from their natural environment, should be patentable has recently provided the setting for the investigation and closer definition of the most basic of the symbolic terms used in patent law, *invention*. Clearly, the various actors (governments, litigants, amici, judges, patentees, etc.) have conflicting goals: on the one hand to encourage industrial development, and on the other to ring-fence natural phenomena to make their use free to all. In this respect, an interesting link can be made between the analysis we undertake and the Chap. 10 of Klaus Hoeyer elsewhere in this volume (Hoeyer 2016). Hoeyer investigates practices of tissue exchange in Europe and identifies two conflicting goals: on one side protecting human bodies and their parts from economic exploitation and commercialisation, on the other side promoting the 'bio-economy' by allowing human body material to become part of market-driven processes of innovation and growth. He seeks an answer to the question "how to understand the co-existence of two competing and conflicting legal frameworks?" With regard to our topic, the patenting of human body material, what is at issue are again two competing aspirations, but this time within a single legal framework, that of patent law. This conflict `between aspirations can perhaps most clearly be illustrated with the words used in Art. 5 of the 'European Biotech Directive', Directive 98/44/EC on the legal protection of biotechnological inventions:

> The human body, at the various stages of its formation and development, and the simple discovery of one of its elements, including the sequence or partial sequence of a gene, *cannot constitute patentable inventions. …*
>
> An element isolated from the human body or otherwise produced by means of a technical process, including the sequence or partial sequence of a gene, *may constitute a patentable invention, even if the structure of that element is identical to that of a natural element.* (European Union 1998, pp. 13–21, emphasis added)

In this single Article, while the human body is excluded from monopolisation through patents, the exclusion for body parts is so qualified as to make them readily available for commercialisation. Clearly, as compared with other areas of biolaw, such as for example the regulation of tissue research discussed by Hoeyer, patent law explicitly leaves the door open to the commercialisation of human body materials, and in this sense appears to offer a poorer symbolisation of human body parts.

Obviously, those representing industry have vastly greater resources available to them than do the NGOs, academics and members of the general public, and their voices have massively more impact. This can be seen for example in the negotiating history of the WTO Agreement on Trade Related aspects of Intellectual Property rights (TRIPS) (World Trade Organization 1994), where the text of the agreement was largely written (ostensibly for the US Government) by attorneys acting for industry, in particular the entertainment, computer and pharmaceutical industries (Sell 2003; Drahos and Braithwaite 2002; Matthews 2002).

That notwithstanding, the general public, NGOs and academics have been particularly active in recent years in submitting *amicus* briefs on questions concerning, for example, the patentability of human embryonic stem cells, plants and animals, essentially biological processes, human DNA, dosage regimes for medical treatment, computer programs, etc. The days are long gone when patent law developed outside the spotlight of general public interest. Indeed, various examples exist of patent cases where the exchanges between the various *amici* (which frequently include political players) and the courts, have clearly influenced the public debate on the issue in question, thus clearly meeting the "communication criterion" identified by van Klink (2016) as a criterion for identifying a law as an example of symbolic legislation in a positive sense, i.e. "communicative legislation". One such example concerns the Myriad patents on breast and ovarian cancer genes (discussed later in this chapter), another concerns the debate in Europe regarding the patentability of human embryonic stem cells (Sterckx 2008; Sterckx and Cockbain 2010).[1]

In a patent application, the applicant defines the monopoly he is asking for, phrased to comply with the relevant regulations. The patent office examiner reviews the application, seeking also to apply those regulations and relying on guidelines, case law and the *travaux préparatoires*. The judges (or appeal board members), often with the help of *amicus curiae* briefs, seek to determine the legally correct meaning of the law. Such briefs may come from industry and patent attorney associations, states, academics, NGOs, companies facing similar problems establishing or denying patentability, and members of the general public. They are supposed to assist the judge/appeal board by *showing how the law should be interpreted*, for example by reference to earlier cases or the *travaux*. However, usually these briefs focus instead on the *wishes or beliefs* of their authors and, as a result, they are often ineffective and simply ignored by the judges.

[1] On a more general note regarding the communication criterion, we should point out that, in Europe, particular "structures for deliberation" are in place "which stimulate the communication between the executive, the judiciary and the addressees of the law" (van Klink 2016), viz. the oral hearings before the Opposition Divisions, Technical Boards of Appeal and Enlarged Board of Appeal of the EPO, all of which are open to the public.

As explained by van Klink (2016), "in many cases symbolic legislation consists of vague norms which are unclear and open to multiple interpretations" and sometimes the legislation even contains "contradictory norms". A prominent question in patent law is the question of what an *invention* is. Patent laws provide for patents to be granted for 'inventions', yet *patent laws do not define what an invention is*, other than negatively (e.g. European patent law provides that 'discoveries' are not inventions). The legislators have left it up to the judiciary to decide. The same is true for 'immoral' and hence unpatentable inventions – the legislators could not decide what would be immoral, and left it to the courts to decide (see Sterckx and Cockbain 2012, chapters 2 and 8).

As far as the central topic of this chapter is concerned, i.e. the way parts of the human body are 'treated' in the context of patent law, as will become clear later in this chapter, the European statute law falls squarely under what van Klink (2016) describes as the *negative* version of symbolic legislation, which he calls 'symbolic legislation' *tout court*.[2] On the one hand, discoveries, e.g. naturally occurring human body materials such as proteins, hormones or genes, or even an extra finger, are *not patentable* according to the EPC. On the other hand, however, when isolated, purified, or otherwise put into a state fit to be used, these things *can be patented* according to the same law, since it is not the naturally-occurring material 'as such'. We will come back to this later. First, we need to provide a bit of information on patents in general.

The *claims* of a patent (or patent application) are a series of numbered sentences, each of which provides a closed definition of the monopoly granted (or, in the case of a patent application, the monopoly sought). They always define either a thing (a product) or a way of doing something (a method or process). Since we are concerned with patent law in relation to human body material, we will focus here on *product claims*.[3] Patents are granted for 'inventions', a term which, as noted earlier, patent laws do not define positively. However, in general, patent laws set out a series of requirements that must be met before a patent may be granted. The most prominent of these 'gatekeeper' requirements are: (1) that the invention is new[4]; (2) that the invention involves an inventive step (or is not obvious); (3) that the invention is useful (or industrially applicable); (4) that the patent applicant has provided sufficient information for others to reproduce the invention; and (5) that the invention is in some way technical in nature. Beyond this, some patent laws specify that certain categories of subject matter may not be patentable, either on the basis that

[2] The same is true for the EPC's exclusions from patentability of methods of medical treatment and of plant and animal 'varieties'. These issues fall outside the scope of this chapter.

[3] However, we must add a note of caution since, at least under European patent law, process claims also provide protection for the products directly obtained by the claimed process.

[4] Novelty, viewed from a patent law perspective, is somewhat counterintuitive for those outside the field and is construed very strictly. For a patent claim to lack novelty, the prior art (i.e. all that was known or available publically before the patent application was filed), must not contain something that falls within the scope of the claim and was reproducible on the basis of publically available teachings. Put differently, to deny patentability for lack of novelty, the prior art must do more than point towards the invention, it must take you by the hand and drag you to it.

they cannot be inventions or that, for political or social reasons, they should not be patented.

In order to demonstrate that the claimed subject matter meets some or all of the requirements mentioned above, *product patent claims to human body material* generally have one (or a combination) of the following formats:

(a) Product X, having the property A.
(b) Product X obtainable by process B.
(c) Product X as deposited in depository C (or derivable therefrom).
(d) Isolated or (partially) purified product X.
(e) Product X in combination with component D.
(f) A (particular) component of product X.
(g) A (particular) derivative of product X.
(h) A sterile composition containing product X.

As will be explained in the following section, since a 1911 US ruling in the so-called adrenalin case, format (d), i.e. a claim to an isolated or (partially) purified product, has been *the format preferred by patent applicants and patent offices.*

## 13.2 The Case That Set the Stage: The Adrenalin Case

In 1900, Japanese chemist Dr Jokichi Takamine sought to patent the hormone epinephrine (adrenalin) which is produced in the adrenal glands of humans and other mammals and which was known (but only in impure form). The claims he sought before the USPTO were in format (d), more precisely in one of his claims:

> A substance possessing the herein-described physiological characteristics and reactions of the [adrenal] glands in a stable and concentrated form, and practically free from inert and associated gland-tissue.

This claim, of course, covers *adrenalin from human tissue as well as other mammalian tissue*. Following an allegation of patent infringement, the case came before Judge Billings Learned Hand of the Circuit Court (i.e. District Court) of the Southern District of New York. Judge Learned Hand, later considered one of the giants of US patent law, decided that format (d), i.e. a claim to an isolated or (partially) purified product, *imparted novelty to the claim.*

In the century following Judge Hand's decision in *Parke-Davis & Co. v. H. K. Mulford Co.* (1911), this approach has been, with a few exceptions, the dominant one in patent law.[5] However, as will be discussed in detail below, in 2009, the Association for Molecular Pathology (AMP) and others questioned the validity of several claims in patents granted to Myriad Genetics (or their licensors, the

[5] In Europe, following the adoption of the European Biotech Directive, the acceptability of this patent claim format has been written into the Regulations of the European Patent Convention, as will be discussed below.

University of Utah) relating to isolated DNA for genes correlated with propensity for breast or ovarian cancer, i.e. BRCA1 and BRCA2. District Court Judge Robert W. Sweet distinguished over *Parke-Davis v Mulford* (the adrenalin decision) and found the claims invalid as being directed to products of nature. In July 2013 the US Supreme Court held in *AMP v Myriad* (2013)[6] that isolated and purified endogenous human DNA was indeed excluded from patentability as a product of nature. But let us start by looking at the *Parke-Davis v. Mulford* decision and its far-reaching effects.

Adrenalin, Adrenaline, or more correctly epinephrine, is a hormone secreted by mammalian adrenal glands. It is widely used medically, for example to treat cardiac arrest by injection into the heart. In 1901, Takamine and his assistant Keizo Uenaka successfully isolated and purified the hormone from the adrenal glands of sheep and oxen. Chemically speaking, epinephrine is a base and, as such, something capable of forming salts with acids. Takamine applied for US patents for the base, its salts, and his process for extracting it from adrenal glands. These patents were assigned to Parke-Davis & Co, and their Adrenaline product rapidly gained market dominance over the earlier adrenal tissue based products. Takamine's royalties amounted to 5 % of Parke-Davis' wholesale price and from this and other business ventures, he earned the equivalent of half a billion US dollars in today's money.[7]

Takamine's US patent applications did not proceed smoothly, at least as far as his claims to the product were concerned. The long-experienced US Patent Office examiner Dr James B Littlewood rejected the claims as being to no more than *an isolated/partially purified form of a natural product*. He supported his rejection by reference to the then leading case *Ex parte Latimer* (1889), a decision of the Commissioner of the US Patent and Trademark Office which related to fibres extracted from pine needles. After many rounds of argument, Examiner Littlewood finally accepted that the product Takamine was claiming was not simply an isolated/purified form of the natural product, and US Patents 730176 and 753177 were issued.

Parke-Davis sued one of the major 'infringers' for infringement of these patents and the case ended up in front of Learned Hand (then a newly minted Federal Judge with less than 2 years' experience and no background in patent law). The question as to whether the isolated and purified adrenalin that was claimed was in fact a *natural product* was not even raised in the case before Judge Hand. Astonishingly, not only was *Ex parte Latimer* not referred to, but it seems clear that Judge Hand was unaware of it. Indeed, he was primarily concerned with whether Takamine's claims lacked *novelty* over a small number of adrenal gland extracts that had been produced.

---

[6] Recently, the Full Court of the Federal Court of Australia in September 2014, has poured scorn upon the US Supreme Court's decision in *AMP v Myriad* (D'Arcy v Myriad Genetics Inc. 2014).

[7] He also significantly funded the Japanese gift of cherry trees to Washington DC, trees which are a feature of Spring in Washington. We are most grateful to Dr Jon Harkness for his help in elucidating the patent history of Takamine, Parke-Davis & Co, and adrenalin (see Harkness 2011).

On the question of whether Takamine was merely claiming a natural product separated from its natural surroundings, Judge Hand indicated that in his original patent application Takamine had claimed the natural product (the 'principle') and that this had been rejected by Examiner Littlewood on the basis of the US Supreme Court decision in *The American Wood Paper Co. v. The Fiber Disintegrating Co.* (1874) that the Examiner interpreted as providing that 'no product is patentable ... which is merely separated by the patentee from its surrounding materials and remains unchanged'.

In a subsequent application, Takamine changed the wording of the product claims so that they 'were not limited to the active principle'. Judge Hand however saw this as opening the claims to potential objections of lack of novelty rather than to the objection that they were directed to an isolated/purified natural product.

Judge Hand found that Takamine's product was a *different* chemical than the naturally occurring one, rather than simply a *purified* version of the natural product, and he was basing this on the belief of the plaintiff's expert. However, he then proceeded to deny the existence of prior case law (as summarized by *Ex parte Latimer*, discussed below) and to *establish the 'isolated/purified' guideline which was to become dominant in the US for more than a 100 years, up to the US Supreme Court decision in 2013 in AMP v Myriad, and which is still dominant in Europe today*:

> But even if it were *merely an extracted product without change, there is no rule that such products are not patentable*. Takamine was the first to make it available for any use by removing it from the other gland-tissue in which it was found, and, while it is of course possible logically to call this a purification of the principle, it became for every practical purpose a new thing commercially and therapeutically. *That was a good ground for a patent.* (*Parke-Davis v. Mulford* 1911, page 103, emphasis added)

## 13.3 The Overlooked Decision: Ex parte Latimer

In *Ex parte Latimer*, then US Commissioner of Patents Benton Jay Hall had considered an appeal from a rejection of a patent application directed to the fibres of the needles of the pine tree *Pinus australis*, which were particularly useful because of their unusual length. Applying the case law developed by the US Supreme Court in *The American Wood Paper Co. v. The Fiber Disintegrating Co.* (1874), and finding that Latimer's process neither produced nor changed the fibres, Commissioner Hall commented:

> *It cannot be said that the applicant in this case has made any discovery, or is entitled to patent the idea, or fact, rather, that fiber can be found in the needles of the Pinus australis*, ... because the mere ascertaining of the character or quality of trees that grow in the forest and the construction of the woody fiber and tissue of which they are composed is not a patentable invention ... any more than to find a new gem or jewel in the earth would entitle the discoverer to patent all gems which should be subsequently found... *Otherwise it would be possible for an element or principle to be secured by patent, and the patentee would obtain the right, to the exclusion of all other men, [to that]... which nature has produced and which nature has intended to be equally for the use of all men.* The result would be that

> … patents might be obtained upon the trees of the forest and the plants of the earth, which would of course be *unreasonable and impossible*. (*Latimer* 1889, pages 125–126, emphasis added)

Commissioner Hall stressed that *freeing something from its natural surroundings* does not, in and of itself, create something new:

> the applicant has done little more than one who gathers the pebbles along the seashore, where the forces of nature have placed them. In the latter case the action of the waves has freed the pebbles from their surroundings or covering as in the former the applicant frees the fiber by his process. (*Latimer* 1889, page 126)

The symbols evoked by Commissioner Hall are of the fruits of nature, the commons, that are open to all. By contrast, in the adrenalin case, Judge Hand was apparently more concerned about the promotion of the bioeconomy, calling the extracted product 'for every practical purpose a new thing commercially and therapeutically' and finding this 'a good ground for a patent' (see above).

Of course, in freeing a natural product from its natural surroundings, chemical bonds may, and usually will, be broken, and new bonds will be formed without altering the relevant (useful) characteristics of the natural product. The argument that such breaking and making of chemical bonds does not cause a natural product to become a new and patentable product, assumed critical importance more than a 100 years later in the *AMP v Myriad* case, to which we now turn.

## 13.4 The Case That Threatened the Bioeconomy: AMP v. Myriad

### *13.4.1 Background*

In 2009, the Association for Molecular Pathology and others sought a declaratory judgement to the effect that certain claims in a bundle of patents licensed to Myriad Genetics Inc. were invalid under 35 USC 101 as being directed to unpatentable *products of nature*. The claims in question related to 'isolated' DNA molecules corresponding to the human genes BRCA1 and BRCA2, genes that, in certain variants, are associated with an increased risk of developing breast and ovarian cancers.

Myriad was, and is, offering diagnostic tests in which a patient's sample is tested for the presence or absence of the undesirable variants, and had been actively dissuading others from performing such tests. The costs of such tests were in the thousands of US dollars, and many potential customers could not afford those costs. The claims included claims to full-length genes, fragments of the genes (e.g. useful as primers and probes in such tests) and cDNA, i.e. DNA from which the non-coding regions have been excised. Being 'isolated' from the rest of the chromosomal DNA, the full-length and fragmentary DNA molecules claimed were of course *chemically different* from native chromosomal DNA in that they terminated with hydrogen atoms and acid groups where, in the chromosomal DNA, further nucleic acid groups would be attached. Thus, *strictly speaking, they were novel molecules, but no more so than would be a leaf or fibre removed from the rest of a plant.*

The case was brought before the Federal District Court of the Southern District of New York, and in 2010 District Judge Robert W Sweet granted summary judgement to the effect that *all of the challenged claims were directed to subject matter which is not patentable under 35 USC 101* (*AMP v. USPTO* 2010). This decision caused immense consternation among the biotech industry and was appealed by Myriad to the Court of Appeals for the Federal Circuit (the 'Federal Circuit'). The appeal was heard by Circuit Judges Alan D Lourie, William C Bryson, and Kimberly A Moore. In the majority opinion (*AMP v. USPTO* 2011), the Federal Circuit overturned Judge Sweet's decision relating to the isolated DNA molecule claims. AMP then appealed to the US Supreme Court which vacated the Federal Circuit's decision and remitted the case to be reconsidered with attention to be given to its then very recent decision in *Mayo v. Prometheus* (2012). In August 2012, the Federal Circuit again overturned Judge Sweet's decision, in substantially the same words (*AMP v. USPTO* 2012), and AMP again appealed to the US Supreme Court.

In June 2013 the Supreme Court duly issued its decision (*AMP v. Myriad* 2013), confirming the *unpatentability* of the claims to 'isolated' full-length and fragmentary DNA molecules, but also confirming the Federal Circuit's decision in relation to the *patentability* of isolated cDNA. The Supreme Court decision has ramifications for all US patents relating to 'isolated' natural products and effectively puts an end to the practice established by the USPTO under *Parke-Davis v. Mulford*.[8] One extraordinary aspect of the appeals was that the USA itself submitted an *amicus* brief arguing against the patentability of isolated fragments of endogenous DNA. Let us now take a look at the language used in Judge Sweet's decision, the Federal Court's second decision, the USA's *amicus* brief, and the Supreme Court's decision.

### *13.4.2 The District Court Decision*

After reviewing earlier relevant case law, Judge Sweet pointed to the question at issue, namely that Supreme Court and other precedents established that purification of a material found in nature was not enough to render the product patentable, it must possess 'markedly different characteristics' than the natural material. Myriad's position had been that the isolated DNA molecules should be assessed for patent-eligibility in the same way as would *any other chemical*. Judge Sweet's however referred to the fact that, unlike normal chemical molecules, DNA is a carrier of information and that it would be erroneous to view DNA as 'no different' from other chemicals that had previously been patented. DNA, he considered, 'serves as the

[8] The USPTO has since changed its guidelines to USPTO Examiners to make this clear (USPTO 2014b).

physical embodiment of laws of nature'. This special nature of DNA was not changed by its isolation and Myriad, in his view, did not:

> establish the existence of differences "in kind" between native and isolated DNA that would establish the subject matter patentability of what is otherwise a product of nature… [T]he isolation of the *BRCA1* and *BRCA2* DNA, while requiring technical skill and considerable labor, was simply the application of techniques well-known to those skilled in the art. … The identification of the *BRCA1* and *BRCA2* gene sequences is unquestionably a valuable scientific achievement for which Myriad deserves recognition, but that is not the same as concluding that it is something for which they are entitled to a *patent*. … (*AMP v. USPTO* 2010)

Thus Judge Sweet considered that the inevitable differences between a gene as found *in situ* and as isolated did not impart a marked difference when the fundamental information-carrying nature of the molecule was unchanged. It was not enough that the claimed molecule was, strictly speaking, a new chemical entity.

### *13.4.3 The Federal Circuit Decision*

The majority opinion of the Federal Circuit was written by Judge Alan Lourie, who has a doctorate in chemistry. Judge William Bryson (who does not) wrote a dissenting opinion. Judge Lourie opened bluntly:

> The isolated DNA molecules before us are not found in nature. They are obtained in the laboratory and are man-made, the product of human ingenuity. While they are prepared from products of nature, so is every other composition of matter. (*AMP v. USPTO* 2012, pages 38–39)

Two earlier US Supreme Court decisions were considered particularly relevant for determining whether something deriving from natural components was itself excluded from patent-eligibility as a product of nature. *Funk v. Kalo* (1948) had considered a mixture of naturally occurring microorganisms to be excluded, while *Diamond v Chakrabarty* (1980) had considered a microorganism, into which a component (a plasmid) from another microorganism had been introduced, not to be excluded, since the resulting microorganism had 'markedly different characteristics from any [bacterium] found in nature'. Judge Lourie thus continued:

> [T]he Supreme Court has drawn a line between compositions that, even if arrayed in useful combinations or harnessed to exploit newly discovered properties, have similar characteristics as in nature, and compositions that human intervention has given "markedly different", or "distinctive"' characteristics. … (*AMP v. USPTO* 2012, page 44)

The claimed isolated DNA molecules did not exist in nature since in nature they were chemically bonded (covalently bonded) to other materials, i.e. they were not molecules as such but parts of larger molecules. When split out from those larger molecules by the breaking of some chemical bonds and the forming of some new ones, they became 'not a purified form of a natural material, but a distinct chemical entity that is obtained by human intervention.'

In other words, for Judge Lourie, breaking covalent bonds, which is implicit in isolating the useful part of a larger naturally occurring structure, was enough. He distinguished over (dissenting) Judge Bryson's analogy of snapping a leaf from a tree:

> With respect, no one could contemplate that snapping a leaf from a tree would be worthy of a patent … Snapping a leaf from a tree is a physical separation, easily done by anyone. Creating a new chemical entity is the work of human transformation, requiring skill, knowledge, and effort. (*AMP v. USPTO* 2012, page 52)

However, snapping a leaf from a tree, *does* in fact involve breaking chemical bonds! As with Judge Hand in *Parke-Davis v Mulford* discussed earlier, Judge Lourie was looking for an 'equitable' reward to be given for a piece of commercially valuable science. He was looking to promoting the bioeconomy rather than looking at the *reasons* for the court-derived exclusion from patentability of products and phenomena of nature, viz. that these should be free to all to use, and property of none (*Funk v. Kalo* 1948).

The *concurring* Federal Circuit Judge, Judge Kimberly Moore, fully agreed with Judge Lourie that *cDNA* should be patent-eligible. On the question of isolated full- or partial-length DNA, however, her position was, put simply, timid:

> If I were deciding this case on a blank canvas, I might conclude that an isolated DNA sequence that includes most or all of a gene is not patentable subject matter. … But we do not decide this case on a blank canvas. (*AMP v. USPTO* 2012, page 14)

Since the US Patent Office had been granting patents on isolated DNA for decades, she was reluctant to upset the settled expectations of patentees and patent applicants when extensive property rights were at issue. In effect, Judge Moore was torn between the aspiration of the US patent law of rejecting patent monopolisation of the natural, as developed in Supreme Court case law, and the aspiration of that same law in terms of stimulating industry, as embedded in the US Constitution. For her, changing the status quo in favour of one over the other required clear congressional guidance.

Judge Bryson, in his *dissent*, was more on a level with the earlier court decisions that had read 35 USC 101 to serve *to exclude from patentability products and processes that existed in nature even if they were not yet known to man*. He stated:

> In its simplest form, the question in this case is whether an individual can obtain patent rights to a human gene. From a common-sense point of view, most observers would answer, "Of course not. Patents are for inventions. A human gene is not an invention." The essence of Myriad's argument in this case is to say that it has not patented a human gene, but something quite different – an *isolated* human gene, which differs from a native gene because the process of extracting it results in changes in its molecular structure (although not in its genetic code). …
>
> [T]he genes are analogous to the "new mineral discovered in the earth" or the "new plant found in the wild" that the Supreme Court referred to in *Chakrabarty* (Diamond v. Chakrabarty 1980). It may be very difficult to extract the newly found mineral or to find, extract, and propagate the newly discovered plant. But that does not make those naturally occurring items the products of invention. … [T]he process of extracting minerals, or taking cuttings from wild plants, like the process of isolating genetic material, can result in some physical or chemical changes to the natural substance. But such changes do not make

> extracted minerals or plant cuttings patentable, and they should not have that effect for isolated genes. … …
>
> [T]here is no magic to a [covalent] chemical bond that requires us to recognize a new product when a [covalent] chemical bond is created or broken, but not when other atomic or molecular forces are altered. … Weaker interatomic forces will be broken when, for example, a dirty diamond is cleaned with water … but that does not make the clean diamond a human-made invention. (*AMP v. USPTO* 2012, pages 3–7)

As can be seen, the parties to and judges in patent cases relating to 'natural products' are avid users of images and analogies to demonstrate their arguments. This is particularly true of Judge Bryson who used the analogy of a *baseball bat* to demonstrate the difference between a product of nature and one of human ingenuity. A baseball bat, made from a single piece of wood by removal of the unwanted parts of the starting block of wood, is clearly not, in itself, a 'natural product', yet the whole of it is substantially the same as it existed in nature save only for the removal of the unwanted parts. The wood constituting the baseball bat existed, substantially unchanged, within the original block of wood, but that block was not a baseball bat, nor did it have the fundamental characteristics of a baseball bat, having that form and those characteristics was achieved by the removal of the unwanted wood following a human design. In contrast, a human gene is not endowed with its fundamental characteristics by the removal of extraneous genetic material – they were there already and are unchanged, fundamentally, by such removal. In Judge Bryson's interpretation, which in our view is the correct one, to be patentable over the natural (over the discovery) requires something more than being strictly novel over the natural. More specifically, Judge Bryson is arguing that being strictly novel does not make something an 'invention'.

### *13.4.4 The Extraordinary USA Amicus Brief*

In *AMP v. Myriad*, the USA, highly unusually, submitted *amicus curiae* briefs, both before the Federal Circuit and before the Supreme Court. The USA also presented its case orally at both these levels. Here we will only comment on the version of the USA's *amicus* brief presented to the Supreme Court. (United States of America 2013)

First, it should be noted that patent applicants tend to pursue claims to all possible aspects of their 'inventions', not all of which have real commercial value. In the *AMP v Myriad* case, it is arguable that the claims of greatest value were those to a method of detecting BRCA1/2 DNA abnormalities and to probes and primers useful in such methods. Elsewhere in the biotech world, however, claims to cDNA (usable to make commercially valuable proteins) might be of serious value. Probes are labelled DNA molecules, and thus not simply fragments of endogenous DNA. Thus the USA, in arguing for the unpatentability *only* of native DNA sequences, was not arguing for a position which would seriously damage the biotech industry except by virtue of the collateral damage that might result from finding isolated naturally occurring biochemicals in general, e.g. plant or bacterial products, to be unpatentable. In effect, the position of the USA was analogous to that set out by European legislators in Art. 5 of the European Biotech Directive – a positive symbolic stance in relation to what its public would see as the 'natural', and hence non-monopolisable, tempered by a negative symbolic stance to shield the commercial interests of the biotech industry.

The campaign against the BRCA patents of Myriad had come to be seen as a feminist, as well as an African-American, issue, and the USA's intervention must be seen in part as political damage-control. The issue, as presented to the public, however, was relatively easy to defuse – you are worried about your genes belonging to industry – no worry, the genes will not be propertised. This politically calming, but commercially harmless, approach did in due course succeed. The ACLU supposedly 'won' its case. (Even if Myriad was left with patent claims just as effective, if not more so, in blocking affordable BRCA1/2 testing in the USA.) Thus with the USA *amicus* briefs one sees the entry of pure politics, rather than commercial, judicial, or patient interests, into the debate on the patentability of natural products. Interestingly, the requests of the political player were met in full.

Again, we should bear in mind that the majority of Myriad's claims to *isolated DNA* were not of real commercial importance transposed to other fields. *cDNA*, however, is different. Where the protein coded for is itself a valuable commercial product, a claim to cDNA is hugely valuable, as imported into another cell line it can drive the production of that protein. Thus, we see the USA arguing for the patent-eligibility of cDNA, even though *the cDNA is clearly obvious if compared with the native DNA since the non-coding regions are obvious and their removal is too*. According to the USA, synthesised DNA, such as cDNA, should be patent-eligible because it does not occur in nature but 'is the product of significant human creativity' (United States of America 2013, page 9).

To a large extent, the USA's *amicus* brief was concerned with the fact that claims to naturally occurring DNA sequences might have the practical effect of *preempting all use of the underlying natural substance*, '[s]ince isolation is a prerequisite to meaningful productive use of native DNA, treating such changes as sufficient to support patent-eligibility would effectively preempt the public's use of the underlying product of nature.' (United States of America 2013, page 11) If a minor structural modification which left a natural substance's operative properties untouched were enough to transform the unpatentable discovery into a patentable invention, then, the USA argued, this might cause the removal of a kidney from the body, making it available for use, also to make the extracted kidney patent-eligible. For any pre-existing natural material to be useful to mankind, some changes might implicitly have to be made to it. For example, for a new plant, found in Brazil, to be useful in the USA, it must be transported to and grown there – the plants grown as a result are not the plants *exactly* as found in Brazil:

> The need to "isolate" a natural substance in order to study or exploit it is hardly unique to DNA. Many natural products – coal beneath the earth, cotton fibers, the stigmas of the saffron flower – must be physically separated, *i.e.*, "isolated" from their environments before becoming useful to mankind. Similarly, many highly reactive elements on the periodic table, such as lithium, boron, and barium, occur in nature only in chemical compounds. (United States of America 2013, pages 24–25)

### *13.4.5 The US Supreme Court Ruling: Who Has Really Won?*

The arguments put forward by the USA were clearly considered convincing by the Supreme Court. The unanimous decision of the Supreme Court was given in 2013 by Justice Clarence Thomas and found that claims to full-length and fragmentary human DNA were *not* patent-eligible under 35 USC 101, but that claims to cDNA *could* be accepted.

Justice Thomas hinted that the determination under 35 USC 101 was not devoid of considerations of novelty and inventiveness. Comparing the case at issue with those in *Diamond v. Chakrabarty* and *Funk v. Kalo*, he commented:

> It is undisputed that Myriad did not *create* or *alter* any of the genetic information encoded in the BRCA1 and BRCA2 genes. The location and order of the nucleotides existed in nature before Myriad found them. Nor did Myriad create or alter the genetic structure of DNA. Instead, Myriad's principal contribution was *uncovering* the precise location and genetic sequence of the BRCA1 and BRCA2 genes… The question is whether this renders the genes patentable … Myriad did not create anything. To be sure, it found an important and useful gene, but separating that gene from its surrounding genetic material *is not an act of invention*. Groundbreaking, innovative, or even brilliant discovery does not by itself satisfy the [35 USC] 101 inquiry. … [The patent claims fall] squarely within the law of nature exception. … Myriad found the location of the BRCA1 and BRCA2 genes, but that discovery, by itself, does not render the BRCA genes "*new* … compositions of matter," [35 USC] 101, that are patent eligible. (*AMP v. Myriad* 2013, pages 12–13, emphasis added)

Justice Thomas, then, was distinguishing between 'discovery' and 'invention', those two words in patent law symbolic of the desire to protect society by leaving the natural free to all to use and of the conflicting desire to stimulate industry by granting patent monopolies. The *AMP v. Myriad* ruling clearly implies that novelty is not enough to transform something discovered in nature into a monopolisable 'invention', an act of invention is required.[9]

## 13.5 The Situation in Europe

### *13.5.1 Background*

Let us now proceed to look at the European situation regarding the patentability of parts of the human body. We should start by pointing out that the dominant patent law in Europe, the European Patent Convention (EPC), is *not* an instrument of the European Union. The European Patent Office is not in any way bound by Directives of the EU, not least because the Member States of the EPC and of the EU overlap, but are not co-extensive.

---

[9] This has been developed further by the US Supreme Court in its more recent decision *Alice v. CLS* (2014), a decision relating to the patentability of computer-implemented inventions.

Having been finalized in 1973, the EPC came into force and the European Patent Office began accepting European patent applications in 1979. However, within the space of a few short years, the European Commission became concerned that Europe was falling behind the US and Japan in terms of biotechnological inventions, which had suddenly become of critical importance with the invention of techniques for DNA multiplication and cutting/splicing. Communications of October 1983 and February 1984, and a White Paper from June 1985 (European Commission 1983), led to the Commission's first proposal for a European Biotech Directive (EBD) in 1988 (European Commission 1988). The European Parliament considered this in 1992, and amended proposals were submitted in 1992 and 1995 (European Commission 1992, 1995a, b). After having rejected the proposed Directive in a plenary vote in 1995, the European Parliament considered a modified proposal again in 1997, and amended proposals (European Commission 1997) were again submitted, resulting in the European Parliament approving the European Biotech Directive in July 1998.

Interestingly, one of the main reasons why the Parliament voted against the proposed Directive in 1995 was said to be the lack of proper consideration of ethical issues, including in particular the patentability of parts of the human body. However, the text that was approved by the European Parliament in 1998 in fact essentially boiled down to the same as the proposal that was rejected in 1995. At first sight, the version that was approved in 1998 might have seemed like a "would-be compromise" whereby "antagonistic groups or parties [were] reconciled", as is sometimes the case for symbolic legislation (van Klink 2016). In reality, however, it seems that the original opponents of the proposed Directive, who changed their vote in 1998, were probably blinded by the lip service that was paid in the text to certain ethical principles, while failing to see that, in the next breath, so to speak, the very same principles were effectively emasculated. To give but one example, let us look again at Art. 5(1) and (2) of the European Biotech Directive:

> The human body, at the various stages of its formation and development, and the simple discovery of one of its elements, including the sequence or partial sequence of a gene, *cannot constitute patentable inventions. …*
>
> An element isolated from the human body or otherwise produced by means of a technical process, including the sequence or partial sequence of a gene, *may constitute a patentable invention, even if the structure of that element is identical to that of a natural element.* (European Union 1998, pp. 13–21, emphasis added)

Despite strenuously arguing that the Directive would not affect the meaning of the exclusions from patentability in the European Patent Convention, the European Commission was blatantly trying to do just that. Indeed, the provisions of the European Biotech Directive were incorporated into the Rules of the EPC in September 1999, just in time to urge the European Patent Office's Enlarged Board of Appeal to recognize them in one of its very worst decisions (*G-1/98 Transgenic plant/NOVARTIS II* 2000).[10] Yet the EBD-derived Rules might have been (and in

[10] In spite of the exclusion from patentability of plant and animal varieties by art. 53(b) EPC, this decision, handed down in December 1999, accepted the patentability of plants and animals as long

fact were, as the Enlarged Board of Appeal has meanwhile, reluctantly, acknowledged) incorrect interpretations of the exclusions from patentability laid down in Art. 52 and 53 EPC. Nevertheless, these incorrect interpretations continue to guide the granting of European patents on parts of the human body.

### *13.5.2 European Patent Law as Symbolic Legislation*

European patent law contains two particular examples of symbolic legislation in relation to human body materials – the exclusion from patentability in Art. 52(2) EPC of *discoveries*, rendered largely symbolic by the qualification in Art. 52(3) EPC to the effect that the exclusion is only of discoveries *'as such'*, and the statement in Art. 53(a) EPC that European patents are not granted for inventions which are *contrary to morality*. Both of these exclusions were tempered by the introduction into the EPC of implementing regulations in September 1999, which tracked the provisions of the European Biotech Directive mentioned earlier.[11] The new Rules made it clear that material discovered and separated, purified or isolated from its natural environment would no longer be a discovery 'as such'. This effectively enshrines in European patent law the isolated/purified test from *Parke-Davis v. Mulford*, the old US adrenalin decision discussed earlier in this chapter.

The part of the EPC of most relevance to the patenting of human body materials is Art. 52(2) EPC, which provides that:

> European patents shall be granted for any inventions, in all fields of technology, provided that they are new, involve an inventive step and are susceptible of industrial application. … The following in particular shall not be regarded as inventions… discoveries … (European Patent Office 2013)

This is qualified by the provision in Art. 52(3) EPC that Art. 52(2) EPC:

> shall exclude the patentability of the subject-matter or activities referred to therein *only* to the extent to which a European patent application or European patent relates to such subject-matter or activities *as such*.' (European Patent Office 2013, emphasis added)

This would seem to approximate to the position long held under US patent law, that products and phenomena of nature are not in themselves patentable, but that *practical applications of them* may be. However, the 'as such' provision of the EPC allows for the interpretation that *adding anything extra to the patent claim escapes the exclusion*. Since any discovery, to be useful, generally requires a bit more (e.g. isolation or purification if a natural product), the 'anything added' interpretation is

as they were claimed at a higher than lowest taxonomic level – thus rodents are potentially patentable, while *Mus mus domesticus* is not (see Sterckx and Cockbain 2012, chapter 7).

[11] In this chapter we do not have the space to explain the reasons why the new Rules tempered the morality clause (Art. 53(a) EPC) (see Sterckx and Cockbain 2012, chapter 8).

clearly required, *but not sufficient, if people are not to be excluded from using the pre-existing but not yet found bounties of nature.*[12]

As mentioned previously, in the field of human body materials, a patentee-friendly interpretation of the exclusion of discoveries was dictated by the European Biotech Directive and written into the Rules of the EPC in 1999 (*without* ratification by the member states): a natural product ceases to be a 'discovery' if isolated or purified. Thus, as will be discussed below, *as a result of this patentee-friendly reinterpretation of the law, the claims to 'isolated' full-length or fragmentary DNA that were rejected by the US Supreme Court in AMP v. Myriad could not be rejected by the EPO as not patent-eligible.*

Few EPO decisions have addressed the exclusion of 'discoveries'. In the following sections we will discuss the most relevant decisions of the EPO's Technical Board of Appeal, and make some general observations on the meaning of 'discoveries' and the interpretation of the exclusion.

### *13.5.3 T-272/95 Relaxin/Howard Florey Institute (2002)*

In December 1983 the Howard Florey Institute of Experimental Physiology and Medicine, based at the University of Melbourne (Australia), filed a patent application relating *inter alia* to DNA fragments encoding a protein precursor to the human hormone relaxin. Relaxin plays a role in facilitating birth since it relaxes or softens ('ripens') the cervix and reshapes the birth canal.

The contribution made by the patent applicants was said to be the production of relaxin in a potentially therapeutically useful form. Echoes of *Parke-Davis v. Mulford*, the adrenalin case discussed earlier, clearly come to mind.

During examination before the European Patent Office, Howard Florey introduced the term '*isolated*' into some of the claims, 'to differentiate the [claimed] gene from the naturally occurring chromosomal gene'. The patent was granted in 1991 (EP-B-112149) and an opposition was filed on behalf of the Green fraction of the European Parliament and the President of that fraction, *inter alia* on the basis of Art. 52(2) (a) EPC, i.e. the exclusion from patentability of discoveries.

According to the opponents, since nobody can invent genes, the claimed human gene was a discovery and not an invention, and thus unpatentable. They argued that isolating and purifying a gene did not change that; the gene performed the same function as it does in the human body.

Howard Florey responded by stating that they had made a surprising discovery that human relaxin was encoded by two separate genes which give rise to separate polypeptide products. The discovery of the H2-relaxin gene was described as

---

[12] Therefore, we would argue that *only subject matter which is both novel and inventive over a discovery* should be patent-eligible. Detailed arguments for this position are provided in Sterckx and Cockbain (2012).

'totally unexpected'. Howard Florey also pointed out that the claimed DNA omitted non-coding regions (i.e. was cDNA).

The Opposition Division of the EPO decided to maintain the patent as granted, and the opponents appealed. Their arguments regarding 'discovery' deserve to be quoted at some length:

> In simple terms, the Proprietor obtained a "code book" from the donors (the genetic material) and "cracked the code" (discovered the number and sequence of human relaxin genes). … In essence what the patentee has done is discovered something about the human body – that there are two relaxin genes and their sequences – and has sought to claim a monopoly in that sequence and proteins derived therefrom. … The gene and the proteins encoded by it are substances which have existed in nature, probably for many thousands of years. In so far as the Proprietor has given anything to the art it is only the discovery of the existence and sequence of the gene and the proteins. There is no element of invention beyond that mere discovery. There is no suggestion that there is any degree of invention in applying the coding information to the making of the relevant proteins. … [T]he patent relates to a discovery as such. (Aglietta et al. 1995, pages 3–4)

Here again we see the argument that 'invention' requires something more than simple novelty over 'discovery'.

The Technical Board of Appeal of the EPO issued its decision in 2002, *referring to the new (EBD-derived) EPC Rules*, and stating that:

> [The claims which] directly or indirectly relate to DNAs encoding the human protein [which] … are described in the patent in suit … as having been obtained by technical processes. They, thus, answer the definition of patentable elements of the human body given in Rule 23(e) (2) EPC. Accordingly, they do not fall within the category of inventions which may not be patented for being discoveries (Article 52(2) (a) EPC). (*T-272/95 Relaxin/ HOWARD FLOREY INSTITUTE* 2002, paras 7–9)

In other words, the *effect of the European Biotech Directive* was clearly far-reaching: *isolation* was itself enough to permit patentability of human body material. The conflict `between the aspiration of patent law to exclude discoveries from patentability was overridden by its aspiration to encourage the bioeconomy by enabling the commercialisation of (human and other) biological materials.

### *13.5.4 T-1213/05 Breast and Ovarian Cancer/University of Utah (2007)*

In 2001 Myriad Genetics was granted a European patent which related *inter alia* to isolated DNA coding for BRCA1. Eight parties opposed: *Sozialdemokratische Partei der Schweiz* (Switzerland); Greenpeace (Germany and Austria); *Institut Curie* (France); *Assistance Publique - Hôpitaux de Paris* (France); *Institut Gustave Roussy* (France); Belgian Society of Human Genetics (as well as thirteen other genetics and cancer societies); Dr Rolf Wilhelms (Germany) and the Netherlands (represented by its Minister of Health).

At the oral hearing in opposition in 2005, two opponents (Greenpeace and Dr Wilhelms) argued that the patent contravened Art. 52(2) EPC since genes are discoveries and thus non-patentable. The central concept underlying the patent application, according to Greenpeace, was the discovery of a link between a gene and a disease. Dr Wilhelms noted that the claimed sequence could be found in nature and was thus a discovery. The Opposition Division rejected this objection, referring to the new EPC Rules and to the Relaxin decision discussed earlier.

Both the patent holder and one of the opponents (the Swiss Social Democratic Party) appealed. Oral proceedings took place in 2007 and the Technical Board of Appeal maintained the patent giving the same reasons as in its decision in *T-272/95 Relaxin/HOWARD FLOREY INSTITUTE.*

### *13.5.5 T-666/05 Mutation/University of Utah (2008)*

In 2001, Myriad Genetics, the *Centre de Recherche du Chul* (Canada) and the Cancer Institute (Tokyo) were granted a European patent application with claims directed to *inter alia* isolated DNA coding for BRCA1 mutations and polymorphisms, and probes for determining whether a BRCA1 mutation indicative of it was present. Six oppositions were filed, by: *Institut Curie*, *Assistance Publique – Hôpitaux de Paris*, *Institut Gustave Roussy*, *Vereniging van Stichtingen Klinische Genetica* (VVSKG, together with the German, Danish and Belgian Society for Genetics, a German Cancer Society and Belgium, represented by its Ministers of Health, Social Affairs and Economic Affairs and Scientific Research), the Netherlands (represented by its Minister of Health), and Greenpeace.

Greenpeace invoked Art. 52(2) (a) EPC, arguing that genes were discoveries rather than inventions. Once again, following the earlier Relaxin decision, the opposition was rejected both by the Opposition Division and the Technical Board of Appeal.

## 13.6 'Discoveries' or 'Products and Phenomena of Nature': What Can It Mean?

Article 1, Section 8 of the US Constitution grants to the US Congress the power to 'promote the Progress of Science and useful Arts, by securing for limited Times to … Inventors the exclusive Right to their … *Discoveries'*. Thus, at first sight, a fundamental difference appears to exist between Europe (where discoveries are excluded from patentability) and the US (where discoveries are apparently regarded as appropriate subjects of patent monopolies). However, at the time the US Constitution was drafted, the meaning of the term 'discovery' was different from its current meaning. What it meant at the time would have been well known to the drafters of the Constitution, not least since it is set out in the then dominant and new

English dictionary (Johnson 1755, page 604)[13]: *the revealing of something that had been (kept) secret.*

In current usage, however, 'discovery' normally means *to find something that was pre-existing and the properties of which were already taking effect at the time of discovery* – one discovers a previously unknown plant or mineral, or one discovers that energy is proportional to mass squared, or indeed that a gene sequence codes for a protein. To the extent that native genes, their variants, and their correlation with disease states are *pre-existing*, identification of the gene and its variants represents a *discovery*, in the current sense of the term.

A *third meaning* of discovery which is highly relevant to patent law relates to situations where *something is identified which had not previously been in operation* (i.e. had not been in effect). Where a product of a marine microorganism, if injected into the human brain, can slow the progress of Alzheimer's, for example, then to use the product to treat Alzheimer's would involve the application of a 'discovery' that had not been previously in effect.

It is interesting to reconsider, in this light, the statement quoted earlier from the US Supreme Court in *Funk v. Kalo* that some discoveries are 'manifestations of laws of nature, free to all men and reserved exclusively to none':

> The qualities of these bacteria, like the heat of the sun, electricity, or the qualities of metals, are part of the storehouse of knowledge of all men. They are manifestations of laws of nature, free to all men and reserved exclusively to none. He who discovers a hitherto unknown phenomenon of nature has no claim to a monopoly of it which the law recognizes. If there is to be invention from such a discovery, it must come from the application of the law of nature to a new and useful end. (Funk v Kalo 1948, p. 130, emphasis added)

What is meant in US patent law by an unpatentable discovery, and what is meant by a discovery in the EPC, we would suggest, is a discovery in the sense of *something already in existence and effect* (e.g. the law of gravity, that energy and matter are interrelated, a newly found mineral, a newly found plant, an algorithm, and other products, phenomena or 'handiwork' of nature). *If such discoveries must be 'free to all men', then so too must be their use in manners which would have been obvious had their existence been known.* Otherwise, their exclusion from patentability would be substantially ineffective as their discoverer could patent all those obvious `uses. Put differently, where a use of a discovery is obvious, then simply proposing that use and seeking to patent it would prevent the discovery from being 'free to all men'. *If the exclusion from patentability of discoveries is to have any teeth, it must extend beyond the discoveries themselves to their obvious modifications and uses.*

## 13.7 Concluding Remarks

In this volume, van Klink (2016) describes two versions of 'symbolic legislation', a negative one and a positive one. We have found patent law to provide instances of both. What is at issue in patent law regarding the patenting of human body materials

[13] Discovery – (a) The act of finding anything hidden; (b) The act of revealing something secret.

are two conflicting aspirations within the same legal framework. This is illustrated by Art. 5 of the European Patent Directive (quoted above) where the signal is given that the monopolisation of human bodies through patents is forbidden, while at the same time the ban on monopolisation of human body materials is so qualified as to make them readily available for commercialisation. Clearly, compared with other areas of biolaw, such as the regulation of human tissue research, patent law is explicitly leaving the door open to commercialisation and so offers a poorer symbolisation of human body materials.

The tension between 'invention' and 'discovery', i.e. the definition of the dividing line between the two, has led in Europe to a (temporary) halt in favour of a boundary that favours commercialisation. By contrast, after a century of practice favouring commerce, the boundary in the US has shifted with the gradual acknowledgement by the US Supreme Court that the transition from what has been discovered to what can be patented must involve an 'act of invention'.

As we hope we have made clear in this chapter, both in the US and in Europe, many "members of the interpretive community" indeed "take seriously their collective responsibility to elaborate the given norms 'in the spirit of the law'." (van Klink 2016) However, a lot of work remains to be done before patent law can come to be regarded as an example of symbolic legislation in a positive sense (communicative legislation). It is particularly deplorable that in Europe, even though the statute law is more explicit and more clear than in the US, and even though structures for deliberation are clearly in place that enable direct communication between the courts and the addressees of the law, the current situation regarding the patenting of parts of the human body (as well as other things that are supposed to be excluded from patentability) is outright alarming.

## References

Agglietta, A., et al. 1995. Statement of grounds of appeal, 26 May 1995, in *T-272/95 Relaxin/ HOWARD FIOREY INSTITUTE* (2002). Available at https://register.epo.org/application?number=EP83307553&lng=en&tab=doclist. Last Accessed 30 Jan 2015.

*Alice Corp. v. CLS Bank International.* 2014. 573 U.S. ____

*Association for Molecular Pathology et al. v. Myriad Genetics, Inc., et al.* 2013. 569 U.S. ____.

*Ass'n for Molecular Pathology v. U.S. Patent & Trademark Office.* 2010. 702 F. Supp. 2d 181.

*Association for Molecular Pathology v. United States Patent and Trademark Office.* 2011. 653 F.3d 1329.

*Association for Molecular Pathology v. United States Patent and Trademark Office.* 2012. 689 F.3d 1303.

*D'Arcy v Myriad Genetics Inc.* 2014. FCAFC 115.

*Diamond v. Chakrabarty.* 1980. 447 U.S. 303.

Drahos, P., and J. Braithwaite. 2002. *Information feudalism: Who owns the knowledge economy?* London: Earthscan Publications.

European Commission. 1983. COM(83)672 final/2, SI(84)144 Annex IV, and COM(85)310.

European Commission. 1988. COM(88)496 final-SYN.

European Commission. 1992. COM(92)589 – final – SYN.

European Commission. 1995a. COM(95)661 final.

European Commission. 1995b. COM(95)661 –final/2.

European Commission. 1997. COM(97)446 final.

European Patent Office. 2013. *European patent convention,* 15th ed. Munich: EPO.

European Union. 1998. *Directive 98/44/EC of the European Parliament and of the Council of 6 July 1998 on the legal protection of biotechnological inventions.* OJ L.213, 30.7.1998, pages 13–21.

*Ex parte Latimer.* 1889. Dec. Comm'r Pat. 123, 46 O.G. 1638.

*Funk Bros. Seed Co. v. Kalo Inoculant Co.* 1948. 333 U.S. 127.

*G-1/98 Transgenic plant/NOVARTIS II.* 2000. Official Journal of the European Patent Office 111–141.

Harkness, J.M. 2011. Dicta on Adrenalin(e). Myriad problems with Learned Hand's product-of-nature. *Journal of the Patent and Trademark Office Society* 93(4): 363–399.

Hoeyer, K. 2016. Bioeconomy, moral friction and symbolic law. In *Symbolic legislation theory and developments in biolaw,* ed. Bart van Klink, Britta van Beers, Lonneke Poort, 161–176. Cham: Springer.

Johnson, S. 1755. *A dictionary of the English language.* Vol. 1, London: Strahan.

*Mayo Collaborative Services v. Prometheus Laboratories, Inc.* 2012. 566 U.S. ____.

Matthews, D. 2002. *Globalising intellectual property rights. The TRIPs agreement.* London: Routledge.

*Morton v. New York Eye Infirmary.* 1862. 17 F. Cas. 879.

*Parke-Davis & Co. v. H. K. Mulford Co.* 1911. 189 Federal Register 95.

Sell, S. 2003. *Private power, public law. The globalization of intellectual property rights.* Cambridge: Cambridge University Press.

Sterckx, S. 2008. The European Patent Convention and the (non-)patentability of human embryonic stem cells – The WARF case. *Intellectual Property Quarterly* 4: 478–495.

Sterckx, S., and J. Cockbain. 2010. Assessing the morality of the commercial exploitation of inventions concerning uses of human embryos and the relevance of moral complicity. *SCRIPT-ed – A Journal of Law, Technology & Society* 7(1): 83–103.

Sterckx, S., and J. Cockbain. 2012. *Exclusions from patentability. How far has the European Patent Office eroded boundaries?* Cambridge: Cambridge University Press.

*T 272/95 Relaxin/HOWARD FLOREY INSTITUTE.* 2002. Decision of European Patent Office Technical Board of Appeal 3.3.4 of 23 October 2002. Available at http://www.epo.org/law practice/case-law-appeals/pdf/t950272eu2.pdf. Last Accessed 30 Jan 2015.

*T-666/05 Mutation/UNIVERSITY OF UTAH.* 2008. Decision of European Patent Office Technical Board of Appeal 3.3.04 of 13 November 2008. Available at http://www.epo.org/law-practice/case-law-appeals/pdf/t050666eu1.pdf. Last Accessed 30 Jan 2015.

*T-1213/05 Breast and ovarian cancer/UNIVERSITY OF UTAH.* 2007. Decision of European Patent Office Technical Board of Appeal 3.3.04 of 27 September 2007. Available at http://www.epo.org/law-practice/case-law-appeals/pdf/t051213eu1.pdf. Last Accessed 30 Jan 2015.

*The American Wood Paper Co. v. The Fiber Disintegrating Co.* 1874. 90 U.S. 566.

United States of America. 2013. Brief for the United States as amicus curiae in support of neither party, submitted to the US Supreme Court in January 2013 in AMP v. Myriad. Available at http://www.americanbar.org/content/dam/aba/publications/supreme_court_preview/briefs-v2/12-398_neither_amcu_us.authcheckdam.pdf. Last Accessed 30 January 2015.

United States Patent and Trademark Office. 2014a. *United States Code Title 35 – Patents.* Available at http://www.uspto.gov/web/offices/pac/mpep/consolidated_laws.pdf. Last Accessed 9 Dec 2014.

United States Patent and Trademark Office. 2014b. *Interim Guidance on Patent Subject Matter Eligibility,* 79 Federal Register 74618–33.

Van Klink, B. 2016. Symbolic legislation. An essentially political concept. In *Symbolic legislation theory and developments in biolaw,* ed. Bart van Klink, Britta van Beers, Lonneke Poort, 19–37. Cham: Springer.

World Trade Organization. 1994. *Trade-Related Aspects of Intellectual Property Rights, Annex 1C of the Marrakesh Agreement Establishing the World Trade Organization, signed in Marrakesh, Morocco on 15 April 1994.* Available at http://www.wto.org/english/docs_e/legal_e/27-trips.pdf. Last Accessed 9 Dec 2014.

# Chapter 14
# Material Uncertainty: Nanomaterials, Regulation and Symbolic Legislation

**Robert G. Lee and Elen Stokes**

## 14.1 Introduction

A symbol is often a graphical representation that takes the form of something physical, tangible and recognisable. It is emblematic in the sense that it is taken to represent something other than the form of the symbol itself – in the way that a dove or an olive branch might be taken to represent peace. A symbol can and must be distinguished from a sign which refers to no more than its depicted object because the task of the symbol is to refer to that which is less well recognised or difficult to determine. Importantly here, what is represented by the symbol is a concept, idea or quality so that a material object conveys the immaterial, the physical form acting as agent to direct our consciousness towards the spiritual or ethereal. This will work only if there is a shared cultural understanding embedded in the symbol. Without this, the communicative power of the symbol is lost and the shorthand fails; clarity of the symbolism gives way to confusion. This suggests that inherent in the symbol is a degree of shared understanding and experience such that the symbol has some ideological quality capable of expressing collective identity (Boyd White 1989).

This chapter explores the symbolic capacities of legislation, in a context in which legislatures and policymakers face complex challenges and are called upon to act. That context is the European Union (EU) regulation of nanotechnology. Applications of nanotechnology promise to transform the world around us, imbuing it with new potential. The technology involves the measurement, manufacture, and manipula-

R.G. Lee (✉)
Law School, Birmingham Law School, University of Birmingham,
Edgbaston, Birmingham B15 2TT, UK
e-mail: R.G.Lee@bham.ac.uk

E. Stokes
Law School, Cardiff Law School, Cardiff University, Museum Avenue,
Cardiff CF10 3AX, UK
e-mail: StokesER@cardiff.ac.uk

 
B. van Klink et al. (eds.), *Symbolic Legislation Theory and Developments in Biolaw*, Legisprudence Library 4, DOI 10.1007/978-3-319-33365-6_14

tion of materials at the nanoscale. The minuscular size of nanomaterials imbues them with specific or enhanced physical and chemical characteristics, compared with equivalent materials in the conventional form. Yet the very materiality of these engineered properties gives rise to a future that is highly immaterial; impossible to discern and shrouded in uncertainty (Lee 2012). As such, this future cannot be consciously shaped, whatever we might wish its qualitative elements to be. But a symbol might serve the here and now, making reference, as with all symbols, to the intangible and giving a name to the nameless. Such a symbol might express collectivism and a shared intention to face, and face down, the future. Legislation may be symbolic in this regard, in that it has meaning beyond its immediate intrinsic significance and creates a sense of unity in confronting contingent futures.

Yet, a future that is uncertain and unpredictable also engenders difficulty in achieving consensus over the appropriate regulatory response (Rappert 1999; Groves 2013).[1] As well as creating fertile ground for conflict`, uncertainty can be perceived as a threat to political authority and its ability to identify and control the possible outcomes. Such uncertainty cannot be objectively addressed and contained by science itself since its very essence is constructed by society as well as science (Latour 1999). This social dimension is culturally seated such that the resolution of uncertainty is no simple matter of greater or better scientific endeavour. Politicians may feel under a pressure to act, but be unwilling or unable to broker any social consensus, and so resort to symbolic legislation. Often this conjures up a certain amount of unease about 'political show business' or 'placebo politics' (Blühdorn 2007), since the focus is on the passage of legislation rather than the content or consequences of those acts (Edelman 1967). Legislation of this sort is described as '*merely* symbolic' (Newig 2007, emphasis added) and 'more symbolic than functional' (Dwyer 1990), to emphasise its apparent superficiality and lack of bite. In these circumstances, 'symbolic' carries pejorative undertones, implying that the legislation is a surrogate for more meaningful or effective action. It is no accident that many of our sources here are taken from literature on environmental legislation which is bemoaned as ineffective (Black 2001) and rarely operating in action as it appears on the face of the statute (Ayres and Braithwaite 1992).

While there are different ways in which legislation may be symbolic, 'symbolism' is more commonly used in this latter sense of closing off opportunities for broader and inclusive dialogue – and, as such, is criticised for its lack of democratic and communicative ambition. Van Klink (Chap. 2 of this collection) conceives of this as a 'negative' representation of symbolic legislation, arising where legislation is directed primarily at political goals. This may include: giving the impression of control even if this has not been or cannot be achieved; improving the status of certain political groups over others; or reaching a workable compromise between different points of view. He contrasts this with a 'positive' approach to symbolic legislation, which, instead of being driven by any narrow political agenda, focuses on enhanced social interaction in the negotiation, interpretation and application of

[1] In the UK, National Foresight programmes have tried to overcome this problem. See e.g. Government Office for Science (2011).

legal norms. This interactive or communicative notion of symbolic legislation is chiefly concerned with the achievement of desirable social ends.

Our aim here is to examine whether, and to what extent, these categorisations of symbolic legislation map onto the EU regulation of nanotechnology. We find that 'negative' elements are unquestionably present, not least in the symbolism of legislation signalling futures under control in various regulatory and institutional contexts. At the same time 'positive' strands are woven into the legislation in the sense that the legislative process opens up space in which those futures can be the subject of debate, which, for now, may be the best that we can hope for in addressing the contingencies of technological development. Along the way, we explore the idea that decisions *not* to legislate can also have a symbolic function. Moreover, we argue that the regulatory environment in which legislation has now been introduced may limit the capacity of that legislation fully to achieve the aspirations of 'positive' symbolism. In the case of nanotechnology, efforts to improve the deliberative and participatory functions of the legislation are ultimately constrained. That many aspects of nanotechnology remain contested and unresolved has meant that legislative responses, even those that aspire to be symbolically 'positive', may struggle to have the desired effect – especially when the dominant legal paradigm continues to exhibit features representative of symbolic legislation in the thinner, 'negative' sense.

## 14.2 The Transformative Potential of Nanotechnologies

'Nanotechnology' is an umbrella term used to describe a group of emerging technologies that allow the design, characterisation, production and application of materials at a tiny scale. Materials are said to be 'nanoscale' if they have one or more dimension in the range one nanometre (one billionth of a metre) to 100 nanometres (BSI 2011: 3). To give a sense of the size, analogies are often drawn with everyday things: for example, a single human hair is usually about 80,000 nanometres (nm) wide and a red blood cell 7,000 nm wide (RS and RAEng 2004: 5).

The development of increasingly sophisticated methods of manipulating matter at the nanoscale has enormous commercial potential. Nanotechnologies enable the production of advanced materials which can have very different properties from the same materials in bulk form. Nanomaterials may be stronger or more chemically reactive, for example, or demonstrate different optical, electrical or magnetic behaviours. Their commercial potential is beginning to be realised across a diverse range of sectors – from healthcare to construction and energy. In medicine, and health the unique properties of nanomaterials are being exploited in medical diagnosis for the early detection of disease, in more targeted clinical therapy and in regenerative medicine for the reconstruction of damaged tissues (Patel et al. 2015: 528).

More trivial examples of the application of nanotechnology in consumer products include nano-textiles, such as crease-free and stain-resistant fabrics coated with nanoparticles, and anti-wrinkle face creams that contain nanoparticles and promise to lift and tighten skin. In those sectors in particular, 'nano' is seen as a powerful marketing brand and as a result it has generated a certain level of publicity, to the point that the label'nano' may be used purely for marketing purposes (Guere 2011). And even though the use of nanoscale gold in cosmetics has virtually nothing to do with (say) the use of polymeric nanoparticles in drug delivery, 'nanotechnology' has become an umbrella term to cover a diverse range of otherwise unrelated applications across different scientific, technological and professional spheres. This homogeneity of language in referring to nanotechnology as single and uniform science is not unimportant in symbolic terms for it encapsulates a shared hope for a brighter, better future heralded by this emergent technology.

In policymaking, this general categorisation of nanotechnology has become a focus of attention primarily because of hypothesised risks associated with certain materials at the nanoscale. It has long been recognised that the physical form of materials is a determining factor of their toxicity. While conventional assumptions about toxicity cannot necessarily be extrapolated to materials at the nanoscale, the very purpose of nanotechnology is that materials have been engineered to give them new form and purpose. Concerns have been raised that the very properties of nanomaterials that make them commercially so attractive could potentially create unforeseen hazards to human health or the environment (RS and RAEng 2004: 35). It is believed that some applications of nanotechnology 'will present risks unlike any that we have encountered before' (Maynard 2008: 6). Not all nanomaterials will pose a greater hazard than their bulk-scale counterparts; however it is conceivable that some nanomaterials will behave, in some cases, in previously unobserved and unfamiliar ways (see, e.g., RCEP 2008; Kendall and Holgate 2012).

This gives rise to a quandary in that, while we can speculate as to the possible effects of nanotechnologies, we are not yet in a position to articulate risks with any degree of precision. Even where there is evidence of a distinctive hazard profile, it may not be clear what this means in terms of risk because of the limited data currently available on the potential exposure routes, exposure levels and toxicity of certain nanomaterials (Maynard 2012). The uncertainty in this area is significant and is typically described in terms of data gaps and the lack of appropriate assessment tools for measuring risk. But uncertainty persists in other ways too, and there is room for further investigation into issues such as the social acceptability of various applications of nanotechnology and of associated incommodities. While more intensive scientific research on toxicology might provide assurance or steer us away from hazard, it is not of itself capable of risk reduction, since it cannot completely navigate the boundaries of ignorance rendering any expectation of complete risk characterisation unattainable. Moreover, other 'facts' about nanomaterials may be more complex and contested than they appear, given that scientific and technical knowledge is itself socially constructed and different players with different stakes in technology-related controversies envision diverse portrayals of 'progress' (Feenberg 2002). These dimensions of uncertainty pose significant challenges for policymakers, especially when it comes to devising regulatory frameworks which claim to rely on 'sound science' as if it were objective and context-free (Jasanoff 1995). The contrast brought about by nanotechnology, is starkly one of a capacity to enable the precise control physical matter at the atomic level while opening up indeterminacy and contingency at a global scale.

## 14.3 Regulatory Responses in the EU

It makes little sense to talk about 'the' EU regulation of nanotechnology as regulatory responses vary, not only among Member States[2] but also – importantly for this chapter – between EU institutions. The purpose here is to examine the regulatory responses of two institutions in particular: those of the Commission, and of the European Parliament. Each of their responses demonstrates a difference in approach to nanotechnology, which reveals the scope for divergence over the appropriate mechanics of regulation and which offers insights into the symbolic quality of the legislation as negotiated by the two institutions. In particular, they direct our attention to whether there is something about the nature of the problem (such as highly uncertain technology futures) or the context (including interactions between EU institutions) that shape the perspectives through which legislation in this area may be regarded as symbolic (Newig 2008: 93–4).

As indicated above, much has been written about symbolic legislation in the context of environmental protection, an area renowned for problems of complexity and high decision stakes that demand responses to threats of potential harm. As such, the term has become associated with 'laws which despite their often ambitious officially declared objectives are designed to remain ecologically ineffective' (Newig 2007: 93). Legislation is said to be symbolic as distinct from functional or instrumental, although it is noted that neither category is exclusive, and that both 'symbolic' and 'functional' represent ideal types rather than factually differentiated categories. It is also the case that symbolic legislation emerges in various forms, and encompasses a broader range of outcomes and activities than is often implied in the literature. For example, Blühdorn develops a typology of different varieties of symbolic politics, which also provides analytical traction for understanding the different ways in which legislation can be symbolic in character. He distinguishes between symbolic politics in the sense of substitute politics on one hand, and the use of symbols in political communication on the other. 'Substitute politics' – and, by extension, substitute legislation – refers to situations where political (or legislative) action does not achieve its apparent purpose. Such action, Blühdorn points out, may be very effective politically, but less so, or not at all, from the perspective of its declared objective (Blühdorn 2007: 256).

By contrast, symbols used in political communication help to 'extend the significance of a statement or action beyond the directly articulated meaning of immediate purpose' (Blühdorn 2007: 255). These symbols may provide tools to aid political

[2] Though to the extent that nanotechnology delivers goods and services which are the subject of free movement principles, the capacity of Member States to take unilateral action in relation to the regulation of such goods and services is necessarily limited.

integration or mobilisation by establishing a shared identity, or alternatively they can offer means of complexity reduction by condensing esoteric or convoluted issues into easily understandable, often visual, images. A more specific categorisation of the symbolic functions of legislation is developed by Van Klink, who also emphasises the broad distinction to be made between ‘thin’ and ‘thick’ accounts of symbolism (Van Klink 2005). He goes on to highlight that, contrary to traditional depictions of symbolic legislation as being devoid of function, ‘thin’ (or ‘negative’) conceptions of symbolic legislation can still have instrumental value. The point being that legislation as an exercise of political power is always symbolic both in the sense of projecting a message, and fashioning a process, however ineffectual the outcomes.

In this chapter we are concerned with the incidence of narrowly reductive and broader communicative forms of symbolic legislation, and the many shades of meaning in between, as illustrated by nanotechnology regulation. We approach the issue from the perspectives of two legislative actors, first examining the European Commission’s stance towards nanotechnology, and secondly presenting the European Parliament’s more recent involvement in the introduction of new, nanotechnology-specific legislation. We show the different ways in which those institutional responses may be construed as symbolic, pointing to examples at different values on the symbolism spectrum. Overall, our finding is that the legislation has been pragmatic and opportunistic in responding to calls for interventions to control nanotechnology without necessarily providing the effective control demanded. It is symbolic of a capacity to control technological futures without offering any tangible tools to do so. On the other hand, one might prefer this approach to one which ignored the calls for control since it is clear that procedural processes surrounding legislation provide a venue and some space in which dialogue might occur and consensus might be built.

## 14.4 The Symbolism of *Not* Legislating

We begin with an account of the European Commission’s stance on nanotechnology regulation, before inquiring into its wider, symbolic significance. The Commission is of the view that new and emerging applications of nanotechnology are already covered by existing regulations, even if those regulations were not designed for such a purpose. This is because existing regulations are broad in remit and are already geared to deal with issues of health and environmental risk. The General Product Safety Directive (Directive 2001/95/EC), for example, imposes an obligation on manufacturers and suppliers to place only safe products on the EU market (Article 3). It covers all products intended for, or likely to be used by, consumers (except products covered by sector-specific legislation), and does not distinguish between products of nano and non-nano form. Sector-specific legislation, such as that on cosmetic products, foods and medicines, contains similarly high standards of safety. Cosmetics, for instance ‘shall be safe for human health when used under normal or reasonably foreseeable conditions of use' (Regulation (EC) No No 1223/2009: Article 3); likewise food 'shall not be placed on the market if it is unsafe' (Regulation (EC) No 178/2002: Article 14).

So existing legislation is said to be comprehensive; it covers general categories of product, and imposes broad-brush conditions of safety without drawing a distinction between products of different technological origin. Consequently, it is assumed to apply nanotechnology-enabled products or substances in the same way as it applies to regular, non-nanotechnology equivalents (Lee and Vaughan 2010).[3] In other words, nanotechnologies can be described as being subject to 'inherited' regulation, passed down through different generations of product development (Stokes 2012). The Commission refers to the range of antecedent obligations, noting that '[v]irtually all product legislation imposes a risk assessment and the adoption of risk management measures' and that '[n]anomaterials are not excluded from this obligation' (European Commission 2008: 6). While the Commission acknowledges that legislation may be in need of updating as new information becomes available, it finds that 'current legislation covers to a large extent risks in relation to nanomaterials and that risks can be dealt with under the current legislative framework' (European Commission 2008: 3). The Commission has in large part maintained this position by choosing not to initiate, or even by resisting, legislative reform in this area.[4]

Just as decisions to enact legislation can symbolise the exercise of political power, so too can decisions *not* to legislate. Committed to a single market designed to promote economic growth (Cecchini 1988), the European Commission is keen, unsurprisingly, on the innovative possibilities of nanotechnology in promoting new trade in goods and services and driving economic activity within the market. In general, the Commission has been guided by technocratic and market-functional aims, such as those of increased high-tech innovation, wealth creation and new product generation. According to its vision of the 'Innovation Union', nanotechnology will contribute to the EU's future trajectory of market advancement and global leadership in the development of key enabling technologies (European Commission 2010). For instance, a report published by the Commission notes that 'the potential of nanotechnology to do good, or at least to make a profit, is clearly immense' (European Commission 2004: 46). It comments on the 'huge commercial potential' of the nanotechnology sector, and makes the claim that 'scientists and businessmen are unanimous: nanotechnology is much more than just a new "hype"' (European

---

[3] A good example of this is provided by control of chemicals under Regulation (EC) No 1907/2006 concerning the Registration, Evaluation, Authorisation and Restriction of Chemicals (REACH) and establishing a European Chemicals Agency [2007] OJ L136/3.

[4] There are limited exceptions to this. For example, the Commission has introduced two implementing acts designed to support the application of existing legislation to nanotechnologies used in the food sector: Regulation (EC) 450/2009 of 29 May 2009 on active and intelligent materials and articles intended to come into contact with food, Preamble 23; and Regulation (EU) of 14 January 2011 on plastic materials and articles intended to come into contact with food, Preamble 23, 28, Arts 9 (2), 13 (4)(b).

Commission 2004: 46). This is itself "hype" insofar as it talks up nanotechnology – the brighter, better future – while closing down other more sceptical dialogues.

Alongside this, and in relation to regulation, the tendency has been to assert that existing regulatory frameworks already do, or can easily be adapted to cover any risk posed by nanomaterials. This assurance is seen as necessary since 'the existence of law quiets and comforts those whose interests and sentiments it embodies' (Gusfield 1968: 58). More than this, however, intervention in the workings of an otherwise free market is seen as antithetical to the underpinning notions of economic liberalisation which drive the single market enterprise. Regulation is seen 'to be too intrusive and stifle market mechanisms, possibly affecting resource allocation and productive efficiency' (Conway, Janod and Nicoletti 2005: 4). This conveniently ignores the reality that the market itself is elaborately politically and legally constructed (Egan 2001) but in denying the case for further regulation and defending the s*tatus quo*: the symbolic politics 'secures the continuation of the system of democratic consumer capitalism and that resolutely obstructs the exploration of any socio-economic alternatives' (Blühdorn 2007: 253).

## 14.5 Opportunities to Legislate and Symbolic Action

The Commission's approach has sparked a backlash from within the Parliament, which depicts the Commission as tardy and unconcerned with the implications of nanotechnology for society, the environment and individual citizens. For instance, the Parliament's Committee on the Environment, Public Health and Food ('the Environment Committee') has said that it:

> [c]onsiders it highly misleading for the Commission to state, in the absence of any nano-specific provisions in Community law, that current legislation covers in principle the relevant risks relating to nanomaterials, when due to the lack of appropriate data and methods to assess the risks relating to nanomaterials it is effectively blind to its risks (European Parliament 2009a: 6).

Unpersuaded by the stance and assurance of the European Commission, the European Parliament's Environment Committee has positively campaigned for the introduction of discrete, nano-specific regulations, citing a need for regulatory amendment to reflect the fact that nanotechnologies may pose new challenges not envisioned by existing legislative measures. The Parliament, in its *Resolution on Regulatory Aspects on Nanomaterials*, notes that the Commission has provided 'only a general legal overview' of current legislation and in doing so has failed to consider the specific characteristics and implications of nanomaterials (European Parliament 2009d: para P). Moreover, the European Parliament '[d]eplores the absence of a proper evaluation of the *de facto* application of the general provisions of Community law in the light of the actual nature of nanomaterials' (European Parliament 2009d: para AA2) and does not accept the Commission's assertion that current legislation adequately addresses the potential risks associated with nanotechnologies (European Parliament 2009d: para AA3). The Parliament goes further, calling on the Commission to review current chemicals regulation with a view to treating all nanomaterials as 'new' substances, so that they are subject to separate methods of assessment and evaluation (European Parliament 2009d: para 11).

Acting on its findings of regulatory deficiency, the European Parliament has driven the process of legislative amendment. The pressure on the Commission by the Parliament has not been to initiate wholesale reform presumably because it doubted this would succeed, given the Commission's enthusiasm for promoting growth in the nanotechnology industry and avoiding unnecessarily restrictive regulation (European Commission 2005: 2). Instead the Parliament has introduced a series of incremental adjustments as and when existing legislation is scheduled for periodic review. So far, the Parliament has successfully incorporated nano-specific requirements into legislation on food additives (Regulation (EC) No 1333/2008, Art 12); food information for consumers (Regulation (EU) No 1169/2011, e.g. Art 18); cosmetic products (Regulation (EC) No 1223/2009, e.g. Art 13 (1) (f)); waste electrical and electronic equipment (Directive 2012/19/EU, Art 8 (4)); restrictions on hazardous substances in electrical and electronic equipment (Directive 2011/65/EU, Recital 16); biocidal products (Regulation (EU) No 528/2012, e.g. Art 19 (1) (f)); and food for infants (Regulation (EU) No 609/2013, Art 9 (2)).

For example, in the case of food additives, the legislation has been amended to require that, where an existing food additive that undergoes a 'significant change in its production methods or in the starting materials used, or there is a change in particle size, for example through nanotechnology', it will be treated as a new market entrant and subject to a separate safety evaluation (Regulation (EC) No 1333/2008: Art 12). Separate assessment and market approval is now also required for biocidal products containing active nanomaterial substances (Regulation (EU) No 528/2012: Arts 4 (4), 19 (1) (f), and 25). Moreover, there are new provisions stating that food, cosmetics and biocidal products containing nanomaterials must carry the label 'nano' (Regulation (EC) No 1223/2009: Art 2 (1) (k); Regulation No 1169/2011: Art 18 (3); Regulation (EU) No 528/2012: Art 58 (3) (d)). Manufacturers and suppliers of cosmetics are under an additional obligation to notify the Commission six months before placing products with nanomaterial ingredients on the market (Regulation (EC) No 1223/2009: Art 16 (3)). Those introducing food for infants or special medical purposes to the EU market must also demonstrate that food with engineered nanomaterial ingredients satisfy the nutritional requirements of, and is suitable for, the persons for whom it is intended, in accordance with generally accepted scientific data (Regulation (EU) No 609/2013: Art 9 (3)). Other provisions, such as that applicable to hazardous substances in electrical and electronic equipment, highlight the importance of taking into account the particular characteristics of nanomaterials in determining whether or not to restrict their market circulation, and encourage the substitution of nanomaterials for 'more environmentally friendly alternatives' (Directive 2011/65/EU: Recital 16).

We will comment later on the nature of these interventions but first there is the question of a wider symbolism of the European Parliament's grasping of the opportunity to legislate. In so doing, the European Parliament has sought to position itself as 'champion' of democracy and of the European citizen in the regulatory debate on nanotechnology. It has done this in several ways, including, for example, the publication of an own-initiative *Resolution* calling on the Commission to take a 'safe, responsible and integrated approach' (European Parliament 2009d: para 4) and to 'pay special attention to the social dimension of the development of nanotechnology' (European Parliament 2009d: para 26). The Parliament has also nailed its colours to the mast by emphasising that the use of nanomaterials 'should respond to the real needs of citizens and that their benefits should be realized in a safe and responsible manner' (European Parliament 2009d: para 1). This is in keeping with the EU consumer's right to choose whether or not to buy the products of new technology, a principle said by the Parliament to be at the core of EU consumer policy (European Parliament 2003: para 23). In Parliamentary debate, it has been noted that consumers 'might like to know whether a food has been produced by the use of nanotechnologies' (European Parliament 2008: Amendment 59), and that they 'need transparent information to be able to make informed choices and purchases' (European Parliament 2008: Amendment 8). As noted by one Member of the European Parliament (MEP): 'the contents must always be evident to the consumer so that anyone who wishes to do so is able to choose to avoid foodstuffs containing nanoparticles or nanomaterials' (European Parliament 2009b).

The language of all of this positions the Parliament as representing the European citizen against the unrepresentative Commission. The narratives in communication between the two institutions signify a power struggle with the Commission by a Parliament wanting to enhance communicative aspects of legislation and these narratives act 'as symbolic vehicles through which cultural creativity is carried out' (Feldman 1990: 809). The opportunistic legislative interventions, pursued by Parliament utilising legislation in areas such as foods and cosmetics that happen to be before it, becomes symbolic of a wider struggle to legitimate political influence. There were, for example, no known engineered nanoparticles in food sold on the European market according to the European Food Safety Authority (EFSA 2009). Similarly the policy area of medical devices is one in which there is no widespread concern regarding nanomaterials, which might well improve the functionality of such devices. Yet in 2012, following this activity by the Parliament, the European Commission requested the Scientific Committee on Emerging and Newly Identified Health Risks (SCENIHR) to deliver an opinion on the safety of medical devices containing nanomaterials (SCENIHR, 2015), in advance of producing draft legislation in this area (European Commission 2012). Thus the Commission were forced to react to re-assert some control over their own legislative agenda in the face of likely nano-related amendments in the Parliament.

## 14.6 Of Labels and Symbols

The question remains whether anything much lies behind the legislative interventions by the Parliament into Commission legislation. On one hand it might seem that these legislative changes are symbolic in the negative sense. For example, at no point do the Parliamentary amendments take a strongly precautionary approach and seek to ban products from the market or place a moratorium on their development. It should be apparent from the foregoing that all of the nano-specific provisions introduced so far focus on improving the flow of information on nanomaterials – whether it be through improved safety assessment, record-keeping of nanomaterial products in the EU market, or the disclosure of nanomaterial ingredients via labelling. The actions by the Parliament might be seen, therefore, to facilitate the market flow of the goods by labelling them (Stokes 2011), providing information about them and thereby enhancing the sovereignty of the European consumer in the market place. In doing so, the communicative aspects of the legislative change serve the market and market actors first and foremost, thereby enhancing 'the social status of groups carrying the affirmed culture and degrades groups carrying that which is condemned as deviant' (Gusfield 1996: 173). On this analysis, Parliament's approach to some extent reproduces the very problems it criticises – 'it conceals the post-ecologist resolve to defend the ecologically exploitative and destructive system of democratic consumer capitalism' (Blühdorn 2007: 269).

Concerns have been raised, more mundanely, about the capacity of 'nano-labels' to convey anything meaningful to consumers. For example, the prospect of nano-labelling has been rejected in other jurisdictions, most notably the United States, because a product's nanomaterial ingredients do not constitute 'material information' about the nature of the product or the consequences of its expected conditions of use (US Food and Drug Administration 2007). Understanding what is meant by the label 'nano' is not necessarily straightforward, not simply because applications of nanotechnology vary quite considerably between different sectors and product-lines, but also because their potential impacts are still matters of conjecture rather than established 'fact'. Confusingly, 'nano' may be adopted as a marketing device to sell products so that, for example, while an i-pod nano may actually utilise some nanotechnology in its screen display this is not the purpose of the label 'nano' which is there to promote the small and lightweight nature of the device. In this way, promotions of 'nano' as desirable may counteract labelling intended as a communication about product content.

There is also the issue that, although newly enacted requirements of the 'nano-label' are designed to communicate the nanomaterial content of a product, it may convey other, unintended messages too. For example, the nano-label might work to inform consumers not only of the presence nanomaterial ingredients but also of the different properties of these ingredients as compared with their bulk-size counterparts. Or, the nano-label could be construed as an indicator of scientific uncertainty or even of increased potential risk. EU policy advisors have been keen to stress that the label 'nano' should not be taken to suggest anything about the product's hazard or risk (SCENIHR 2015: 31). It remains possible, however, that 'products using the term "nano" might be misunderstood by consumers as a warning' (European Council 2009: 2), even though this is not the stated policy aim (European Parliament 2009c). 'Nano', therefore, is capable of sending mixed messages and, given these uncertainties, misgivings are often expressed about the capacity of nano-labels to help consumers to make more informed choices. If this is so, then the European Parliament's intervention, and the attempts to control nanotechnology in a manner which serves little instrumental purpose, suggests a process of 'symbolic re-assurance' (Edelman 1967). Worse still, it would appear to assert that highly uncertain and contingent futures attaching to innovation are capable of being discerned and disciplined. This is high symbolism for the legislation may be seen as capturing that which is truly beyond us.

## 14.7 Creating Space for Dialogue?

The interventions by the European Parliament, even if they lack immediate utility, might be seen in a more positive light, if seen as part of wider drive to improve the societal knowledge base and thereby open up communicative spaces. By quite simple legislative steps, described above, the Parliament forces a Commission response in the form of the Commission's own definition of a nanomaterial (European Commission 2011) and, albeit at a rather technical level, debate is engendered about the efficacy of competing definitions. The European Parliament has also sought to make up for what it considers to be a lack of detailed engagement with nanotechnology on the part of the Commission. Commenting on the Commission's treatment of nanotechnology regulation, one MEP concluded that:

> What is absent, above all, is the willingness to consider concerns other than safety risks, not least the issue of whether or not new technologies are desirable, or issues to do with people's convictions about life in general. The benefits and possible adverse effects must first of all be considered, in order to prevent choices being made solely on the basis of economic value while the technology is still at an early stage in its development. (European Parliament 2006)

In this regard, the Parliament has been publicly and proactively committed to raising policy awareness of the broader, societal issues at stake – not only in the context of nanotechnology but also as regards other emerging technologies. For example, it notes that the 'likely convergence of nanotechnology with biotechnology, biology, cognitive sciences and information technology raises serious questions relating to ethics, safety, security and respect for fundamental rights that need to be analysed' (European Parliament 2009d: Preamble).

In this more positive sense, the legislation does become a symbol around which people can unite. Risk governance literature has consistently emphasised the value of communication and inclusion in gaining the confidence necessary to re-assure. This may involve an integrated approach which draws upon a wide range of relevant knowledge and experience, including lay knowledge, and which is borne out of repeated reflection in the face of and concerning ignorance, uncertainty, ambiguity and complexity. The question is not the legislative question of 'how safe is safe enough' but a much wider and open question of 'how much uncertainty is the collective willing to exchange for some benefit (s)?' (Van Asselt and Renn 2011)

The rationality behind such analytic-deliberative approaches is to develop the potential to counter shortfalls in the different framings of risk (Macnaghten 2010). Only by the inclusion of multiple stakeholders (recognising throughout that the public at large is the ultimate stakeholder) throughout the process of risk governance will a robust and fair outcome be achieved while enhancing social learning which might prove invaluable in the risk governance effort (Webler, Kastenholz and Renn 1995). The legislative agenda of the European Parliament may be opportunistic rather than comprehensive. It may seize opportunity to attach controls to nanomaterials on the market in a manner which endorses the legitimacy of the market for such materials. Rather than any early and upstream debate on these issues, its controls are downstream interventions often at the point of contact between citizen and nanomaterial.

The danger of such 'downstream' approaches is a ready acceptance of the actuality of the technology only thereafter seeking to the control of risks. It ignores the possibility of points of departure on the entire question of the market circulation of the product or of points of divergence on issues of concern. Deliberative processes have frequently revealed fundamental concerns and divergent views about the very need for the technology while challenging the motives behind its deployment and raising questions of benefit sharing and reward (Doubleday 2007).

In spite of calls for greater upstream deliberation and public engagement, the risk regulation of nanotechnology has taken the form of downstream devices such as labelling. It remains to be seen whether the attaching of labels has the effect of closing down rather than opening up the issues. It might have such a dumbing down effect if the impression is that the label has somehow disclosed the necessary information and sealed off more vital discussion about science, values and what society expects from technology-based innovation. It is possible, however, that the informational provisions promoted by the European Parliament are a step on the way to a more open debate on these issues. This might be the case if, in the institutional skirmish on regulation of nanotechnology, the Parliament can seek to trump the technocratic vision of the Commission with a genuine resort to democratic appeal. Whether this is so depends not so much on whether people might be thought to be denied choice in the narrow sense of the ability to make meaningful decisions on the back of a label but on a much wider question of whether the informational provisions fostered by the Parliament can lead to open public discourse and democratic debate. If the only inclusivity of this approach is to allow the European citizen to join the ranks of those shopping for nanotechnology-enhanced products, then little will have been achieved and, in the light of the low instrumental value of the interventions pursued, the legislation adopted is symbolic of little more than a constitutional tussle with the Commission.

Yet if one asks the crucial question of whether one would want or would reject the legislative changes promoted by the Parliament, then they should be supported, if only because, without a greater flow of information, it is hard to see how the space for public participation and better processes of deliberative engagement can be created. However imperfect, it is hard to see that the legislation is 'merely' symbolic because around that symbol there is room to congregate and deliberate on the futures to which we may aspire.

## 14.8 Conclusion

In this chapter, we have tried to show that the regulation of nanotechnology inevitably is represented in symbols – often of promise but with undertones of menace. Legislating to regulate nanomaterials is inevitably symbolic of a capacity to control not just these materials but the prospects to which they will give rise. This may well be illusory; raising the question of what it is we hope for in legislation, especially when that legislation represents a political mêlée in which power is at issue and where regulatory reform becomes a battleground to facilitate or curtail dominant political choices in favour of the technology. The answer to that question may lie in the realisation that choices not to legislate may be no less symbolic. Legislation with the capacity to inform and engender debate may constitute a symbol around which we could develop a shared understanding that allows us to face contingency with greater collective confidence.

## References

Ayres, I., and J. Braithwaite. 1992. *Responsive regulation. Transcending the deregulation debate*. Oxford: Oxford University Press.

Black, J. 2001. Decentring regulation. Understanding the role of regulation and self-regulation in a 'post-Regulatory' world. *Current Legal Problems* 54(1): 103–147.

Blühdorn, I. 2007. Sustaining the unsustainable. Symbolic politics and the politics of simulation. *Environmental Politics* 16(2): 251–275.

Boyd White, J. 1989. *Heracles' Bow. Essays on the rhetoric and poetics of the law*. Madison: University of Wisconsin Press.

BSI (British Standards Institution). 2011. Vocabulary – Nanoparticles, Publicly Available Specification 71. London: BSI.

Cecchini, P. 1988. *The European challenge, 1992. The benefits of a single market*. Aldershot: Gower.

Conway, P., V. Janod, and G. Nicoletti. 2005. *Product market regulation in OECD countries: 1998 to 2003*. Paris: Organisation for Economic Co-operation and Development.

Directive 2001/95/EC on general product safety, as amended [2002] OJ L11/4.

Directive 2011/65/EU on the restriction of the use of certain hazardous substances in electrical and electronic equipment [2011] OJ L174/88.

Doubleday R. 2007. Risk, public engagement and reflexivity. Alternative framings of the public dimensions of nanotechnology. *Health, Risk and Society* 9(2): 211–227.

Dwyer, J.P. 1990. The pathology of symbolic legislation. *Ecology Law Quarterly* 17(2): 233–316.

Edelman, M. 1967. *The symbolic uses of politics*. Urbana: University of Illinois Press.
Egan, M. 2001. *Constructing a European market. Standards, regulation and governance*. Oxford: Oxford University Press.
European Commission. 2004. *Nanotechnology. Innovation for tomorrow's world*. Brussels: European Commission Directorate-General for Research.
European Commission. 2005. Communication on nanosciences and nanotechnologies. An action plan for Europe 2005–2009. COM 243 final.
European Commission. 2008. Communication on regulatory aspects of nanomaterials. COM 366 final.
European Commission. 2010. Communication on Europe 2020 Flagship Initiative Innovation Union. COM 546 final.
European Commission. 2012. Proposal for a regulation of the European Parliament and of the Council on medical devices, and amending Directive 2001/83/EC, Regulation (EC) No 178/2002 and Regulation (EC) No 1223/2009. COM 542 final.
European Parliament. 2003. Resolution on the commission communication to the European Parliament, the Council, the Economic and Social Committee and the Committee of the regions on 'consumer policy strategy 2002–2006'. P5_TA(2003) 0100.
European Parliament. 2006. Member of the European Parliament (MEP) Bastiaan Belder on behalf of the Freedom and Democracy Group. European Parliament debate No 4 of 28 September 2006.
European Parliament. 2008. Report on the proposal for the regulation of novel foods. A6-0512/2008.
European Parliament. 2009a. Motion for a European Parliament resolution on regulatory aspects of nanomaterials. A6-0255/2009.
European Parliament. 2009b. MEP Åsa Westlund on behalf of the Party of European socialists. European parliament debate No 16 of 24 March 2009.
European Parliament. 2009c. MEP Dagmar Roth-Behrendt (Rapporteur). European Parliament debate No 15 of 23 March 2009.
European Parliament. 2009d. Resolution of 24 April 2009 on regulatory aspects of nanomaterials, P6 TA(2009)0328.
Feldman, S.P. 1990. Stories as cultural creativity. On the relation between symbolism and politics in organizational change. *Human Relations* 43(9): 809–828.
Feenberg, A. 2002. *Transforming technology. A critical theory revisited*. Oxford: Oxford University Press.
Government Office for Science. 2011. *Foresight. Migration and global environmental change*. London: Government Office for Science.
Groves, C. 2013. Four scenarios for nanotechnologies in the UK, 2011–2020. *Technology Analysis and Strategic Management* 25(5): 507–526.
Guere, G.P. 2011. Labeling nano-enabled consumer products. *Nano Today* 6: 117–121.
Gusfield, J.R. 1968. On legislating morals. The symbolic process of designating deviance. *California Law Review* 56(1): 54–73.
Gusfield, J.R. 1996. *Contested meanings. The construction of alcohol problems*. Madison: University of Wisconsin Press.
Jasanoff, S. 1995. Procedural choices in regulatory science. *Technology in Society* 17(3): 279–293.
Kendall, M., and S. Holgate. 2012. Health impact and toxicological effects of nanomaterials in the lung. *Respirology* 17(5): 743–758.
Latour, B. 1999. *Pandora's hope. Essays on the reality of science studies*. Cambridge, MA: Harvard University Press.
Lee, R.G. 2012. Look at mother nature on the run in the 21st century. Responsibility, research and innovation. *Transnational Environmental Law* 1(1): 105–117.
Lee, R.G., and S. Vaughan. 2010. REACHing down. Nanomaterials and chemical safety in the European Union. *Law Innovation and Technology* 2(2): 193–217.
Macnaghten, P. 2010. Researching technoscientific concerns in the making. Narrative structures, public responses and emerging nanotechnologies. *Environment and Planning A* 42(1): 23–37.

Maynard, A.D. 2008. *Nanotechnology. A research strategy for addressing risk*. Washington, DC: Woodrow Wilson International Center for Scholars.
Maynard, R.L. 2012. Nano-technology and nano-toxicology. *Emerging Health Threats Journal* 5: 17508.
Newig, J. 2007. Symbolic environmental legislation and societal self-deception. *Environmental Politics* 16(2): 276–296.
Newig, J. 2008. Symbolic environmental legislation and societal self-deception. In *The politics of unsustainability. Eco-politics in the post-ecologist era*, ed. I. Blühdorn and I. Welsh. London: Routledge.
Patel, S., R. Nanda, and S. Sahoo. 2015. Nanotechnology in healthcare. Applications and challenges. *Medicinal Chemistry* 5(21): 528–533.
Rappert, B. 1999. Rationalising the future? Foresight in science and technology policy co-ordination. *Futures* 31(6): 527–545.
RCEP (Royal Commission on Environmental Pollution). 2008. *Novel materials in the environment. The case of nanotechnology. Twenty-seventh report, Cm 7468*. London: RCEP.
Regulation (EC) No 178/2002 laying down the general principles and requirements of food law, establishing the European food safety authority and laying down the procedures in matters of food. 2002. OJ L31/1.
Regulation (EC) No 1333/2008 on food additives. 2008. OJ L354/16.
Regulation (EC) No 1223/2009 on cosmetic products. 2009. OJ L342/59.
Regulation (EU) No 1169/2011 on the provision of food information to consumers. 2011. OJ L304/18
RS and RAEng (Royal Society and Royal Academy of Engineering). 2004. *Nanoscience and nanotechnologies. Opportunities and uncertainties*. London: RS.
Stokes, E. 2011. You are what you eat. Market citizens and the right to know about nano foods. *Journal of Human Rights and the Environment* 2(2): 178–200.
Stokes, E. 2012. Nanotechnology and the products of inherited regulation. *Journal of Law and Society* 39(1): 93–112.
US Food and Drug Administration. 2007. *Nanotechnology. A Report of the US FDA nanotechnology task force*. Rockville: FDA..
Van Asselt, M., and O. Renn. 2011. Risk governance. *Journal of Risk Research* 14(4): 431–449.
Van Klink, B. 2005. An effective-historical view on the symbolic working of law. In *Social and symbolic effects of legislation under the rule of law*, ed. N. Zeegers et al. New York: Edwin Mellen Press.
Webler, T., H. Kastenholz, and O. Renn. 1995. Public participation in impact assessment. A social learning perspective. *Environmental Impact Assessment Review* 15(5): 443–463.

# Chapter 15
# The Democratic Legitimacy of Interactive Legislation of the European Union Concerning Human Embryo Research

**Nicolle Zeegers**

## 15.1 Introduction

In May 2014 the European Commission refused the request of nearly two million citizens, mobilized by the European Citizens' Initiative ('ECI') *One of us*, to submit a legislative proposal in order to ban all human stem cell research.[1] The reason given was that Member States and the European Parliament had only recently discussed and decided the EU policy in this area and had agreed to conditionally fund human embryonic stem cell research. However reasonable the latter may sound, the refusal of such a successful ECI, which was only the second one the Commission decided on, can be regarded as highly remarkable considering the fact that the ECI is meant to be a tool for giving ordinary citizens access to the political agenda of the EU in order to improve its democratic legitimacy. Could the recent closure of a long lasting and complex decision making procedure be a legitimate reason for the Commission to refuse the wish of so many contenders to start all over again?

---

A draft version of this chapter was discussed at the authors' conference in Amsterdam. I would like to thank the conference participants, and especially Baerbel Dorbeck-Jung, Lisette ten Haaf, Lonneke Poort and Wibren van der Burg for their helpful comments. I thank Adam McCann for his corrections of the English.

[1] *One of us* proposes an amendment to Article 16 of the *Proposal of a Regulation of the European Parliament and Council that establishes a framework program for research and innovation (2014–2020) – Horizon 2020-* COM (2011) 809 final edition. Section 3 of this article, that concerns the ethical principles, stipulates what research areas are not funded and the proposed amendment is to add here under d: *research activities that destroy human embryos, including those aimed at obtaining stem cells, and research involving the use of human embryonic stem cells in subsequent steps to obtain them* (Proof of legal act on official website).

N. Zeegers (✉)
Faculty of Law, University of Groningen,
Oude Kijk in 't Jatstraat 26, Groningen 9712 EK, the Netherlands
e-mail: n.e.h.m.zeegers@rug.nl



B. van Klink et al. (eds.), *Symbolic Legislation Theory and Developments in Biolaw*, Legisprudence Library 4, DOI 10.1007/978-3-319-33365-6_15

What is at stake here is how to determine the norms that should guide EU funding of research involving the use of human embryos considering the clear differences that exist between European Member States' and interest groups' positions concerning the moral status of the early human embryo. The mobilization of a considerable number of European citizens against the existing EU policy by the ECI *One of us* highlights the deepness of the disagreement concerning this issue. This has been the motive for addressing the democratic legitimacy of the decision making process in the EU concerning norms for research with human embryos in this contribution.

This decision making will be approached as a case of interactive legislation at the EU level in order to answer the questions of whether the norms for funding research with human embryos have been established in a democratically legitimate way and, more importantly, what are the criteria to measure such legitimacy in interactive legislation? As such this contribution focuses on symbolic or interactive legislation as legislative strategy to discuss biotechnological matters (Denoted as part three in the introduction to this book): it analyses and evaluates how such legislative strategy helps the EU to cope with differences in EU member states' moral stances on the use of early human embryos in research. By focusing on symbolic legislation as a legislative strategy, this contribution acknowledges that a dominant function of law in the case discussed is to express values in the political sphere (see Chap. 2, this volume).[2] In the first section, I will introduce the case of norm formulation concerning research with human embryos as part of the EU research funding programme and explain why this can be regarded as interactive legislation. Subsequently, the role of ECIs will be addressed and it will be made clear that the yardstick for legitimacy in this case of EU decision making must be predominantly inferred from the idea of deliberative democracy. In the third section, this yardstick will be applied and in the fourth section, a conclusion will be drawn about how such legitimacy could be improved.

## 15.2 The EU Case of Legislating Research with Human Embryos and the Interactive Approach

The European Union's involvement with the use (and destruction) of human embryos in research is not direct regulation, as this is not within its competence, but in setting out the rules for EU funding of such research. The authority to legally regulate the use of human embryos in research lies with the national legislatures (Plomer 2010).[3] As can be inferred from their national laws; fundamental differences exist in the moral and ontological status accorded to the early human embryo

---

[2] However, different to Van Klink and following Dahl and Lindblom (1953), political in this contribution refers to (efforts to) resolve inevitable conflicts in a peaceful manner.

[3] The European Convention on Human Rights in *Evans v UK* has left the protection granted to human embryos to the discretion of member states in recognition of the diversity of national moral cultures on this question.

in the Member States of the EU. In some member states the human embryo has the full right to protection from the moment of its conception, such as in Germany and Ireland, whereas in other Member States human embryos that not have reached the development stage of 14 days can be, and are indeed, used in research. In 1990, the German and the UK government were the first two governments in the EU to enact laws addressing this issue. With respect to the content, these countries' laws are almost opposites. The German *Embryo Protection Act* (1990) forbids the use of human embryos in research, defining this embryo as 'the fertilized, human egg capable of development, from the time of conception onwards as well as any totipotent cell taken from an embryo that, given the right conditions, could develop into an individual'. The UK *Human Fertilisation and Embryology Act* (1990) (HFE Act) takes a fundamentally different approach (Lee and Morgan 2001).[4] With the 'no, unless' formulation in this Act, flexibility is built into the legal regime in such way that research is conditionally allowed, even with embryos that are specially created for the research.[5]

For the UK and other Member States of the European Union that allow the use of human embryos for research, such as France, the Netherlands, Belgium and Spain, the life-saving possibilities that human embryonic stem cells offer, outweigh the respect that should be paid to human embryos younger than 14 days. Said jurisdictions do not deem the early human embryo unworthy of any protection but apply the gradualist approach by departing from the rule that in any case only (material derived from) embryos that have not yet reached the development stage of 14 days can be used in research.[6] The (minority of) Member States that hold the principle that human embryos should legally be treated as persons from the moment of conception reject the use of human embryos in any case. Although such absolutism does not seem to leave any room at all for the formulation of compromises, the

---

[4] Although some of the rules in this Act are also straightforward prohibitions, for example, the prohibition on the development of an embryo in vitro beyond 14 days. Human Fertilisation and Embryology Act 1990, schedule 2, para. 3 (2) and (6).

[5] With the Act, the HFE Authority – as a statutory body – was set up and made responsible for licensing the use of embryos in research. In order to get the approval of the HFE Authority, researchers who wish to use embryos in their research must submit a research protocol that makes clear that the research is "necessary or desirable" for one or more of the purposes of the *HFE Act*. Desirability is assessed in terms of the contribution to scientific knowledge or human health that can be expected from the research. 'Necessary' means that creating such embryos (instead of using other sources of stem cells) is necessary for the research. In addition to being convinced on the latter point, the Authority cannot issue a license unless it is satisfied that the creation and/or use of embryos are necessary for the research.

[6] In addition, the use of human embryos in research is only allowed under the following four conditions: Firstly, such research has to comply with the necessity principle which means that the results expected could not be reached otherwise. Secondly, the research must serve the research aim of increasing scientific knowledge in basic research or medical knowledge for the development of diagnostic, preventative or therapeutic methods to be applied to humans. Thirdly, there must be informed consent of the donors of the embryos in accordance with national law and the donor's personal data have to be protected; Fourthly, research proposals will be subject to scientific and ethical review.

opposite is true. In Germany, for example, such compromise was evident from the moment scientific developments led to the expectation of great biomedical advances to be made in embryonic stem cell research, such as finding therapies for irreversible organ and tissue failure. Because of these expectations, Germany, at the beginning of the twenty-first century, felt the need to make possible the use of embryonic stem cells in research. The definition of the embryo in their *Embryo Protection Act* (1990) did not preclude the possibility to import embryonic stem cells from abroad.[7] In 2002, such importation was made possible with the Stem Cell Act.[8] Austria and Denmark have chosen an approach similar to Germany in allowing importation of embryonic stem cell lines while internally prohibiting their procurement from human embryos (Salter 2007).

The tension between principles and promises, whether in terms of positive effects on citizens' future health or in terms of economic revenues, is clearly something all Member States of the EU have tried to accommodate by choosing the 'right' policy concerning the use of human embryos in research. For those adhering to a gradualist approach concerning the moral status of the human embryo, compromises may be more obvious'. However, also adherents of the other approach combine their principle of absolute protection with allowing the import of human stem cells from abroad or with allowing the use of therapies resulting from research that has included the destruction of human embryos. So whereas compromises for them at first sight seem to be impossible and out of the question, at closer look compromises turn out to be conceivable. In any case, at the EU level, the different actors have the task of finding common ground in the existing norms concerning the use of human embryos in research in the Member States. Such common ground is necessary to formulate an EU policy on the funding of such research that a majority can agree on.

Why may the decision making process concerning the norms for EU research funding with human embryos be approached as interactive legislation? The day-to-day decision making in the EU is done by a range of actors, Member States, EU institutions as well as interest groups. The lack of a clear locus of power, such as the

[7] This is because human embryonic stem cells that already have been harvested are pluripotent and not totipotent and by this definition do not fall within the ambit of the Act. Pluripotent cells are cells capable of giving rise to every type of cell found in the human body, whilst totipotent cells are also capable of developing into cells needed for human development, including extra-embryonic tissues, for example the placenta. The definition of embryo adopted by the Embryo Protection Act encompasses totipotent, but not pluripotent, cells.

[8] This Act stipulates what stem cell lines could be allowed to be imported: Firstly, these have to be derived before 1 January 2002 in the country of origin according to legal provisions applicable in that country (the cut-off date criterion); secondly, were obtained from embryos produced by medically-assisted in vitro fertilisation in order to induce pregnancy, but were no longer used for this purpose and, thirdly, no compensation or other benefit in money has been granted for donation of ESC (embryonic stem cells). The setting of the cut-off was meant to make sure only stem cells of supernumerary embryos that already exist at the moment of enactment can be imported in order to avoid an indirect pressure on foreign IVF-clinics to produce more human embryos in future than before in order to have more supernumerary embryos left which could provide in German scientists' need for stem cell lines.

executive or the legislator, makes such EU decision making different from decision making by national governments. As Bomberg et al. (2012) contend, Member States, EU institutions and interest groups must bargain and share power in order to reach an agreement acceptable to all actors or at least most of them. Of course, there is a clear procedure of decision making for legislation: The European Commission drafts and tables the proposals for legislation and subsequently the Council (Member States represented) and the European Parliament examine the proposal, make amendments to it and are the ones who finally have to approve it. However, debating different viewpoints and negotiating compromises to a considerable extent is done in, mostly informal, communications preceding and adjacent to the formal procedures. This debate and such negotiation concerning the ethical boundaries of the use of embryos in research takes place between representatives of the Commission, Member States and interest groups positioned in a more or less horizontally relation to each other. Because of the importance of horizontal, interactive processes in this process of norm development it fits into Van der Burg's description of interactive legislation (Van der Burg 2014: 145) and Poort's description of a communicative approach to law (see Chap. 5, this volume). Also, a second characteristic of 'interactive legislation' is present as the formulation and the implementation of the norms in the context of conditions for funding are more or less merged into one continuous process (see Van der Burg's chapter in this volume). A third characteristic of 'interactive legislation' that can be discerned is that the emphasis in this case of legislation is more on the codification of norms that have emerged in the societies of the Member States, or more precisely the common denominator in these different norms, than on the modification of norms.

## 15.3 Deliberative Democracy and ECIs

Biotechnological issues such as the use of human embryos for research bring with them specific challenges for legislators, also at the level of the European Union. Firstly, no widespread agreement exists on the fundamental ethical issue of when human life begins and in connection with this what are the rights of the early human embryo. Secondly, the developments in biomedical technology are constantly opening up new possibilities, such as cures and therapies for disease and genetic defects. In the decision making process concerning the use of the human embryo in research, the rights to protection of the early human embryo have to be weighed up each time anew against such new possibilities. Because of these challenges, law making concerning the issue is complicated and cannot be done without the advice of scientific experts. But what about ordinary citizens, what role would citizen participation have to play in processes of law making that regard such complex matters?

Robert Dahl, an important political scientist on pluralism, with regard to having citizens participate in legislation processes concerning complex, technical, matters has pointed at the relevance of such citizens' competence (Dahl 1961).[9] After all the resulting decisions in addition to being democratically supported need to be rational in the sense of presenting the best solution to a problem (Dahl 1970; Blokland 2011: 215). Dahl in such cases would rather accept that the policy would be formulated by a minority, in this case predominantly including experts in embryonic stem cells, because they would at least be knowledgeable concerning the technical details of the issue.[10] However, according to the adversaries of Dahl's work, his neglect of the importance of citizens' participation in policy making is exactly the weak point of Dahl's idea of democratic policy making (Munnichs 2000: 28; Held 1987).

Undoubtedly, in our current society public participation in policy making concerning biotechnology has become more popular as well as more urgent than Dahl in his time could conceive. Relevant differences in practical circumstances between current society and the society at Dahl's time are, firstly, the higher level of information and education of current citizens; secondly, the general acknowledgement that experts' knowledge is not purely objective nor fallible and, thirdly, that by entering a dialogue with its citizens about biotechnology, a government can build trust in the institutions that regulate it (Bovenkerk 2011: 95).

Current scholars in pluralist theory, such as Bruce Ackerman, Gutmann and Thompson, have anticipated such changes by incorporating the idea of deliberative democracy in their work (Gutmann and Thompson 2004). Deliberative democracy is a form of democracy in which deliberation is central to decision making. Mere voting is not the primary source of legitimacy for the law but authentic deliberation is.[11] Deliberative democracy has also been presented as a solution for the downside of the big role interest groups in modern society have acquired in law making. As Marxist and neo-corporatist adversaries of Dahl's interest group democracy expected and warned against, such form of democracy has been accompanied by power concentration because of the big economic power of multinational corporations, respectively, the privileged access of some interest groups, such as employee and employer organizations, to government. The involvement of other interest groups and unorganized citizens in the legislative decision making would present a counterforce against such concentration of power.

How does the citizen participation advocated in deliberative democracy relate to the legitimacy of government decisions in representative democracy? Procedures of representative democracy are built around finding a compromise between political parties. The legitimacy of this compromise is related to the number of seats political parties have in parliament, which at its turn is decided by popular election. So, the number of seats each political party occupies in parliament is an indicator for the relative weight each viewpoint has in the representative democratic decision making process as well as the compromise resulting from this process. After all, parliament has to adopt such compromise as the end decision.

---

[9] Robert Dahl is an important theorist on pluralism who also did empirical research into whether and how pluralism works in practice.

[10] Provided that in other policy areas the policy would be formulated by other minorities.

[11] Such deliberation can result in decisions based either on consensus or a majority.

The procedures of deliberative democracy are built around a search for consensus, instead of compromise. EU decision making as such can already be characterized as consensual in the sense described by Sudbery as 'the Commission seeking optimal consensual outcomes' holds 'long informal discussions with a large number of actors before proposals are even made public'(Sudbery 2003: 81). The danger of such consensual style is that it hides conflicts 'in viewpoints and stifles debate. Poort (2012, 2013 and this volume) has shown that in several cases of national 'deliberative democratic' decision making in the field of biotechnology this search for consensus has led to decisions that rested on the exclusion of controversy and diverging viewpoints and through this exclusion formed an impediment to further norm development. According the original idea of 'deliberative democracy', a decision that is based on consensus would be the ideal result of decision making. However, Poort concludes that conceiving of consensus as the ideal outcome or orientating aim of decision making is not helpful. Instead, Poort wants the adherents of deliberative democracy to embrace the ethos of controversies. With the latter ethos it would be acknowledged that the outcome of 'deliberative democratic' decision making most of the time is a 'temporary political achievement' and that this outcome rests on a compromise. Otherwise, by denying that diverging viewpoints still exist at the moment a decision is taken, according to Poort, we would run the risk, firstly, that the supporters of this diverging view point will radicalize and turn their back at the decision and the procedures behind it. Secondly, Poort fears the rationality of decisions would be undermined by the exclusion of such viewpoints, as taking all viewpoints in consideration leads to a clearer picture of what is at stake and a better definition of the problem.

Poort rightly points at the dangerous consequences of the exclusion and denial of certain viewpoints. This brings us to two legitimacy problems that 'deliberative democratic decision making' has in comparison with 'representative democratic decision making'. The first problem is the lack of an indicator for the relative weight that should be given to the different viewpoints that have been brought into the process of deliberation. As explained above the involvement of a parliament in the decision making would solve this problem. Therefore it would be preferable to use 'deliberative democratic' decision making only as an extension of the decision making that takes place in the representative democratic institutions.[12] As in the case of EU decision making addressed in this article decisions are adopted by a parliament in the end – European parliament – we can leave this event here. A second legitimacy problem revolves around the question of how long (new) viewpoints should be brought into the process before a decision can be taken. The advocates of 'deliberative democracy' want to open up the debate for as many actors as possible but in order to reach an agreement it would be necessary at some point in time to draw a line (Van Klink 2014: 25). In other words: inclusiveness as requirement on account

[12] In fact, also in the cases Poort (2012, 2013) analysed the deliberative procedures run through in addition to and not instead of representative democratic procedures. In the Dutch context the CAB, for example, sends it recommendations to the Minister who decides. This offers opportunity to parliament to ask questions concerning these decisions.

of the finiteness of time brings a dilemma that touches the central question of this article: the legitimacy of EU decision making concerning the ethical norms guiding the funding of the use of embryos in research. The fact that the ECI *One of us* required to open up the debate concerning the possibility of such funding begs the question of what line can be drawn in regard with admitting (new) viewpoints and start the decision making all over again. Before elaborating on how to solve this dilemma in Sect. 15.3.2, in Sect. 15.3.1 the ECI *One of us* and the Commission's reply will be described and given context by elaborating on the function and role of ECIs more generally.

### *15.3.1 The European Citizens' Initiative 'One of Us'*

The European Citizens' Initiative is meant to give citizens as (a group of) individuals access to the political agenda of the EU.[13] It is a legal tool created for citizens to call on the European Commission to propose legislation on matters where the Commission has competence to do this.[14] A Citizen's Initiative has to be supported by at least one million EU citizens, coming from at least seven different Member States. In addition for each of these Member States, a minimum number of signatories is required.[15] The Commission is obligated, within 3 months after the initiative has been submitted, to communicate its legal and political conclusions on a Citizens' Initiative as well as the action it intends to take and the reasons for doing this.[16] However, as the Commission is free to decide not to take any action at all, not each Citizens' Initiative that is in accordance with the rules for submission will automatically evolve into an item on the EU agenda. Considering the extent of support that has to be organized for an ECI to be submitted, it is in fact predominantly an instrument for interest groups to gain access for a specific issue to the EU agenda by mobilizing the support of citizens across the Member States. In the case of the initiative *One of us* the interest groups involved consist of religious groups belonging to the Catholic, Orthodox and Protestant churches. The *Brüstle judgment* of the

[13]The European Commission has the monopoly of legislative Initiative. The European Parliament, the European Council as well as a quarter of the member states can ask the commission for a law proposal. With the ECI also citizens under certain conditions can ask the commission to initiate such law proposal.

[14]In article 1 of *Regulation (EU) No 211/2011* this is formulated as follows: "citizens' initiative" means an initiative submitted to the Commission in accordance with this Regulation, inviting this commission, within the framework of its powers, to submit any appropriate proposal on matters where citizens consider that a legal act of the Union is required for the purpose of implementing the Treaties.' The *Regulation*, containing the rules and procedures governing the ECI, was adopted by the EP and the Council of the EU on 16 February 2011. The Citizen's Initiatives could be launched from April 1, 2012.

[15]The minimum number of signatories per member state is the number of Members of the European Parliament delegated by the state multiplied by 750.

[16]*Regulation (EU) No 211/201*, article 10 (1) (c).

Court of Justice of the European Union has encouraged the initiators because they consider this 'a weighty argument allowing the Court to grant legal protection to the human prenatal life' (Puppinck 2013). The initiators of the ECI pursued the goal of demanding the EU to end the financing of activities which presuppose the destruction of human embryos. *One of us* has proven to be successful in gathering almost two million signatories, more than enough for the European Citizens' Initiative to be submitted to the European Commission. Still the Commission, as it announced in its reply to the ECI *One of us*, decided not to submit a legislative proposal concerning the request made.[17] According to Máire Geoghegan-Quin, European Commissioner for *Research, Innovation and Science* the Member States and the European Parliament had agreed to conditionally fund human embryonic stem cell research because 'Embryonic stem cells are unique and offer potential for life-saving treatments, with clinical trials already underway'. Because these actors just recently had decided this policy, after long processes of deliberation, the Commission did not want to submit a legislative proposal that would reopen debate and restart the decision making process all over again.

### 15.3.2 The Deliberative Approach

Could this refusal of the Commission be assessed as democratically legitimate according to the deliberative approach? This question brings us back to the dilemma raised by the requirement of inclusiveness in this approach. A procedure for solving this dilemma can be inferred from Poort's *Ethos of controversies* as will be explained here. According to the deliberative approach, as many viewpoints as possible should be addressed in the decision making. Poort, for good reasons, advocates an *Ethos of Controversies* (instead of an ethos of consensus) to be applied in cases of norm development concerning ethically controversial research (Poort 2012, 2013 and this volume). This *Ethos of Controversies* would, firstly, structure the discussion around a focus on the variety of the differences in viewpoints, concerns and preferences instead of on the commonalities. Secondly, in this discussion the various viewpoints should be confronted with each other in order to have the actors explain, think through, and reconsider their viewpoints. However, Poort also admits that the end decisions have to be made and that at that point some extent of disagreement will probably still exist. Therefore, thirdly, she asks for awareness of conflicts 'that still exist after the decision has been made. Acknowledgement of such conflict and the fact that decisions have a political character would stimulate a more open debate during further norm development.

This author's approach is primarily focussed on the question of how to keep the debate open for as many viewpoints as possible. However, at the same time she admits that decisions have to be taken at a point where disagreement still exists.

---

[17] European Citizens' Initiative: European Commission replies to 'One of us', press release, 28 May 2014, Brussels.

So apparently this is the point where the actors adhering to a viewpoint that they feel is not satisfactorily represented in the decision have to accept their loss, at least temporarily. When is that point reached and is the Commission, as they did with respect to *One of Us*' request, allowed to close off the decision making process? Although Poort does not address this question directly, the criteria for when this would be legitimate can be inferred from her ethos of controversies (Poort 2013 and this volume). Firstly, the viewpoint presented has been addressed already in the decision making and, secondly, the decision makers at the end the decision making process have acknowledged that the compromise reached only was a 'temporary political achievement'.

## 15.4 Analysis of the EU Decision Making

Has the viewpoint that 'any destruction of human embryo should be banned from funding because it is to be treated as a person from the moment of conception' been included in the decision making processes concerning the norms for research funding with human embryos? In order to answer this question the decision making processes concerning the *Sixth and Seventh Framework* for research and technology, in 2002 respectively 2006, and the more recent decision making process concerning *Horizon 2020* have been analysed.

Unanimity has existed all along concerning two activities to be excluded from EU funding: human cloning for reproductive purposes and the modification of the genetic heritage of human beings which could make such changes heritable (germ line gene therapy).

Divergence in viewpoints is most manifest concerning the question whether research activities intended to create human embryos solely for the purpose of research or for the purpose of stem cell procurement should be funded.[18] On 7 September 2000, the European Parliament passed a resolution opposed to such creation. As Salter contends the emotive and categorical terms in which the debate in parliament was couched, showed that the opponents of human embryo stem cell research were not prepared to negotiate or compromise (Salter 2007). According to this author, the *European Group on Ethics in Science and New Technology* (EGE) subsequently, in the arena of ethical advice, has made several efforts to distinguish 'ethical components' in order to create room for negotiation and the formulation of compromise.[19] The notion that the embryo's status depends on its source and for example would be different for 'embryos that are specially created for research' than for 'supernumerary embryos'(created for the treatment of infertility but no

[18] Including the creation of human embryos by means of somatic cell nuclear transfer (commonly referred to as therapeutic cloning).

[19] The *European Group on Ethics in Science and New Technology* is a group of experts who advises the Commission, the European Council and the European Parliament on issues concerning science and technology.

longer needed to fulfil parent's wish to have children) would stem from this group's report concerning human stem cell research (EGE 2000). Salter, pointing in addition at the opinion EGE formulated concerning the patentability of cells obtained from embryos, strongly suggests EGE's 'ethical refinement' was meant to pave the way for the formulation of compromises concerning the issue of the use of human embryos in research (EGE 2002; Salter 2007:280). With these efforts, EGE would respond to demands of scientists and industries for more regulatory protection of their investments in human embryonic stem cell research.

With respect to the decision making process concerning the *Sixth Framework Programme* this search for compromises also became visible in the formal procedures. During the First Reading of the Commission's proposal for this programme in Parliament, adherents of the viewpoint of absolute protection of the human embryo had made it clear they would strongly oppose the possibility of the EU funding of the use of supernumerary or spare embryos. In June 2002, the Parliament voted through the overarching *Sixth Framework Programme* but moved the issue of what they labelled *the ethical conditions for human embryo and human embryonic stem cells research* to the decision making procedure of the relevant *Specific Research Programme*. By this move, ethical conditions, among which the question whether spare embryos should be exempted or not, were turned into a 'technical matter' to be decided upon by the Council without the agreement of Parliament. Under pressure of Austria, Germany, Ireland and Italy, the Council decided for a moratorium on the EU funding of human embryo and human embryonic stem cells research until December 2003. The Council asked the Commission first to propose further guidelines concerning the ethical conditions that should guide Community funding of such research and these guidelines were to be produced by December 2003.

In June 2003, the Commission presented its proposal concerning such guidelines. As Salter explains, with this proposal the Commission tried to accommodate the critics of the use of human embryos as such by formulating a compromise based on the selection of embryos on the basis of criteria of its source and the date of its creation.[20] The Commission proposed a guideline that reads as follows: only research conducted upon supernumerary embryos created prior to 27 June 2002 and stem cells derived from those embryos will be eligible for funding (European Commission European Commission 2003).[21] By the restriction to supernumerary embryos the Commission proved to have listened to the amendments that earlier had

[20] Another manner in which the EU contributes to optimizing research, while at the same time making sure embryonic stem cell lines will only be created if necessary, is the funding of a European registry for human embryonic stem cell lines. On 29 March 2007 it was announced in press that the European Commission had agreed to fund such registry in their Research Framework Programme. The main objective of this registry is to provide comprehensive information about all human embryonic stem cells lines available in Europe.

[21] The commission motivates this proposal as follows: 'In order to allay fears that Community funding might indirectly encourage the production of embryos by in vitro fertilisation (IVF) over and above the number required and to send out a political signal, the Commission is proposing that only supernumerary embryos created before 27 June 2002 (date of adoption of the 6[th] Framework Programme) can be used.'

been brought forward during the First Reading in Parliament (European Parliament 2001).[22] However, by adding a cut-off date, reminiscent of the rather strict German legislation concerning the import of human embryonic stem cell lines, it incited protests from international scientists who warned against the application of this date of embryo creation criterion because of its impact on the freedom and quality of their research (Research Europe Research Europe 2003). The European Parliament seems to have agreed with this scientific view and, as under the terms of the consultation procedure, removed the 27 June 2002 restriction from the ethical guidelines as well as enlarged the embryos source criterion to include human embryos produced by spontaneous or therapeutic abortion and supernumerary embryos from IVF treatment (without further restriction) (European Parliament 2003). However, the Commission although receiving a favourable opinion from the European Parliament did not succeed in subsequently getting the proposal accepted by the Council of 3 December 2003. In practice, this resulted in no research with newly obtained human embryonic stem cells being funded under the *Sixth Framework Programme* (EGE 2007: 15 and 26). Under the *Seventh Framework Programme* for research, technological development and demonstration (2007–2013) such obtainment of human embryonic stem cells also was excluded. The Council agreement was based on a Commission Declaration stating that

> The European Commission will continue with the current practice and will not submit to the Regulatory Committee proposals for projects which include research activities which destroy human embryos, including for the procurement of stem cells. The exclusion of funding of this step of research will not prevent Community funding of subsequent steps involving human embryonic stem cells. (Decision No 2006/1982/EC: 20).

On 18 December 2006 the *Seventh Framework Programme*, including the above Commission Declaration, afterwards denoted as *the 2006 compromise*, was adopted in co-decision of the Council and the European Parliament. EU funding in fact has been restricted to the use of already banked or isolated stem cells. The *2006 compromise* became contested again in the decision making process concerning Horizon *2020*, the framework program for research and innovation (2014–2020). At the one hand there were organizations that wanted to widen the possibilities. The European Humanist Federation (2013), for example, wanted the EU 'when appropriate and where permitted' also to allow European funding for the creation of human embryos for research purposes or for the procurement of stem cells. This would require an amendment of Article 19.[23] On the other hand, several MEP's in the European Parliament's plenary debate on Horizon 2020, expressed their support to *One of us*

---

[22] The compromise was that the Sixth *Framework Programme* would fund 'research on "supernumerary" early-stage (i.e., up to 14 days)'by which were meant 'embryos genuinely created for the treatment of infertility so as to increase the success rate of IVF but no longer needed for that purpose and when destined for destruction'. European Parliament (2001).

[23] The European Humanist Federation (EHF) unites more than 50 humanist and secularist organisations from about 20 European countries and promotes a secular Europe, defending equal treatment of everyone regardless of religion or belief, fighting religious conservatism and privilege in Europe and at the EU level.

seeking to end the financing of activities that presuppose the destruction of human embryos.[24] In December 2013, the framework program was adopted by the Council (European Union 2013). In fact, *Horizon 2020* has finally maintained the 7th framework position in which the EU only funds research activities involving existing human embryonic stem cell lines.[25]

To conclude: The analysis of the EU decision making concerning the norms for research funding with human embryos shows that the viewpoint of absolute protection of the human embryo has been amply taken into consideration. At some point efforts have been made to downplay the controversy between its adherents and the proponents of human stem cell research. For example, in 2002 the ethical guidelines were turned into a 'technical matter' and moved away from the European Parliament to another decision making procedure of which the EP was excluded. However, the proponents of absolute protection among the Member States have made sure its reappearance at centre stage. The Council under this pressure set a moratorium on the EU funding of the use of embryos in research until the end of 2003 and subsequently the Council has made sure that no research with newly obtained human embryonic stem cells would be funded.

## 15.5 Conclusion and Discussion

In this article an argument has been developed concerning how to determine whether ethical norms for EU research funding have been established in a democratically legitimate way. The case study specifically concerned the funding of research involving the use of human embryos. However, considering the divergence in cultural norms that exists across the European Union, the argument's relevance may be wider.

The norm development concerned fits into the interactive approach to legislation, at least to a considerable extent. Therefore its democratic legitimacy should be assessed by asking two questions that are inferred from Poort's *Ethos of Controversies*: firstly, have all the existing viewpoints concerning the norms been addressed in the decision making and, secondly, did the decision makers at the end of the decision making process acknowledge that the compromise reached was only a 'temporary political achievement'.

The answer to the first question with respect to the case study is positive. The analysis above shows that the viewpoint of the supporters of *One of us* had been taken into consideration all along the EU decision making processes. Therefore the refusal of the European Commission to submit the legal proposal of *One of us* to a considerable extent can be regarded legitimate.

---

[24] Franz Obermayr (NI, AT) Konrad Szymański, Ewald Satdler and Miroslaw Piotrowsk (ECR, Poland) 20 November 2013.

[25] MEPs, such as Carvalho, rapporteur of the Horizon 2020 specific programme, have declared this in the hearing concerning *One of Us*.

What about the second question, the acknowledgement by the decision makers that the outcome is 'a temporary political achievement' and rests on a compromise? The communication of the Commission about the EU funding policy concerning the use of human embryos in research shows that room for improvement exists on this point. Apparently, the promises for public health that human embryonic stem cells offer, are seen as the decisive reason to conditionally allow the funding of the use of such cells in research. However, would it not be fair to the dissenting minority in this light to also address the option of reconsideration in the case where such promises will not be fulfilled? Of course researchers should be given enough time to find out whether the research works out the way they expect. But really acknowledging the temporary political character of the decision would come down to specifying points in time where results can be assessed and a change of policy can be discussed. In other words, where not only the political debate will be opened up again but also a new decision making process could start. A new technological development would in any case be a good reason for re-starting the policy debate. A development that for instance would promise to find a cure for other human defects would incite a demand for a wider facilitation (read funding) of the use of human embryos in research. As alluded to in Sect. 15.2, such developments are exactly the challenges policy makers in the field of biotechnology have to cope with.

This would leave us with two conditions under which the EU, in the light of democratic legitimacy, should be flexible enough to re-open the decision making process concerning the funding of research with human embryos: Firstly, when it is no longer conceivable that the promises of this research for public health will be fulfilled within a certain time limit and, secondly, when new technological developments make still other improvements of public health possible.

## References


Blokland, Hans. 2011. *Pluralism, democracy and political knowledge. Robert A. Dahl and his critics on modern politics*. Farnham: Ashgate Publishing Limited.

Bomberg, Elizabeth, et al. 2012. Introduction. In *The European union. How does it work?* ed. Elizabeth Bomber et al., 3–18. Oxford: Oxford University Press.

Bovenkerk, Bernice. 2011. *The biotechnology debate. Democracy in the face of intractable disagreement*. Dordrecht: Springer.

Burg, Van der. 2014. *The dynamics of law and morality*. Farnham: Ashgate Publishing Limited.

Dahl, Robert, and Charles Lindblom. 1953. *Politics, economics and welfare* . New York: Harper & Row.

Dahl, Robert. 1961. *Who governs? Democracy and power in an American City, New Haven*. New Haven: Yale University Press.

Dahl, Robert. 1970. *After the revolution? Authority in a good society*. New Haven: Yale University Press.

Decision 2002/834/EC of the European Parliament and of the Council of 30 September 2002 ***c** oncerning the Sixth Framework for research and technology, Official Journal L* 294, 29/10/2002.

Decision 2006/1982/EC of the European Parliament and of the Council of 18 December 2006 concerning the Seventh Framework Programme of the European Community for research, technological development and demonstration activities (2007–2013).

European Commission. 2003. Proposal from DG of the European Commission, 26 November 2003. (15385/03), p. 2.

European Group on Ethics in Science and New Technology. 2000. *Ethical aspects of human stem cell research*. Opinion no. 15 to the European Commission.

European Group on Ethics in Science and New Technology. 2002. *Ethical aspects of patenting involving human stem cells*. Opinion no. 16 to the European Commission.

European Group on Ethics in Science and New Technologies. 2007. *Recommendations on the ethical review of hESC FP7 research projects*, Opinion no. 22 to the European Commission.

European Parliament. 2001. Proposal for a European Parliament and Council decision concerning the multiannual framework programme 2002–06 of the European Community for research, technological development and demonstration activities aimed at contributing towards the creation of the European Research Area, COM(2001) 94, 14 November.

European Parliament. 2003. Resolution on the proposal for a Council Decision amending Decision 2002/834/EC on the specific programme for research, technological development and demonstration. Integrating and strengthening the European Research Area (2002–06), COM (2003)390, 19 November.

European Humanist Federation. 2013. Keep dogma out of European Research. Position paper on the Framework Programme for Research and Innovation. "Horizon 2020" (2014–2020). http://humanistfederation.eu/wp-content/uploads/2012/05/2.-Horizon-2020-EHFs-amendments-proposals3.pdf. Accessed 10 Apr 2014.

European Union. 2013. Regulation No 1291/2013 of the European Parliament and Council of 11 December that establishes Horizon 2020 – The framework program for research and innovation (2014–2020).

Gutmann, Amy, and Dennis Thompson. 2004. *Why deliberative democracy*. Princeton: Princeton University Press.

Held, David. 1987. *Models of democracy*. Cambridge: Polity Press.

Klink, Bart van. 2014. Symboolwetgeving. De opkomst, ondergang en wederopstanding van een begrip. *Regelmaat* 29: 5–26

Lee, R., and D. Morgan. 2001. *Human fertilisation and embryology. Regulating the reproductive revolution*. London: Blackstone Press Limited.

Munnichs, Geert. 2000. *Publiek ongenoegen en politieke geloofwaardigheid. Democratische legitimiteit in een ontzuilde samenleving*. Assen: Van Gorkum.

Plomer, Aurora. 2010. Human Dignity, Human Rights and Article 6(1) of the Biotech Directive. In *Embryonic stem cell patents. European law and ethics*, ed. Plomer Aurora and Torremans Paul. Oxford: Oxford University Press.

Poort, Lonneke. 2012. An ethos of controversies. A critical analysis of the interactive legislative approach. *Legisprudence* 6: 35–55.

Poort, Lonneke. 2013. *Consensus & controversies in animal biotechnology. An interactive legislative approach to animal biotechnology in Denmark, Switzerland, and the Netherlands*. Den Haag: Eleven International Publishing.

Puppinck, G. 2013. Synthetic analysis of the EJC case C-34/10 Oliver Brüstle v Greenpeace eV and its ethical consequences. European Center for Law and Justice. http://eclj.org/Resources. Accessed 10 Feb 2014.

Research Europe. 2003. EU stem cell guidelines raise supply issues. Research Europe, July 31, p. 4.

Salter, Brian. 2007. Bioethics, politics and the moral economy of human embryonic stem cell science. The case of the European Union's sixth framework programme. *New Genetics and Society* 26: 269–288.

Sudbery, Imogen. 2003. Bridging the legitimacy gap in the EU. Can civil society help to bring the union closer to its citizens? *Collegium* 26: 75–95.

# Chapter 16
# Changing Expectations of Experts: The Symbolic Role of Ethics Committees

**Lonneke Poort and Bernice Bovenkerk**

## 16.1 Introduction

Rapid developments in medical and biotechnological research demonstrate the enormous influence new technologies can have on our daily life.[1] Preconception analysis can be used to identify genetic diseases in an early stage. Stem cell researchers from Cardiff University won the Nobel prize in 2007 for improving diagnosis of cancer and contributing to the development of better treatments.[2] At the same time, the use of embryos for stem cell research is highly controversial as it touches upon the very concept of life. It raises the question to what extent we want to identify the genetic diseases of our unborn children. Moreover, the risks and consequences are not yet fully understood.

The lack of knowledge about risks and consequences on the one hand and the controversies about the concept of life on the other, turn these medical and biotechnological issues into a complex policy problem. Addressing these issues on a political level is difficult as these issues are so intricately bound to fundamental values. Decision making will, therefore, always involve favouring one position over the

[1] For example, the creation of in vitro meat could drastically change the way we produce our protein. http://nos.nl/artikel/536816-veestapel-weg-door-kweekvlees.html, last accessed 5 August 2013. See also Van der Weele (2014).

[2] M.R. Capecchi, M.J. Evans, and Oliver Smitties, nobel prize for medicine 2007, http://www.nobelprize.org/nobel_prizes/medicine/laureates/2007/index.html, last accessed July 2013.

L. Poort (✉)
Department of Sociology, Theory and Methodology, Erasmus University Rotterdam, Erasmus School of Law, Burgemeester Oudlaan 50, 3062 PA Rotterdam, the Netherlands
e-mail: lonnekepoort@gmail.com

B. Bovenkerk
Philosophy Group, Wageningen University, Leeuwenborch, Hollandseweg 1, Wageningen 6706 KN, the Netherlands
e-mail: b.bovenkerk@uu.nl



B. van Klink et al. (eds.), *Symbolic Legislation Theory and Developments in Biolaw*, Legisprudence Library 4, DOI 10.1007/978-3-319-33365-6_16

other. As these issues concern fundamental values, decision making always involves heavy political conflicts'. Governments have the tendency to avoid these conflicts. In some areas where conceptions of the good life are concerned, such as citizens' practicing their religious beliefs or sexual conduct between consenting adults, governments can generally avoid political conflicts by making sure that each person can make her own decisions. In cases such as our treatment of animals or novel technologies this will not work, as the choices that some persons make will affect the lives of others and the lives of animals. Not making a decision is as politically loaded as making a decision. In other words, policy decisions have to be made regarding these complex policy problems and governments need to take a stance.

One way for governments to deal with this tension between avoiding heavy political conflicts 'and the need for decision making, is to install expert committees for advising in decision making or even deferring decision making to. After all, by relying on the experts' advice, governments do not explicitly need to take a stance. Expert committees fit well within liberal democratic pluralism with its emphasis on procedural solutions for social controversies (Moreno 2009). Expert committees can, therefore, be regarded as a way of depoliticizing contentious problems, aiming to reconcile antagonistic groups in society (Bovenkerk 2012). Furthermore, installing expert committees expresses that policy makers take the issue seriously. To that extent, the instalment of expert committees has a symbolic function.

One type of expert committee, which is often installed to address complex policy problems, is the ethics committee. The constitution and role of ethics committees differ in practice. In general, we can identify an ethics committee as an interdisciplinary group of experts. Ethics committees' members can be experts on science, society, law, religion, and/or ethics. The committees are appointed either to give advice on the moral impact of contentious issues or to stimulate and feed debate. In practice, ethics committees may even have a decisive role, raising the question of what the basis of their authority is.[3]

In this paper, we focus on the role of ethics committees by putting the role of ethicists as experts up for discussion. Two decades of discussion in science and technology literature has concluded that the distinction between facts and values is generally an artificial one. In theory the technocratic paradigm of decision making has, consequently, been discredited. However, we show that in practice, even in ethics committees, the normative complexity of the contentious issues is still discussed in a technocratic way. We do so by drawing on the findings of our qualitative research into biotechnology and animal experimentation ethics committees in several different countries (Sect. 16.3). This qualitative comparative research pinpoints the misunderstandings that exist about experts in general and about normative experts in particular. To that extent, we will give two main criticisms of the committee system: (1) they have an underlying tendency to regard science as neutral on the one hand and the status of ethical judgments as either wholly subjective or completely objective in the sense of authoritative on the other hand; (2) following from this, the expectations of ethics committees tend to be both overestimated and

[3] For example the Dutch Committee on Animal Biotechnology, see Bovenkerk and Poort (2008).

underestimated, regarding their part in public debate and their part in decision making. Consequently, their symbolic value in practice often lies more in window-dressing that ethical concerns are addressed than in a clear communicative role.

To explain the mismatch of expectations of ethical experts in practice, we build upon the discussions in social science on expert involvement (Sect. 16.2). In these discussions, the role of (scientific) experts in decision making has already been extensively reflected on and they, therefore, provide a good starting-point for our analysis. The insights offered by the social science perspectives offer a good explanation for the general idea on expert involvement in policy making, and shows that there is a difference between scientific and normative expertise. In these studies, democraticizing expertise is argued for as an alternative. In our opinion, democraticizing expertise cannot provide an answer to the complexity of moral disagreement in biotechnological and medical technological issues. We, therefore, carry out a more extensive reflection on the role of normative expertise.

The main issue that we intend to address in this chapter is, however, not the science and technology perspective on expertise. Our focus is more on practical reality where the clear distinction between facts and values is muddled. In practice, people expect ethics committees to deliver similar results and operate along similar lines as more scientifically centred expert committees who have a more decisive role. Even though, by appointing ethics committees, a clear distinction is made between these two different types of committees, it remains unclear where the particular mandate of ethics committees lies (Sect. 13.3). Are they meant to solve moral dilemmas or simply to give input into public discussion? In other words, what is expected of the committees by decision-makers?

Our argument is that we cannot expect similar outcomes from ethical experts as we expect from scientific experts. Furthermore, the pitfalls of expert involvement cannot be solved solely by involving lay people. Ethics committees are no substitute for public debate or vice versa. So we need to ask what constitutes moral expertise and to what extent this kind of expertise can be relevant for decision making. In other words, what can decision-makers expect from both the ethical expert and ethics committees?

The aim of this chapter is to find an alternative understanding of the symbolic value of ethics committees and of the ethical expert in the field of medical- and biotechnologies. Ethics committees do not only serve a symbolic function in the negative sense of being ineffective ways of reaching the goal of window dressing (see the Chap. 2 in this volume by Van Klink). In our view, discounting of normative perspectives, in both theory and practice, stands in the way of defining the true communicative function ethics committees can play (Sect. 14.4). In a more positive understanding of symbolic legislation, ethics committees serve an important role in stimulating debate and societal interaction. From a legal theory perspective, as followed in this volume on symbolic dimensions of biolaw, the symbolic functions of law-making can be interpreted as functions of communication in which vocabulary is developed for further communication (Van Klink 1998). Furthermore, the vocabulary developed in the communicative framework can contribute to making concrete the values that the law intends to express and protect. We will argue that ethicists can have a particular kind of ethical expertise and this expertise lies in its communicative function. To give room to this communicative function, it is necessary to change expectations.

## 16.2 Theoretical Reflection on Expert Involvement

In this section, we describe the main ideas that have emanated from social science studies on expert involvement. These ideas oscillate between scientific expertise on the one hand and lay people and stakeholder involvement on the other hand. In the last decades we see a development from the dominant technocratic position to one of democratizing expertise.

An early position in discussing expertise is in terms of providers of 'neutral' scientific knowledge. This position assumes that scientific facts can be used to determine correct or 'good' policy. In general, the facts about technology are considered unproblematic and undisputable. Facts can, therefore, legitimate decision making substantially and can be used as a basis for law-making. Experts are not seen as co-producers of legal standards and thus not as law-makers themselves, but as reporting to decision-makers. A leading thought in this position is that 'truth speaks power'. This position is also referred to as a technocratic understanding of scientific expertise. In practice this is reflected in technocratic ways of decision making, which we will illustrate in the next section.

Since the last decade, in Science and Technology Studies (STS), it has become controversial to strongly rely on scientists and to legitimate decisions built primarily on scientific expertise (Nowotny 2003: p. 151). Instead, new understandings of the distinction between lay people and experts were defined and put up for discussion. The leading question in these criticisms was 'how is scientific consensus formed?' (Collins and Evans 2002: p. 241). The strong relation between politicians and scientists conflicts 'with leading principles in governance, such as transparency and public access to deliberation and assessment procedures (Nowotny 2003: p. 154–155). As it only has to report to decision-makers, expertise is not open for contestation, which conflicts with these governance principles. According to Nowotny this understanding of the expert is controversial as it does not lead to 'socially robust knowledge'. She questions whether scientific facts are in reality unproblematic and undisputable. For knowledge to be more 'socially robust' it should be challenged by a larger community. Valid knowledge does not only derive from the lab, but is also valid on democratic grounds. This view emphasizes the social construction of knowledge. Consequently, lay persons or symbolic users should be included too (Nowotny et al. 2001; Nowotny 2003: p. 155).

Jasanoff (2003: p. 159–160) in this context, points out that expertise is not so much a tool to find the facts. Instead, expertise is also made in culture and in political processes of decision making: expertise is a product of politics and culture. Jasanoff suggests that expertise in this view is politics by different means. Whereas expertise tends to be considered universally valid, in political practice, expertise is used to strengthen political views. Politicians in her view draw on experts' recommendations in an instrumental way by cherry picking only those parts of the recommendations that are useful for their ideas.

All these discussions deconstruct the technocratic understanding of expertise (Collins and Evans 2002).[4] In STS, additionally, attempts are made to *re*construct the understanding of expertise and of knowledge by calls for on the one hand democratizing science and on the other hand politicizing it.

By democraticizing scientific expertise, the distinction between the public and the experts is blurred. Science is considered a crucial source of knowledge, but so are other forms of knowledge. Citizens are seen as both critics and producers of knowledge. The quality of knowledge depends on its contestation by means of, for example, extended peer review and open expert controversy (Tickner and Wright 2003: p. 217–218). Experts and citizens should function as co-producers of policy and eventually of decision making. Legitimacy is not only found in the quality of knowledge, but also in inclusiveness. An example of a tool used to democratize expertise are the consensus conferences, which are often organised in Denmark as well as in the Netherlands on all kinds of controversial issues such as animal biotechnology and cloning. In these consensus conferences the public can actually be involved and potentially influence decision making (Joss 1999).

Another attempt to reconstruct knowledge is made by Collins and Evans (2002) who explore the question 'why science should be granted legitimacy because of the kind of knowledge it is' (Collins and Evans 2002: p. 241). Collins and Evans do acknowledge that science is somehow different from lay perspectives, but do not claim that science is the dominant or only position on which scientific consensus must be based.

These attempts all seem to break through the traditional distinction between lay people and expertise. On the one hand, experts are no longer considered advisors to policy-makers, but instead co-producers. On the other hand, citizens are equally considered experts, bringing in relevant knowledge or providing a kind of counter expertise (Wynne 1996).

In general, reflection in STS has shown that the positivistic image does not give an accurate account of the relationship between science and experiential knowledge. Furthermore, for example, Jasanoff questions the relationship between science and ethics by pinpointing the mistaken picture of science as completely objective on the one hand and ethics as purely subjective, as a non-expertise (Jasanoff 2003: p. 158). In reality, many value judgments may enter in scientific inquiry, from the moment of problem definition, to the moment of selection of theoretical concepts, of putting forward a hypothesis, of selection of method, of interpretation of test results, of application of the test results, and finally to the way in which the knowledge is dispersed (see for example Longino 1990; Putnam: in Nussbaum, Sen, and Sugden 1993). In the philosophy of science a traditional

---

[4] Another example can be found in the analysis of the development of standards in the areas of global food safety. Alessandra Arcuri (2014) describes that decision making in this field is mainly left up to technocratic regime by relying too much on scientific experts only.

assumption has been made that even though some non-epistemic values can legitimately enter the scientific domain (for example, when choosing a subject matter to investigate or in order to limit acceptable research methods, like in the case of research with human subjects), science should otherwise remain free of non-epistemic values and value-free science is a desirable and achievable goal (Douglas 2000). As Douglas (2000: p. 565) shows, however, 'non-epistemic values [do] have a legitimate role to play in the internal stages of science'.[5] Moreover, the ideal of science as completely free of non-epistemic values implicitly takes the practice of carrying out randomized controlled double blind trials, as is common in medical experiments, as the standard of evidence for all scientific endeavour. Many problems cannot be researched using such an approach, however (Slob and Staman 2012). According to Sarewitz (2004: p. 386) the application of a positivistic image of science to the political domain, particularly in the context of environmental controversies, is problematic: 'the notion that science is a source of facts and theories about reality that can and should settle disputes and guide political action' is misguided and ignores 20 years of literature in the constructivist tradition.[6]

Sarewitz' thesis is that the natural world is complex and can, therefore, be analysed and framed differently within different scientific disciplines, sometimes leading to opposite conclusions drawn from the same set of facts. Consequently, more scientific research carried out on a particular complex phenomenon, will usually lead to more disagreement - a condition Sarewitz calls 'an excess of objectivity' (Sarewitz 2004: p. 389). The increasing disagreement is caused by the use of different frameworks that the experts from different disciplines may use; consequently, experts come to different conclusions based on the same data (see Sect. 3.3).

These extensive reflections on the mistaken picture of the relation between science and experiential knowledge and between science and values have only had limited influence on the understanding and expectations of ethics committees. In decision-making practice, the dominant position still follows a rather technocratic understanding of expertise. This position is reflected in the so-called concept of evidence-based policy. Politicians rely heavily on experts' recommendations when drafting their policies. This policy strategy strongly builds on the idea that societal

[5] Scientific reasoning can often not even do without it, because there is always a risk of error in different stages of research and what error one is willing to accept in the end depends on a non-epistemic value choice. Douglas (2000) argues that there are many stages in scientific endeavour where there is 'inductive risk', in other words the possibility of false positive or false negative outcomes, or the possibility that the outcomes lead the researcher to accept his/her hypothesis while the hypothesis is in reality false, or the other way around, that he/she rejects a hypothesis that is in reality right. The consequences of false positives are often overregulation (for example when the toxic properties of a chemical are measured) and the consequence of false negatives underregulation (and a possible health risk to the population) and thus they have a consequence in the real world. This means that the scientist in question will have to make a value judgment about what is an acceptable outcome in the case of inductive risk. In other words, non-epistemic values do and should enter the internal scientific domain.

[6] While Sarewitz' writing could be subsumed under the critical stream of the modern model, as he attacks the fact/value distinction, the implications of his work in our view do not fit within the modern model, but belong instead in the model that we want to propose.

problems can be neutralised by the application of objective knowledge.[7] In other words, the technocratic understanding of expertise, also concerning ethics committees, still seems to prevail. In the next section, we will illustrate this claim.

## 16.3 Ethics Committees

### *16.3.1 Case Studies*

The case studies we refer to are biotechnology ethics committees in the Netherlands and Switzerland and animal experimentation committees in the Netherlands and Sweden. We do not go into the functioning of these committees in great detail, as we have already done so in earlier publications (Bovenkerk and Poort 2008; Bovenkerk 2012; Poort 2013; Poort et al. 2013). The biotechnology ethics committees we have studied are the Dutch Committee for Animal Biotechnology (CAB) and the Swiss Ethics Committee on Non Human Gene Technology (ECNH). Both committees were installed to advice their governments on the use of animal and/or plant biotechnology. Both had a mixed composition, made up of for example ethicists, lawyers, molecular biologists, animal welfare experts, medical experts, and social scientists. No lay people were involved in these committees. While both committees gave advice to their respective decision-making authorities it was eventually these authorities who were responsible for licensing.

Furthermore, we draw on research carried out into Swedish and Dutch animal experimentation committees (AEC's). In the Netherlands, these committees are often institution bound and give advice to the license holder about specific animal experiments that researchers want to carry out. On the basis of this advice the license holder decides whether or not the experiments can be carried out. Usually the advice does not lead to a rejection of a particular experiment, but specific conditions (for example about numbers or species of animals used or methodological set-up) are often set before the researchers can go ahead. In Sweden, seven local animal ethics committees examine applications for all non-human animal experiments. The body responsible for the apparatus is the Ministry of Agriculture. The ethical evaluation is mandatory for all projects including animal experimentation, and the Swedish committees play a decisive role. Each of the seven committees consists of scientific experts and laypersons. The scientific experts come from universities and pharmaceutical companies; this category also includes animal technicians and veterinarians. The laypersons represent local political parties, animal welfare and animal rights organizations. One major difference – for our purposes – between the Swedish and Dutch AEC's is that the former are composed of both experts and lay persons, whereas the latter solely contain experts, both ethical and scientific.

[7] See M Slob and J Staman (2012) *Policy and the evidence beast*. The Hague: Rathenau Institute.

Our criticism in the next section is not directed towards the functioning of committee members as such, but rather towards their mandate and the expectations that policy makers and the public have of them. We conclude that these expectations lead to a legalistic climate and incompatible roles of the committees.

### 16.3.2 Ethics Committees in Perspective

In this section, we address two main criticisms of the committee system: (1) the supposed value neutrality of science on the one hand and the status of ethical judgments as either wholly subjective or completely objective in the sense of authoritative; (2) following from this, the role of ethics committees is often overestimated, both regarding their part in public debate and their part in decision making. We illustrate these criticism with the examples of the animal ethics and animal experimentation committees in Sweden, Switzerland, and the Netherlands, introduced in the previous section.

Ethics committees are assigned to either stimulate public debate or to advise on decision making related to moral issues. Installing ethics committees expresses that decision-makers do take the moral dimensions of the issue at stake seriously. Ethics committees therefore have a symbolic function in decision making around issues in the field of new medical and biotechnological developments. However, the expectations of 'ethical expertise' reveal a mistaken picture of this symbolic function and their role in at least three ways. First of all, ethical experts are expected to provide the morally right answer. Second, due to this expectation, their advisory role turns into a more decisive one. And by doing so, their context of moral reasoning risks becoming a legal one. Third, their role in stimulating public debate is often mistaken as one in which members of the ethics committees represent the moral concerns of 'the public'. The function of ethics committees seems to become one of window dressing. Let us explain that.

In the advisory role, the members of the ethics committees are mistakenly considered to have a prerogative on moral values and in that capacity being able to provide the morally right answer. And of course, due to experiential knowledge, ethical experts do develop more substantial knowledge about what the morally right answer could be. Their opinion may even be more reliable than those of a lay person as the ethical experts have instruments to explain their arguments. However, their value judgment in itself is not worth more than those of other actors with experiential knowledge. As fundamental values are ultimately beyond argumentation, ethical experts do not have claim to holding better values. Moreover, their expertise is not one of providing the morally right answer, but providing the instruments to think through value judgments. If we were to understand the quality of ethical expertise as lying in the ability to provide the morally rights answer, the role of ethics committees risks becoming one of decision making. Having this role, ethics committees are 'seduced' to provide one clear 'morally right' answer. However, to come to a clear answer built on agreement or consensus, fundamental moral considerations are often silenced and debate foreclosed (Poort et al. 2013: p. 3–4). In the Swedish animal ethics committees (AEC's) ethical evaluation is often restricted to the more technocratic issues regarding how to decrease animal suffering by technological solutions. Restriction to more technocratic issues can be explained by the strong scientific discourse dominating the committees. Furthermore, in practice, the committees were expected to provide clear answers that the Minister could build on and it is simply easier to reach a consensus on the technocratic issues. The ethical issues about the justification of animal suffering, on the other hand, were difficult to 'solve' dealing with the diverse perspectives of the committee members (Poort et al. 2013: p. 4). However, while the committees could more easily agree on technical issues, consensus in committees does not make the outcome morally sound (Moreno 2009: p. 482). Consequently, in practice, the ethics committees do not function fundamentally different than scientific expert committees involved in decision making. Their 'label' as being an ethics committee is purely symbolic in terms of window dressing. A similar problem is encountered regarding Dutch AEC's. While officially their primary task is that of weighing the costs of the experiment to the animals against the benefits for humans or animals, in practice discussion in these committees focuses on technical and methodological issues, such as implementation of alternatives (refinement, reduction, and replacement), type of animal used and specific test set-up. Not much discussion is spent on the ethical justification of the goal of the research or the weighing of this goal vis-à-vis animal death and suffering (Stafleu 1994; Swart et al. 2004). Members of AEC's acknowledge that their values are not better than anyone else's and therefore they feel that it is not up to them to make a moral judgment about the worth of the goal of research. At the background, a view of moral subjectivism seems to be present within the committees, while at the same time their opinions tend to be viewed as morally authoritative.[8] A positive recommendation is often regarded as a moral 'stamp of approval', or a symbolic guarantee that the behaviour of scientists in the study is morally right; however, as Verweij et al. (2000) argue, a positive recommendation by an ethics committee should not be taken as a substitute for moral responsibility of the researchers.

In the Netherlands, the CAB officially plays an advisory role. It is the Minister of Agriculture, Nature and Food Quality[9] who decides whether or not to grant the license for using animals in biotechnological procedures. In reality, the CAB plays the decisive role as the Minister always follows the Committees' advise. This role changes the discourse for the CAB and limits the deliberations to individual cases without reaching a higher aggregation level. One strategy that the CAB developed for dealing with this problem is the strategy of reframing the issue (Paula 2008). As the issue was reduced to more practical and technical matters, this strategy was intended to smooth over moral conflicts'. Regarding the more practical and technical matters, the CAB could reach a unanimous decision in most cases. As a result,

[8] Another problem, as Verweij et al. (2000, p. 347) suggest is that 'the possibility of a negative judgment about a research project often pits scientists and committees toward a struggle for power – which is the death sentence for any moral deliberation'.

[9] Now subsumed under the Ministry for Economic Affairs.

the moral status of animals was no longer discussed by the committees. Bovenkerk questions whether the 90 % rate of unanimous decisions reflects true consensus or whether this 'means that the more divisive moral problems were excluded from Committee discussions?' (Bovenkerk 2012). The fact that the CAB in practice has a last say in licensing decisions, has created a situation in which the CAB's decisions were contested in court, for example, by researchers or by interest groups. As a response, the committee already anticipated possible objections in their advice. This way a legalistic climate has arisen; instead of being allowed a more open and comprehensive moral discussion, the committee members were forced to adopt a legalistic moral framework (Paula 2008; Bovenkerk and Poort 2008). These examples show that the framework for moral reasoning has become a legal or political one, resulting in, for the sake of agreement, downplaying and simplifying the complex moral issues into more technocratic issues. In other words, the communicative function of ethics committees is limited and does not contribute to making the values that the decision-makers want to establish, concrete.

In their role of stimulating debate, ethics committees risk being considered representatives of the public. For example, the Swiss ECNH is installed to advise the government on the moral impact of non human gene technology as well as to stimulate public debate. However, the committee tends to be used by government to contain and substitute public debate instead of stimulating it. The latter would not be problematic, were it not that the semblance of consensus presented by the committee functions to exclude certain voices and therefore functions to contain public debate (Poort 2013: chapter 6). The composition of the committees is not necessarily built on representation, and neither are their frameworks for reasoning publicly accountable. Again, also in public debate, the role of ethics committees risks to become one of window dressing in which the instalments of the ethics-committees becomes an excuse for limiting public input.

These examples illustrate the overestimation as well as the limits of ethics committees. An overestimation, because the lack of clarity on the function that ethics committees actually do have, can lead to groundless expectations. The resulting lack of transparency may lead to an ambiguous role in depoliticizing decision making and replacing public debate as well as to a shift from an advisory role into a decisive role.

The limits also relate to the lack of transparency and clarity of the function that ethics committees do actually have. The role of the CAB was officially one of advising and stimulating debate, but in practice became a more decisive one. The latter restricted a more fundamental debate in the committee as reasoning was suddenly restricted by a legalistic context (Bovenkerk and Poort 2008; Paula 2008). This stood in the way of a more open debate in which members could bring to bear diverse and sometimes opposing viewpoints. However, these difficulties could be expected, considering the incentives and the framework for reasoning of the CAB and ECNH. Their terms of reference combine advising on the moral impact with a role of stimulating public debate, but these roles are not easily reconciled. Both tasks require different discourses for reasoning that are mutually contradictory. While in advising the committee is forced to take up a clear viewpoint and if necessarily to defend it legally, stimulating debate requires pondering a variety of viewpoints and arguments.

In STS, an argument for democraticizing expertise are presented as a tool to address these pitfalls related to public debate. Our critique is, however, directed at the expectations on the CAB and ECNH's role in stimulating broader and fundamental debate. It would be helpful to limit these expectations and create room within other arenas for more fundamental debate.

### *16.3.3 How Should We Understand Ethical Expertise?*

The misunderstanding of the role of ethics committees may be explained by the standard understanding of the role of experts within a positivistic image of science. In the same way that scientific experts are expected to bring in universally valid or professionally certified knowledge, ethical experts are expected to bring in universally valid or professionally certified moral judgements. Approaching ethical expertise in a similar way as scientific expertise leads to an overestimation of the experts' role. As has become clear, the narrative of scientific experts as being providers of clear and universally valid facts can be questioned in itself, but surely, should not be equated with the narrative of the ethical experts. In the previous section we presented several examples from decision-making practice that illustrated that in practice the scientific experts' narrative is still followed. In STS this narrative has already been challenged and criticized theoretically, but in everyday practice, scientific experts are still seen as providers of clear facts. Moreover, ethical experts are approached in a similar way. In this section, we argue, first of all, that you cannot approach ethical expertise in the same way as scientific expertise. Second, we argue, that in line with insight of STS, there should be more room for values and scientists should make their methods and value claims more explicit.

In Sect. 16.3.2, we have noted problems with the instalment and functioning of ethics committees. In our view an important cause of these problems is the tendency to expect the same type of outcomes from an ethics committee as from a technical or scientific expert committee. However, the subject matter of both types of committee diverges; where one is dealing principally with value judgments, the other is dealing with scientific facts or technical solutions. While we acknowledge that facts and values cannot be completely separated, it is important to note that the goals of the two types of committee differ; they are meant to generate a different type of outcome. Ethics committees are often regarded as any other technical or scientific committee, while in the end they are asked to make a value judgment. This creates a problem especially when committees are used to legitimize political decisions. A politician can say 'I have consulted the scientific experts and they believe that this is the best solution'. But can the politician also say 'I have consulted the ethical experts and they believe this is the best solution?'. This raises the question whether one can be a ethical expert. The relevance of this question is well described by Julian Baggini (2010: p. 17): 'Ethical experts are distinctive in the limits of their authority. We usually defer to most experts as the last word on their subjects. Not so the ethical expert. Hardly anyone ever thinks that something is right because an ethical expert says it is'. The question about ethical expertise is raised because of the fact that while moral philosophers deal with matters of value, an ethical expert's values are not necessarily better than those of a lay person. Values are not open to being scientifically researched and tested.

Still, in a different sense, ethicists could be regarded as ethical experts. Nussbaum (2002), following Socrates and Aristotle, argues that even though ordinary people have a certain moral competence, they often hold inconsistent views and their views are not always well-informed. Ethical experts can point out these inconsistencies, they have a better grasp of fallacies and of logical reasoning. Moreover, they have studied literature of earlier philosophers who have grappled with similar questions, which stops them from making the same mistakes as these earlier philosophers. Finally, moral philosophers have knowledge of ethical theories and they can consider judgments against the background of these theories. In short, by studying the theories and judgments of philosophers before them, they can take a broader viewpoint from which to analyse a particular problem. Nevertheless, Nussbaum warns against hierarchically ranking ethical experts above other members of the public, as this would not show proper respect for the equality of other citizens in a democracy with regard to moral and political issues. Ethical experts, then, should not have the last word on moral matters. One reason is that agents are responsible for their own actions, and completely deferring a moral decision to ethical experts takes away this responsibility (Verweij et al. 2000). The ultimate goal of ethics committees should be to enable researchers to make their own reflected decisions and to contribute to communicative and expressive functions of decision making.

The two reasons that Nussbaum and Verweij et al. mention for not giving ethical experts the last word, are both in the end practical in nature, however. They do not imply the impossibility of being an expert on moral matters per se. Other commentators do not look as favourably on the idea of the ethical expert, suggesting that the ethical expert will simply posit her own personal opinion as authoritative (p.e. Crosthwaite 2005). This viewpoint, however, mistakenly equates ethical experts with ethical scholars. An ethical scholar, or moral philosopher, can argue for a specific viewpoint from within a moral theory. He or she does will take a substantive stance on the question what moral theory or meta-ethical position is the right or most valid one. This will lead to a personal, albeit well reflected, opinion. An ethical expert on the other hand, is expected to take a more pluralistic standpoint and analyse the subject matter from within several theoretical perspectives. The specific role of the ethical expert consists in clarifying and analysing debates, making implicit value assumptions explicit, pointing out logical inconsistencies, and moderating debate, rather than taking a personal stance. The critique of ethical experts makes two further problematic assumptions, namely (1) that an ethical expert's role would be to make the final decision on a moral issue and (2) that an ethical expert will operate as a single individual. Baggini (2010: p. 26) instead thinks that only collectives could reach substantive decisions together. This is because committees can view issues from more different perspectives than individuals and because ethics is in part a matter of balancing claims and interests and different people in a committee can represent these different claims and interests.[10] Nevertheless, even a collective is not necessarily representative of all opinions and values held in society, and the composition of an ethics committee can greatly influence outcomes of its deliberations. We must remember that the members of ethics committees are not elected individuals, nor do they usually represent a certain segment of society. Besides, the members of ethics committees usually don't all have a background in moral philosophy; ethics committees tend to be composed of experts from different academic fields.

Perhaps it is something in the nature of moral judgments and the process of moral judgment formation that means that one group simply cannot make decisions for the moral agent. We do not hold a subjectivist view of ethics, which states that in the end moral judgments are always subjective and cannot be criticized from outside of the viewpoint of the moral agent. We do think it is possible to claim that one position is better from an ethical point of view than another position, for example because it is better reflected on and coheres more with our considered judgments and with empirical evidence. Still, this does not mean that there is only one right position to take in a moral dilemma. With Sarewitz, we can wonder however, if we could not say the same about the expert decisions of empirical scientists. When a scientific expert reaches a conclusion this is often presented as authoritative; yet scientists from other disciplines or paradigms could reach different conclusions, based on the same reality. While empirical scientists may have a higher claim to objectivity, it seems to us that the difference between the authoritative character of an ethical judgment and a scientific judgment is a gradual rather than an absolute one, particularly when we are dealing with unstructured problems, where facts and values are more entangled in the first place. This means that the argument that an ethics committee cannot make the final moral judgment for anyone else, also holds for other expert committees. In the end, a decision has to be made by the moral agent, in this case the government or policy maker who has been given a mandate by the public, and this agent has to weigh both ethical arguments and empirical findings. An additional argument for reserving the final decision making for the politician or policy maker is that policy decisions have to be made coherent with decisions in other policy fields; this is something that you cannot expect from scientists or ethicists.

Due to the diversity of disciplinary lenses through which a problem can be approached, it is possible to find a scientifically legitimate set of facts to support each different value-based view in an environmental controversy. Some scientific (sub)disciplines fit better with certain values and interests and others with the opposite values and interests, leading to a situation where the same phenomenon is approached from such different angles that different research groups do not only yield opposite conclusions, but also do not even appear to address the same problem. For example, it has been shown that the risk perception of scientists regarding

[10] It should be noted, however, that Baggini appears to be thinking about ethics committees completely composed of ethical experts, which is generally not the case in practice.

nuclear waste disposal is influenced by their scientific discipline and even by the question whether they are employed by a public or a private institution (Barke and Jenkins-Smith 1993). Life scientists tend to perceive greater risks from nuclear waste disposal and oppose putting unconsenting persons at risk more strongly than chemists, physicists, and engineers. This phenomenon is especially at play in environmental problems, as different disciplines tend to operate from conflicting scientific views of nature. This is one reason why the debate about plant biotechnology is so intractable: molecular biologists look at the molecular level and argue that GMOs are substantially equivalent to their non-modified relatives, while ecologists look at the role of GMOs in the environment, with all its connections and unpredictabilities (Sarewitz 2004; see also Bovenkerk 2012). For this reason, Sarewitz (2004: p. 392) concludes that 'stripping out conflicts 'of interest and ideological commitments to look at "what the science is really telling us" can be a meaningless exercise.....disciplinary perspective itself can be viewed as a sort of conflict of interest that can never be evaded'.

This analysis discredits the technocratic approach that still seems dominant in decision making, in which uncertainties about environmental or technological problems can be reduced simply by the application of more scientific research and the involvement of more scientific experts in the formation of policy. Rather, more research and the involvement of a greater variety of disciplines is unlikely to create one coherent picture. There is no one best methodology or approach to deal with complex policy problems, but a variety of problem definitions examined through different and conflicting value frameworks and disciplines. The foregoing has consequences for the role of experts in the solution of complex policy problems. Expert knowledge always needs to be regarded in its proper context; experts need to make the value framework behind their methods and claims explicit. Moreover, experts should not have the last word in policy decisions; their voice is but one among other voices. As Sarewitz (2004: p. 400) concludes, policy decisions need to be made on the political level, with a discussion about relevant values and politicians can no longer 'hide behind scientific controversy'.

## 16.4 Changing Expectations

While science is not objective in the sense of value neutral, ethical judgments are, at the same time, not wholly subjective, as is often believed. Except for extreme moral relativists, ethicists generally believe that objective principles or norms exist, and if not they at least believe in intersubjectively reached valid judgments. This is not necessarily to say that ethical theorizing and scientific inquiry have the same structure. Our point is merely that the role of values, amongst which moral values, in science and the role of ethical discussion have been underestimated and that this has had consequences for the way the role of scientific expertise has been framed. If values enter scientific inquiry at diverse points, it becomes important for scientists to be aware of these values and make them explicit, so that these can be critically assessed along with the usual assessments of the scientific methodology used. This requires a greater transparency on the part of scientists.

In this section, we want to show to what extent the role of ethics committees is overestimated and, at the same time, underestimated. Therefore, we pinpoint first of all the expectations that decision makers can have ethical experts and, secondly, of ethics committees and moral experts.

Above we have argued that the values of an ethical expert are not necessarily better than anyone else's. Nevertheless, one remark must be made here: when operating as an ethical expert for a while, the ethical expert may gain 'experiential knowledge'. To that extent, the ethical expert may additionally become a moral expert with substantive knowledge about the moral complexity of the issue at stake. Furthermore, in light of his quality as an ethical expert, the ethical expert may be expected to have a higher reliability of his moral judgments. We could argue here, that, therefore, the ethical experts may have 'better' moral judgments which the decision-maker can build upon. The issue here is, however, that we question whether the experiential knowledge of the ethical expert legitimates a higher authority on moral judgments. The ethical expert as a moral expert will follow a particular value framework which need to be explicated. In this quality, the ethical expert does not differ from other moral agents or from scientific experts, for that matter. Besides, other players and actors have this experiential knowledge too. Nonetheless, the ethical expert may be more capable in reflecting upon his value assumptions. But, if the ethical expert would do a good job, all involved actors would become moral experts with capabilities for making value frameworks explicit. The quality of being an ethical expert does not make his value assumptions in itself 'better' than those of other actors. Like those actors, the ethical expert has no authority for making moral judgments. The government has the decision making authority; the ethical expert has a different role which will be explained in context of the second pillar.

Secondly, the expectations decision-makers have of ethics committees need to be reconsidered. A first pillar in our argument concerns the relation between scientific and normative expertise. Our position within the discussion about expert involvement is that facts and values are not easily separated in most fields of science and that objective ideologically neutral scientific facts are a myth. Inspired by Sarewitz (2004), we challenge the ideal of evidence based decision making. We do not want to replace the ideal of value-free science with fact-free science. Rather, we think it is important to make value judgments in research explicit, so that the legitimacy of these choices can be discussed. Also, the implicit value assumptions of lay 'experiential' knowledge should be brought to light, something which was also challenged in the idea of democraticizing expertise. Formal *and* informal experts should confront each other with their value assumptions and make their value frameworks explicit. To that extent, ethics committees can have a role of communicator in which the ethical experts in these committees provide the communicative frameworks in which decision makers and other stakeholders can reflect upon their value assumptions and make those assumptions more explicit.

To that extent, the ethical experts within ethics committees can have a role of communicator in which the ethical experts provide the communicative frameworks in which decision makers and other stakeholders can reflect upon their value assumptions and make those assumptions more explicit. The ethics committee as such may still have an advisory role, building on these communicative frameworks, but we should beware of the risks of granting ethics committees too decisive a role. The risk is that while every member of the committee has his or her own disciplinary lense – and thus partial viewpoint - the focus on consensus formation that an advisory role demands, will cover up dissensus and will limit the legitimacy of different viewpoints (see Poort 2013: chapter 10).

This brings us to the second pillar in our argument, which is the need for context sensitivity. What do we mean by this? We think that the context in which a complex policy problem is developed and tends to be addressed is relevant for our understanding of the relation between experts, science, and policy. Determining the context includes analysis of the specific roles that different players can play. In our understanding each player has a different role to play. We want to emphasise that it is important that these roles are specified as otherwise these roles can become blurred and false expectations are created, as we saw, for example in the case of animal experimentation committees. By clearly stipulating the roles and contexts of each different type of expertise, the risks of overvaluing expertise may be overcome (Castle and Culver 2013, p. 40). Our main argument is, thus, that one has to see the role of the different players in the right context. In this context, we should ask: What can we expect of the different players? What kind of expertise is referred to? What are the limits of the various roles and expertises?

What these roles are, is up for discussion depending on the context. Connecting to Collins and Evans' (2002) idea on reconstruction of expertise: when defining the role of experts we should wonder what kind of knowledge claims can be made by the experts? How to legitimate these claims? In the report 'Policy and the evidence beast', the Dutch Rathenau Institute makes specific recommendations to this effect (Slob and Staman 2012). As for scientists, they argue that it takes a specific kind of expertise to translate science into policy. If a scientist wants to advice policy-makers she should realise that she thereby assumes a different role than that of the academic scientist. These choices have institutional consequences. Just as we cannot equate the ethical expert with the moral philosopher, the roles of scientific researcher and scientific expert are two different things.. Being a scientific researcher does not make you a scientific expert yet. The same thing goes for ethicists: a moral philosopher and an ethical expert fulfil two different functions. While a moral philosopher can make strong claims from within a particular normative framework, an ethical expert needs to take a more value-pluralistic stance. We should note, however, that this expertise does not transfer the role of policy-maker to them; they should be regarded strictly as advisors. While we press that scientific and ethical experts make their value assumptions explicit, this does not give them a green card to push their own ideological views; they cannot become the vehicle of NGOs without losing credibility. Experts should be prepared to not only make their value assumptions explicit, but also to put them up for debate publicly. If a decision is made that is contrary to their own personal convictions, they should resign to that fact. It is not their place to push a specific political agenda. In the case of lack of knowledge or dissensus, experts should not be tempted to take a clear position. Rather, they should explain the basis of this scientific dissensus. In this context the role of open expert controversy is invaluable (see Beck 1992). They should leave it up to the politicians to be the final arbiter and to weigh the uncertainties and different positions. After all, they have to relate their decisions to other policy decisions as well (to reduce inconsistency with other policy areas) and moreover, they have been elected by the public and therefore have more decision-making legitimacy than unelected experts.

We think ethical experts should be given the role of moderator of the public debate and, therefore, ethical experts would have a communicative function. What do we mean by moderation? Firstly, ethicists can structure the discussion, simply because they tend to have a helicopter view. They are trained in analysing ethical debates and have knowledge of ethical theories. They can draw on their study of past ethical debates and bring to bear insight into normative frameworks that are being referred to in debates. The experts are also trained in recognizing and signalling implicit value assumptions, clarifying concepts, thinking logically, and pointing out inconsistencies. Secondly, they can offer background knowledge that can feed into the debate. The role of the ethics committee as a whole can be a bit further reaching. They can advise politicians, because they can clarify the different positions in the debate. This way they can help politicians reach their own judgments and take their own moral responsibility, without making the decisions for them.

## 16.5 Conclusion

In this paper, we have criticized the general understanding of the role of experts in decision making. We have pinpointed the misunderstanding of the positivistic image of experts in terms of scientific inquiry as well as of the expectations politicians (and the public) have of experts. We did so by taking as a reference point general discussions about expert involvement in social science studies in which the technocratic understanding is questioned. Furthermore, we presented a critical reflection of the role of ethical experts. Based on these reflections and the outcomes of our case studies, we have drawn further-going conclusions regarding the positivistic image of scientific expertise by challenging the fact-value distinction. To start with, we have pointed out that facts are not value-free at all. Second, we have illustrated that normative expertise is essentially different from scientific expertise. Seeing normative experts in a similar light as scientific experts, resulted in false expectations and overvaluing the role of the various experts. It is therefore necessary to explicate the roles the various players have and what is to be expected from them.

At the same time, the role of ethical experts are underestimated. Whereas in STS an argument for democraticizing expertise is made, we have suggested an alternative understanding of the symbolic function that ethics committees can have: a communicative function. The communicative function that ethics committees can have can best be explained as one of moderator of debate. Our arguments builds on two pillars. First, there is a need to be aware that facts and values are not easily separated in most fields of science. Facts are not value-free and always depend on either the disciplinary lense of the experts or even their personal value-framework or world-view. In light of this understanding, there is a need to analyse and foreground the role of normative experts. At the same time, we cannot expect similar outcomes from these experts as we do from scientific experts. This relates to the second pillar: the need for context sensitivity. With context sensitivity we refer to the context in which a policy problem is raised as well as to how the relation between experts, science and policy is understood. The context defines the roles that the different actors play. In order to overcome the risk of overvaluing the role of experts or having false expectations it is valuable that the roles of the different players are defined and specified. Seeing the expert in the right context is required in order to make sure we understand her role correctly.

## References

Arcuri, Alessandra. 2014. The coproduction of the global regulatory regime for food safety standards and the limits of a technocratic ethos. *EUI Working Papers, RSCAS* 97: 1–17.

Baggini, Julian. 2010. No one is an ethical expert. Long live ethical expertise. In *Ethiek in discussie. Praktijkvoorbeelden van ethische expertise*, ed. M. Van den Hoven, L. Van der Scheer, and D. Willems, 15–30. Gorcum: Van Assen.

Barke, Richard, and Hank Jenkins-Smith. 1993. Politics and scientific expertise. Scientists, risk perception, and nuclear waste policy. *Risk Analysis* 13(4): 425–439.

Bovenkerk, Bernice. 2012. *The biotechnology debate. Democracy in the face of intractable disagreement*. Dordrecht: Springer.

Bovenkerk, Bernice, and Lonneke M. Poort. 2008. The role of ethics committees in public debate. *International Journal of Applied Philosophy* 22(10): 19–35.

Castle, David, and Keith Culver. 2013. Getting to 'No'. The method of contested exchange. *Science and Public Policy* 40: 34–42.

Collins, Harry, and Robert Evans. 2002. The third wave of science studies. *Social Studies of Science* 32: 235–296.

Crosthwaithe, Jan. 2005. Moral expertise. A problem in the professional ethics of professional ethicists. *Bioethics* 9: 361–379.

Douglas, Heather. 2000. Inductive risk and values in science. *Philosophy of Science* 67(4): 559–579.

Jasanoff, Sheila. 2003. Accountability. (No?) Accounting of expertise. *Science and Public Policy* 30: 157–162.

Joss, Simon. 1999. Public participation in science and technology policy- and decision-making – Empheral phenomen or lasting change? *Science and Public Policy* 26(5): 290–293.

Longino, Helen. 1990. *Science as social knowledge. Values and objectivity in scientific inquiry*. Princeton: Princeton University Press.

Moreno, Jonathan D. 2009. Ethics committees and ethics consultants. A companion to bioethics. *Malden: Wiley-Blackwell* 2009: 475–484.

Nowotny, Helga. 2003. Dilemma of expertise. Democratising expertise and socially robust knowledge. *Science and Public Policy* 30: 151–156.

Nowotny, Helga, Peter Scott, and Michael T. Gibbons. 2001. *Rethinking science, knowledge and the public in an age of uncertainty*. Cambridge: Polity Press.

Nussbaum, Martha C. 2002. Moral expertise? Constitutional narratives and philosophical argument. *Metaphilosophy* 33(5): 502–520.

Paula, Lino E. 2008. *Ethics committees, public debate and regulation. An evaluation of policy instrument in bioethics governance*. Amsterdam: Athena Institute.

Poort, Lonneke. 2013. *Consensus & controversies in animal biotechnology. An interactive legislative approach to animal biotechnology in Denmark, Switzerland, and the Netherlands*. Den Haag: Eleven International Publishing.

Poort, Lonneke M., Tora Holmberg, and Malin Ideland. 2013. Bringing in the controversy. Re-politicizing the de-politicized strategy of ethics committees. *Life Sciences, Society and Policy* 9(11): 1–14.

Putnam, Robert. 1993. Objectivity of the science-ethics distinction. In *The quality of life*, ed. M. Nussbaum and A.K. Sen, 143–157. Oxford: Clarendon.

Sarewitz, Daniel. 2004. How science makes environmental controversies worse. *Environmental Science & Policy* 7: 385–403.

Slob, Marjan, and Jan Staman. 2012. *Policy and the evidence beast*. The Hague: Rathenau Institute.

Stafleu, Frans. 1994. *The ethical acceptability of animal experiments as judged by researchers*. Faculty of Animal Science: Utrecht University.

Swart, Jac A.A., H. Hub Zwart, and Wolters Jan (eds.). 2004. *DEC's in discussie. De beoordeling van dierproeven in Nederland*. Budel: Damon.

Tickner, Joel A., and Sarah Wright. 2003. US perspective. The precautionary principle and democratizing expertise. A US perspective. *Science and Public Policy* 30: 213–218.

Ulrich, Beck. 1992. Risk society. Towards a new modernity. New Delhi: Sage. (Translated from the German Risikogesellschaft 1986.)

Van der Weele, C. 2014. In vitro meat. In *Encyclopedia of food and agricultural ethics*, ed. P.B. Thompson and D.M. Kaplan, 1219–1225. Dordrecht: Springer.

Van Klink, Bart. 1998. *De wet als symbool. Over wettelijke communicatie en de Wet gelijke behandeling van mannen en vrouwen bij de arbeid*. Deventer: W.E.J. Tjeenk Willink.

Verweij, Marcel, Frans W.A. Brom, and Alex Huibers. 2000. Do's and don'ts for ethics committees. Practical lessons learned in the Netherlands. *International Perspectives* 12(4): 344–357.

Wynne, Brian. 1996. Misunderstood misunderstandings. Social identities and public uptake of science. In *Misunderstanding science? The public reconstruction of science and technology*, ed. A. Irwin and B. Wynne, 19–46. Cambridge: Cambridge University Press.

# Chapter 17
# Law as a Symbolic Order: Some Concluding Remarks

**Britta van Beers, Bart van Klink, and Lonneke Poort**

## 17.1 Introduction

The idea that the legal order can also be understood as a symbolic order has been explored in this volume from various disciplinary perspectives, ranging from symbolic legislation theory to bioethics, from biolaw to EU law, and from sociology to anthropology. In the contributions to part I of this volume various aspects of the general theory of symbolic legislation were discussed. Subsequentely, in parts II and III, the symbolic dimensions of law were analyzed and examined in an area of law in which the notion of the symbolic has acquired a central role: biolaw. In these concluding remarks, we offer a reflection on how the notion of the symbolic has been discussed and developed in the various contributions to this volume. Our aim is to bring together the volume's first theoretical part with the subsequent two more applied parts about biolaw.

A recurring thought in this volume is that the notion of the symbolic is fundamental to come to an understanding of the meaning and the functioning of law. How fundamental the notion of the symbolic is to law, can be illustrated by the fact that law's symbolic dimensions already surface in what is perhaps the prime act of any legal order: the insertion of new generations into the community of legal subjects. As Pessers writes in her chapter, the subjection of the individual to the legal order, and the individual's subsequent transformation into a legal subject – this highly

B. van Beers (✉) • B. van Klink
Department of Legal Theory and Legal History, Vrije Universiteit Amsterdam,
De Boelelaan 1105, 1081 HV Amsterdam, the Netherlands
e-mail: b.c.van.beers@vu.nl; b.van.klink@vu.nl

L. Poort
Department of Sociology, Theory and Methodology, Erasmus University Rotterdam,
Erasmus School of Law, Burgemeester Oudlaan 50, 3062 PA Rotterdam, the Netherlands
e-mail: lonnekepoort@gmail.com

 
B. van Klink et al. (eds.), *Symbolic Legislation Theory and Developments in Biolaw*, Legisprudence Library 4, DOI 10.1007/978-3-319-33365-6_17

imaginary depiction of the human as a free, equal and dignified being – can be regarded as the individual's birth in the symbolic order of law. From this perspective, the very first and constitutive act of the legal order can be qualified as a symbolic act.

Nevertheless, the notion of the symbolic remains elusive. This can also be illustrated through the symbolic act of legal personification. Indeed, the continuing legal-ethical debates on the meaning of legal subjectivity and human dignity, especially in the field of biolaw, underline how the symbolic dimensions of law are an area of controversy and heated debate.

In its effective history, the symbolic has acquired many and heterogeneous meanings. In one of its most basic meanings, a symbol is understood as 'a visible sign of something invisible' (Merriam-Webster dictionary). Correspondingly, the notion of the legal order as a symbolic order suggests a deeper layer of meaning to law than the immediately visible contents of, for example, a statute or a judicial decision. Yet more can and should be said about the meaning of the symbolic for reflection on the functioning of law.

A second common understanding of the notion of symbol is 'something that stands for or suggests something else by reason of relationship, association, convention, or accidental resemblance' (Merriam-Webster dictionary). Within the debates on the symbolic dimensions of law this 'something else', to which the symbolic refers, takes on the following shape: law is more than a series of commands backed up by sanctions; law is also part of a larger narrative, and embodies certain values, imageries and representations. In Ricoeur's words, the notion of the symbolic encompasses 'within a single emblematic notion the different ways in which language can give figure to obligation: as an imperative, to be sure, an injunction, but also as counsel, advice, shared customs, founding narratives', etc. (Ricoeur 2007: 84; for a further discussion of Ricoeur's approach, see Chap. 11 in this volume).

## 17.2 Negative and Positive Concepts of Symbolic Legislation

This split between traditional, instrumentalist approaches and narrative, symbolic approaches to the functions and effects of law also gives rise to two opposite yet interrelated readings of symbolic law: a negative and a positive concept (for general reflection on these twin concepts, see Chap. 2). The negative approach emphasizes the shortcomings of symbolic legislation. Symbolic legislation is then perceived as essentially toothless, ineffective legislation that is promulgated not so much to achieve the manifest goals but latent political goals, such as simulation of power in situations of crisis. A positive understanding of symbolic legislation takes an opposite stance and emphasizes the shortcomings of a traditional, instrumentalist understanding of legislation. It stresses the importance of symbolic expression of values, communication and interaction as strategies to promote the law's efficacy. Symbolic legislation is then perceived as a means to escape from the constraints of an instrumentalist framework and as a way to come to a fuller account of law.

Within the legal governance of technological developments, many instances can be found to illustrate symbolic law in its negative dimension. A very clear example is the manner in which the principle of non-commercialization currently functions within EU biolaw. According to this principle, the human body and its parts cannot be used, as such, for financial gain (see, for example, Article 3 par. 2 EU Charter of Fundamental Rights). Nevertheless, this principle has not been able to prevent the rise of a market in human body, a market which has been even recognized as part of the EU internal market and is, as such, regulated and facilitated by EU directives and regulations (see Chap. 10).

Accordingly, Sterckx & Cockbain's chapter (Chap. 13) lays bare the contradictory mobilization of the principle of non-commercialization in the field of EU patent law. Article 5 of the European Patent Directive gives the signal 'that the monopolization of human bodies through patents is forbidden, while at the same time the ban on the monopolization of human body materials is so qualified as to make them readily available for commercialisation' (see also Chap. 10). A similar tension can be found in the EU governance of nanotechnologies. As Lee & Stokes argue in their contribution (Chap. 14), the legislation in this field has been 'pragmatic and opportunistic in responding to calls for interventions to control nanotechnology without necessarily providing the effective control demanded'.

However, as will be discussed in the next section, within biolegal regulation many instances of the positive concept of symbolic legislation can also be recognized. As both Sterckx & Cockbain and Lee & Stokes acknowledge, it is undeniable that the ineffective laws which they criticize can also be qualified as symbolic legislation in the positive sense. EU patent law expresses certain deeply held convictions on the value and status of the human body, even if these have not been consistently thought through and applied; similarly, EU laws on nanotechnologies, despite fundamental shortcomings, may pave the way for further debate, which can be regarded as an important symbolic function. This suggests that the negative and positive understandings of symbolic legislation are not mutually exclusive categories.

Indeed, there are several reasons to question the distinction between negative and positive understandings of biolaw. Hoeyer's analysis of the legal regulation of human tissue markets, for example, offers ground for a 'third way' in these debates. Laws that aim to protect against commercialization and commodification of the human body may at first sight seem to be examples of symbolic legislation in the negative sense, as the market in human tissues is currently thriving. Hoeyer's fieldwork in and around the agencies regulating and procuring human tissues reveals however that the symbolic laws in this area do influence biomedical practice. The performative effects of the symbolic laws in this field are perhaps different than what may be expected, but they are no less real (see Chap. 10).

More generally, Van Klink (Chap. 2) wants to demonstrate that the choice for either a negative or positive understanding of symbolic legislation depends not so much on the legislation itself, but rather on the political presuppositions of the researcher. If a researcher is critical towards state power in general, s/he will be more inclined to distrust the motives of the legislature. From a external critical-sociological perspective, symbolic legislation appears to be a mere instrument to gain or maintain power and preserve the political and legal order as it is. If a scholar, on the other hand, believes in the existing order and the officials supporting this order, s/he may conceive of symbolic legislation as a means to promote its underlying values and ideals in a more democratic and responsive way. From an internal communicative or interactive perspective, symbolic legislation does not equal bad or ineffective legislation; it constitutes instead an alternative legislative strategy that achieves its goals, not by means of coercion primarily but through communication and interaction.

Schwitters (Chap. 4) puts the fundamental distinction between symbolic legislation (in the positive sense) and instrumental legislation in perspective in a different way. In his view, symbolic legislation depends, as much as traditional instrumental legislation, on the coercive power of the state and the formal procedural legitimacy. Moreover, as Lembcke argues in this volume (Chap. 6), symbolic legislation may become instrumental, for instance when it helps to solve or mitigate political conflicts in society.

With the term 'anthropotechnology', German philosopher Peter Sloterdijk, in his well-known essay *Regeln für den Menschenpark*, exposed how traditional anthropotechnologies, such as education, reading and philosophy, are now being replaced by technologies that intervene with the genetic and biological aspects of human life. Interestingly, Pessers argues that a similar development is taking place with regard to law as an anthropotechnology: the humanizing function of law is coming under pressure from the emerging anthropotechnologies of genetics and biomedicine. This current tendency is not without risk. As Pessers writes, 'when the symbolic tissue is destroyed, the social fabric also threatens to dissolve'. Similarly, De Dijn is highly critical about the tendency within contemporary, 'revisionist' schools of ethical and legal thought to cut their ties with the symbolic categories and concepts of the lived world, and replace these with more 'rational' or 'scientific' modes of reasoning. According to De Dijn, the danger is that human rights and human dignity are then reinterpreted and adapted to merely facilitate unreflected technological progress and market expansion.

## 17.3 The Symbolic Functions of Biolaw

A positive reading of the role of symbolic laws, which pays more attention to the symbolic effects of biolaw within the governance of new technologies than to biolaw's inability to affect biomedical practice, makes it possible to come to a deeper understanding of the complex interplay between biomedical law and biomedical realities. In the contributions to this volume, various symbolic functions of law have been mentioned and elucidated, which are of special importance to biomedical regulation.

A first symbolic function of biomedical laws that recurs in several chapters is what could be called law's *expressive* function (see Chaps. 2, 3 and 11 in this volume): law is used as a vehicle to express certain important collective values and aspirations. In connection, as Schwitters argues, the law may also signal what is appropriate behavior. For example, within the recitals of international biolegal conventions and declarations, human dignity and respect for human life are often mentioned as the central values to be upheld. Similarly, an important function of much biomedical legislation is to bring to expression the special status of human body materials and human embryos, and to distinguish these from ordinary objects of property law. As these status questions are fraught with moral and political controversy, the expressive function of biomedical legislation should not be regarded as mere window dressing, but can be a vital contribution to the socio-cultural process of coming to terms with new technologies. Both Zeegers' chapter about the European debates on the special status of the human embryo (Chap. 15) and Herring's chapter about the scholarly discussions on the special status of the human body (Chap. 8), offer telling illustrations of the struggles that may accompany law's expressive function. Moreover, even if the depiction of the human body or the human embryo may be quite implicit within certain biomedical laws, that depiction may still far-reaching consequences. As Herring, for example, argues, current biomedical laws paint a rather problematic picture of the human body 'which reinforces a particular set of values concerning the body', namely a rather individualized and property oriented approach. In his contribution, Priban is very skeptical about law's capacity to promote social cohesion by giving expression to moral values. In his view, law cannot claim general moral authority in a functionally differentiated society (see further below).

A second symbolic function of law, that is closely connected to the previous one, is the way in which biomedical laws may set the stage for further debate. From that perspective, the law can be said to offer a communicative framework which can facilitate public debate and stimulate further development and interpretation of certain norms and values (see Chaps. 2, 3, and 5). Van Klink refers to this function as the *constitutive* function, as it constitutes an interpretative community. However, Van der Burg and Poort use the term 'communicative function', because they consider the notion of constitutive function to be too much state-centered. In their view, it is not the legislature that creates an interpretative community; instead, through legislation, it connects to already existing patterns of communication and interaction in society. As biomedical technologies confront societies with radically new questions and uncertainties, and give rise to heated public debates, the communicative or constitutive function of biolaw can become of vital importance by offering the tools for further public and political discussion, such as certain normative frameworks and distinctions.

A third symbolic function of biomedical laws, which emerges in several contributions to this volume, is the way in which the language of law is part of and contributes to the foundational categories and distinctions through which we make sense of the world around us. This third symbolic function of law, which Van Klink labels as an *epistemic* one, and which several other authors regard as a specific part of the *communicative* function (see, for instance, Chaps. 5 and 11), has emerged in biolegal contexts as a consequence of the fact that the hybrid products of biomedical technologies seem to defy traditional categories and distinctions. Biomedical hybrids, such as frozen embryos or human tissue engineered products, seem to question the foundational distinctions between the natural and the artificial, subject and object, man and machine and life and death (see Chaps. 9, 11 and 12). As De Dijn writes, new technologies and their accompanying markets 'actively contribute to the liquefaction or hybridization of fundamental symbolic categories (such as life and death, male and female, man and animal), and to the transgression of the boundaries between them'.

It could be said, that under these circumstances, 'the language of law is of increasing importance in the collective symbolization of novel biotechnological entities' (Chap. 11). Nevertheless, the question remains as to how the legal symbolization of these hybrid entities is to take place. To what extent should biolaw facilitate the technological erosion of existing symbolic categories by replacing these with new categories? And to what extent should biolaw resist that tendency, and instead serve to protect and maintain the existing symbolic order (see De Dijn in his concluding remarks, Chap. 9)?

The last symbolic function of biolaw, which is discussed in the chapters of Pessers, De Dijn and Van Beers, is what could be called its *humanizing* or *anthropological* function. According to Pessers (Chap. 12), law can be regarded as an *antropotechnology* in itself, that is, a technology which contributes in vital ways to the humanization of human beings. Indeed, through their subjection to the law, and being recognized as legal subjects, individuals are, in a way, tamed. As classic social contract theory makes clear, the installation of a legal order, and the conferral of legal subjectivity, enables individuals to leave the state of nature in which man is a wolf to man (*homo homini lupus est*), and to be inserted from birth 'in a reciprocal relationship of rights and duties towards others'. More generally, the humanizing or anthropological function of law manifests itself through its role in the symbolic mediation of the biological facts of life. That is, the symbolic values and rituals surrounding the biological events of birth, death and reproduction are reflected and reinforced by certain legal concepts and constructions. For example, legal arrangements in the context of marriage, kinship and parenthood contribute to and maintain a larger anthropological and institutional narrative which permeates the stories we tell about ourselves and each other (see also Chap. 11).

## 17.4 The Challenge of Normative Pluralism

The etymology of the word 'symbol' from the Greek word *symbolon* (assimilated from *syn*, 'together', and *bole*, 'to throw') suggests that a symbolic order is something which is necessarily shared (see Chaps. 7, 11, and 14). However, in a pluralist, postmodern society this collective aspect of the symbolic order is under constant pressure. According to Priban (Chap. 7), law lacks the capacity as well as the moral authority to give symbolic expression to fundamental values in society. In a differentiated society, law is just one of the subsystems and therefore it cannot make claims (moral or other) which are accepted by or acceptable to the other subsystems. Lembcke (Chap. 6), on the other hand, does not rule out the possibility that symbolic legislation may contribute to the state's authority, if it succeeds in convincing citizens to follow the law voluntarily. However, if law is used as merely an instrument of coercion, authority is lost and transformed into power.

Indeed, the challenge of a pluralistic society for the symbolic dimensions of biolaw recurs in various contributions to this volume. In De Dijn's contribution (Chap. 9), the main tension is (as said) between broad ethics, which takes its starting point in lived experience, and revisionist ethics, which takes a more instrumentalist and pragmatic approach. In Herring' contribution (Chap. 8), relational and communal accounts of the human body collide with individualized, property-based approaches. Furthermore, in Hoeyer's and Sterckx & Cockbain's chapters (Chaps. 10 and 13 respectively), market-based approaches to the human body are contrasted with non-market based ones.

Since the EU is characterized by a wide variety of moral and religious traditions, the friction between these contesting symbolic orders can impede the process of harmonization. As various chapters in this volume illustrate, one of the ways in which EU institutions have responded to the challenge of pluralism in the field of biolaw and bioethics in Europe, is through technocratic solutions, such as specific strategies of risk regulation. As also Lee & Stokes point out in their analysis of EU risk regulation of nanotechnologies, these risk strategies are often unable to take into account the broader uncertainties caused by emerging technologies, such as uncertainty on these technologies' social acceptability (see Chap. 14) or the uncertainties caused by these technologies on a symbolic level (see Chaps. 9, 11, and 12). Moreover, it should be emphasized that scientific and technological knowledge is itself socially constructed (see Chaps. 11, 14, and 16).

The technocratic tendency within governance of bioethical matters is also criticized by Poort & Bovenkerk (Chap. 16). As they argue, governments install expert committees in order to avoid political conflicts'. By relying on the expert's advise, they do not have to take difficult decisions in highly controversial matters. According to them, expert committees fit well within liberal democratic pluralism with its emphasis on procedural solutions for social controversies. In their contribution, Poort & Bovenkerk focus in particular on ethics committees in the field of medical and biotechnologies. From the start, they stress that ethics committees can never replace the public debate. In their view, people have to change their expectations about what ethicists can and cannot do. An ethical commission can never be the final authority when it comes to morally sensitive issues. Ultimately, political decisions have to be taken by the government, after intense public debate. However, ethicists can fulfill a useful symbolic function by providing the vocabulary by means of which people can discuss controversial matters. The communicative (or epistemic) function that ethics committees can have, can best be characterized as a moderator of the debate.

Facing the challenge of pluralism, Van der Burg and Poort both promote an interactive legislative approach in their contributions to this volume (Chaps. 3 and 5 respectively). In the interactive approach, the legislature is not in the center of the legislative process; it is only one of the possible driving forces (not necessarily the most important one) behind the creation of new legal norms. The creation of law is a complex and on-going social process in which many actors interact with each other on a horizontal level. For instance, it was not the legislature that developed legislation on embryo research, but the scientific community in dialogue with other stakeholders, among which patient organizations and the general public. In embryo legislation, the expressive and communicative function are very important. It does not consist of detailed norms and instructions, but it offers basic principles and general rules. It confirms the intrinsic value of embryos without providing a clear definition thereof. Usually, it is left to committees in which various stakeholders are represented, to give meaning to this value in the context of a specific case. It is the aim of the interactive approach to include as many organizations and people in the legislative process, so that many different views can be heard.

In her chapter, Poort similarly stresses the importance of on-going norm development and a broad public debate. That may, however, conflict' with the law's aim to provide legal security and closure. According to Poort, theories of legislation tend to focus too much on achieving consensus. When it comes to morally sensitive issues – for instance in the field of biotechnology –, it is often not possible to achieve consensus. Therefore the legislature should, as Poort argues, adopt an ethos of controversies. An ethos of controversies structures the decision-making process in complex matters when aiming for consensus is premature. Poort proposes a two-track approach consisting of a legal track, in which legally binding decisions are taken concerning the law's content, and a moral track, in which the moral debate continues also after the decision has been made. The combination of these two tracks is meant to secure that norm development can continue whenever it is needed, while legal conflicts' can also be brought to an end. Following this approach, there is not necessarily a tension between ending conflicts and stimulating dynamics. In other words, through the use of general clauses, symbolic legislation is able to offer closure and, at the same time, it leaves room to normative pluralism.

Zeegers' chapter (Chap. 15) thinks Poort's analysis through in the context of EU policies regarding biomedical research funding. She discusses whether ethical norms for EU research funding have been established in a democratically legitimate way. Building on Poort's notion of ethos of controversies, she assesses democratic legitimacy on the basis of the following questions. Firstly, have all the existing viewpoints concerning the norms been addressed in the decision making? Secondly, did the decision makers at the end of the decision making process acknowledge that the compromise reached was only a 'temporary political achievement'? Zeegers focuses in particular on the funding of research involving the use of human embryos. However, her case study has a broader relevance given the divergence in cultural (and ethical) norms that exists in the EU. According to her, there are two conditions under which the EU should be flexible enough to re-open the decision-making process: to begin with, when it is no longer conceivable that the promises of this research for public health will be fulfilled within a certain period of time; and, subsequently, when new technological developments make still other improvements of public health possible.

In conclusion, we see that, within the interactive approach to legislation, normative pluralism is acknowledged by including as many viewpoints as possible in the decision-making process as well as by postponing the moment of legal closure. The debate on the law's content can always be re-opened, so that other viewpoints that have been excluded so far can be taken into account.

The discussion on the symbolic value and symbolic functions of law in general, and biolaw in particular, will no doubt continue in the years to come. Reflection on the symbolic dimensions of (bio)law is likely to become more important as the number of ethical and technological challenges which require new legislative strategies will continue to grow. Moreover, these challenges each cause new uncertainties on a symbolic level through their incessant interrogation of existing symbolic categories and values. From that perspective, these recent developments have served to expose the extent to which the legal order is, in various ways, also a symbolic order. Therefore, even if the notion of the symbolic remains elusive, the need for a better understanding of law's symbolic dimensions has perhaps never been more pressing than in today's society. We hope that this volume can contribute to that goal.

# Index




B. van Klink et al. (eds.), *Symbolic Legislation Theory and Developments in Biolaw*, Legisprudence Library 4, DOI 10.1007/978-3-319-33365-6

MIX
Papier aus verantwortungsvollen Quellen
Paper from responsible sources
FSC® C105338

FSC
www.fsc.org

If you have any concerns about our products,
you can contact us on
**ProductSafety@springernature.com**

In case Publisher is established outside the EU,
the EU authorized representative is:
**Springer Nature Customer Service Center GmbH**
**Europaplatz 3, 69115 Heidelberg, Germany**

Printed by Libri Plureos GmbH
in Hamburg, Germany